Studies in Computational Intelligence

Volume 744

Series editor

Janusz Kacprzyk, Polish Academy of Sciences, Warsaw, Poland
e-mail: kacprzyk@ibspan.waw.pl

The series "Studies in Computational Intelligence" (SCI) publishes new developments and advances in the various areas of computational intelligence—quickly and with a high quality. The intent is to cover the theory, applications, and design methods of computational intelligence, as embedded in the fields of engineering, computer science, physics and life sciences, as well as the methodologies behind them. The series contains monographs, lecture notes and edited volumes in computational intelligence spanning the areas of neural networks, connectionist systems, genetic algorithms, evolutionary computation, artificial intelligence, cellular automata, self-organizing systems, soft computing, fuzzy systems, and hybrid intelligent systems. Of particular value to both the contributors and the readership are the short publication timeframe and the world-wide distribution, which enable both wide and rapid dissemination of research output.

More information about this series at http://www.springer.com/series/7092

Xin-She Yang

Editor

Nature-Inspired Algorithms and Applied Optimization

 Springer

Editor
Xin-She Yang
School of Science and Technology
Middlesex University
London
UK

ISSN 1860-949X   ISSN 1860-9503 (electronic)
Studies in Computational Intelligence
ISBN 978-3-319-88465-3   ISBN 978-3-319-67669-2 (eBook)
https://doi.org/10.1007/978-3-319-67669-2

Printed on acid-free paper

This Springer imprint is published by Springer Nature
The registered company is Springer International Publishing AG
The registered company address is: Gewerbestrasse 11, 6330 Cham, Switzerland

Preface

Nature-inspired algorithms, especially those based on swarm intelligence, have been successfully applied to solve a variety of optimization problems in real-world applications, and thus their popularity has also increased significantly in recent years. The applications of nature-inspired optimization algorithms have been very diverse, from engineering optimization to feature selection and from scheduling to vehicle routing. Consequently, significant progress has been made with several thousand new research papers published in these areas in the past few years.

This edited book reviews and summarizes the state-of-the-art developments in nature-inspired algorithms with an emphasis on applied optimization in real-world applications. The algorithms covered in this book includes ant colony optimization, bat algorithm, cuckoo search, directional bat algorithm, differential evolution, firefly algorithm, flower pollination algorithm, genetic algorithm, particle swarm optimization, simulated annealing and others. The application topics include classification, feature selection, computational geometry curve-fitting, economic load dispatch, knapsack problems, mass damper tuning, modelling to generate alternatives, hypercomplex representations, vehicle routing with time windows, wireless networks, wireless butterfly networks and others.

In addition, some rigorous theoretical analyses of nature-inspired algorithms have also been presented. An overview of mathematical tools used for analyzing nature-inspired algorithms is presented to provide an informal but relatively comprehensive summary. In addition, no free lunch theorems are reviewed in the context of metaheuristic optimization, and a convergence analysis of the cuckoo search algorithm has been carried out using Markov chain theory. All these can form a solid foundation for the in-depth understanding of the working mechanisms for such powerful algorithms.

It is worth pointing out that the developments in nature-inspired computing are so rapid that it is estimated that there are more than 150 algorithms and variants in the current literature. Thus, it is not possible and not our intention to review all of them. Instead, we have focused on the diversity and different characteristics of algorithmic structures and their capabilities in solving a wider range of problems in various disciplines.

Despite the success and popularity of nature-inspired algorithms, there are still some questions and issues that require further research. In addition to the lack of a rigorous mathematical framework for analyzing such algorithms, an important area of research is parameter tuning and parameter control. As almost all algorithms have algorithm-dependent parameters, their settings will largely influence the performance of the algorithm under consideration. However, how to efficiently tune an algorithm and vary/control its parameters is still unresolved. At the same time, it is also difficult to achieve a fine balance of exploration and exploitation for a given algorithm and a given set of problems. Furthermore, though no free lunch theorems hold for averaged performance for all problem sets, free lunches can potentially exist for a finite set of problems. After all, for a given type of problems, some algorithms (especially those uses the landscape-specific knowledge of the problem of interest) are more effective than others. Therefore, how to incorporate problem-specific knowledge effectively requires further studies.

Though there are many case studies in real-world applications, the scales of such applications are relatively moderate, and the number of design variables is typically about a few dozens to a few hundred. In reality, many applications can have thousands or even millions of design variables, such large-scale problems can be very challenging to solve because they are usually computationally expensive. It is not quite clear how to scale up the present techniques to tackle large-scale, computationally extensive optimization problems. Therefore, there is a strong need to review carefully the state-of-the-art developments concerning bio-inspired computation, swarm intelligence and optimization techniques in general so as to identify important research challenges, to inspire further research and to encourage innovative approaches that can ultimately help to develop effective tools for solving hard optimization problems in real-world applications.

This book is a timely attempt to achieve such objectives with emphasis on applied optimization. As a timely snapshot of the latest developments, this book will be interested by students, researchers and professionals in many disciplines, and can thus serve as an ideal reference for graduates and researchers in computer science, evolutionary computing, machine learning, computational intelligence and engineering, as well as engineers in various disciplines and industrial applications.

I would like to thank the reviewers for their constructive comments on the manuscripts of all the chapters during the peer-review process. I also would like to thank the editors, especially Drs. Thomas Ditzinger and Ravi Vengadachalam, and staff at Springer for their help and professionalism.

London, UK   Xin-She Yang
August 2017

Contents

Contributors

Zaid Abdi Alkareem Alyasseri School of Computer Sciences, Universiti Sains Malaysia (USM), Pulau Pinang, Malaysia; ECE Department - Faculty of Engineering, University of Kufa, Najaf, Iraq

Mohammed Azmi Al-Betar Department of Information Technology, Al-Huson University College, Al-Balqa Applied University, Irbid, Al-Huson, Jordan

Mohammed A. Awadallah Department of Computer Science, Al-Aqsa University, Gaza, Palestine

Gebrail Bekdaş Department of Civil Engineering, Istanbul University, Avcılar, Istanbul, Turkey

Roberto Carballedo Deusto Institute of Technology (DeustoTech), University of Deusto, Bilbao, Spain

Asma Chakri Industrial Mechanics Laboratory, Department of Mechanical Engineering, University Badji Mokhtar of Annaba (UBMA), Annaba, Algeria

Javier Del Ser TECNALIA, Derio, Spain; University of the Basque Country (UPV/EHU), Bilbao, Spain; Basque Center for Applied Mathematics (BCAM), Bilbao, Spain

E. Emary Faculty of Computers and Information, Cairo University, Giza, Egypt

Iztok Fister Jr. Faculty of Electrical Engineering and Computer Science, University of Maribor, Maribor, Slovenia

A.H. Gandomi School of Business, Stevens Institute of Technology, Hoboken, NJ, USA

Akemi Gálvez Faculty of Sciences, Department of Information Science, Toho University, Funabashi, Japan; Department of Applied Mathematics and Computational Sciences, University of Cantabria, Santander, Spain

Xing-Shi He College of Science, Xi'an Polytechnic University, Xi'an, China

J. Michael Herrmann The University of Edinburgh, Edinburgh, Scotland

Andrés Iglesias Faculty of Sciences, Department of Information Science, Toho University, Funabashi, Japan; Department of Applied Mathematics and Computational Sciences, University of Cantabria, Santander, Spain

T. Jayabarathi School of Electrical Engineering, VIT University, Vellore, India

Thomas Joyce The University of Edinburgh, Edinburgh, Scotland

Aylin Ece Kayabekir Department of Civil Engineering, Istanbul University, Avcılar, Istanbul, Turkey

Ahamad Tajudin Khader School of Computer Sciences, Universiti Sains Malaysia (USM), Pulau Pinang, Malaysia

Pedro Lopez-Garcia Deusto Institute of Technology (DeustoTech), University of Deusto, Bilbao, Spain

Carlos Loucera Department of Communications Engineering, Universidad de Cantabria, Santander, Spain

Sinan Melih Nigdeli Department of Civil Engineering, Istanbul University, Avcılar, Istanbul, Turkey

Eneko Osaba Deusto Institute of Technology (DeustoTech), University of Deusto, Bilbao, Spain

João Paulo Papa School of Sciences, São Paulo State University, Bauru, Brazil

T. Raghunathan School of Electrical Engineering, VIT University, Vellore, India

Haroun Ragueb Energy and Mechanical Engineering Laboratory, Department of Mechanical Engineering, Faculty of Engineering Sciences, University M'hamed Bougara of Boumerdes (UMBB) Avenue of Independence, Boumerdes, Algeria

Gustavo Henrique de Rosa Department of Computing, São Paulo State University, Bauru, Brazil

Quoc-Tuan Vien Middlesex University, London, UK

Fan Wang College of Science, Xi'an Polytechnic University, Xi'an, People's Republic of China

Yan Wang College of Science, Xi'an Polytechnic University, Xi'an, People's Republic of China

Xin-She Yang School of Science and Technology, Middlesex University, London, UK

Julian Scott Yeomans OMIS Area, Schulich School of Business, York University, Toronto, ON, Canada

Hossam M. Zawbaa Faculty of Computers and Information, Beni-Suef University, Beni Suef, Egypt

Mathematical Analysis of Nature-Inspired Algorithms

Xin-She Yang

Abstract Nature-inspired algorithms are a class of effective tools for solving optimization problems and these algorithms have good properties such as simplicity, flexibility and high efficiency. Despite their popularity in practice, a mathematical framework is yet to be developed to analyze these algorithms theoretically. This work intends to analyze nature-inspired algorithms both qualitatively and quantitatively. We briefly outline the links between self-organization and algorithms, and then analyze algorithms using Markov chain theory, dynamic system and other methods. This can serve as a basis for building a multidisciplinary framework for algorithm analysis.

Keywords Algorithm · Bat algorithm · Cuckoo search · Differential evolution · Firefly algorithm · Flower pollination algorithm · Particle swarm optimization · Metaheuristics · Nature-inspired computation · Optimization · Self-organization · Swarm intelligence

1 Introduction

Optimization is important in many disciplines from engineering designs to business scheduling. Most such optimization problems require sophisticated optimization tools to solve, and there are a diverse spectrum of algorithms used in the literature, from traditional gradient-based algorithms and simplex methods to evolutionary algorithms and nature-inspired metaheuristic algorithms [7, 37]. In recent years, nature-inspired algorithms have become widely used for dealing with highly nonlinear problems and tough optimization problems [20, 33–35, 37]. Most of such nature-inspired algorithms are based on swarm intelligence, intending to mimic the diverse characteristics in natural systems.

Though the literature in this area is quite vast, however, most studies are about various applications of algorithms. There is little literature on theoretical analysis

X.-S. Yang (✉)
School of Science and Technology, Middlesex University, London NW4 4BT, UK
e-mail: x.yang@mdx.ac.uk; xy227@cam.ac.uk

© Springer International Publishing AG 2018
X.-S. Yang (ed.), *Nature-Inspired Algorithms and Applied Optimization*,
Studies in Computational Intelligence 744, https://doi.org/10.1007/978-3-319-67669-2_1

of these algorithms. In fact, there is a significant gap between theory and practice. Most nature-inspired metaheuristic algorithms have successful applications in practice, but their theoretical analysis lags far behind. Apart from a few limited results about the convergence and stability concerning particle genetic algorithms [28], particle swarm optimization [8] and others, no theoretical analysis has been carried out about many other algorithms. It is often the case that we know these algorithms can work well in practice, but we rarely understand why they work. As a result, the applications can be a heuristic process itself and there is little information on how to improve them. Such lack of understanding may hinder the development of effective algorithms and some researchers even cast doubt on certain metaheuristics. Therefore, there is a strong need to do more rigorous mathematical analysis of nature-inspired algorithms.

Therefore, this book chapter will first introduce the fundamentals of algorithms and optimization in Sect. 2, followed by the outlines of all major nature-inspired algorithms in Sect. 3. Then, in Sect. 4, the emphasis will be on the analysis of main characteristics of optimization algorithms and their links to self-organization. Section 5 provides some preliminary framework for analyzing these algorithms mathematically. Section 6 concludes with some discussions and open problems.

2 Algorithm, Optimization and Metaheuristics

Optimization problems tend to be nonlinear with complex objective landscapes. The algorithms used for solving optimization can be traditional algorithms such as gradient-based methods and quadratic programming, evolutionary algorithms, heuristic or metaheuristic algorithms and various hybrid techniques.

2.1 The Essence of an Algorithm

An algorithm is a computational procedure. For example, Newton's method for finding the roots of a polynomial $p(x) = 0$ can be written as

$$x_{t+1} = x_t - \frac{p(x_t)}{p'(x_t)}, \tag{1}$$

where x_t is the approximation at iteration t, and $p'(x)$ is the first derivative of $p(x)$. This procedure typically starts with an initial guess x_0. In most cases, as along as $p' \neq 0$ and x_0 is not too far away, this algorithm can work very well. But if x_0 is too far away from the true solution $x_* = \lim_{t \to \infty} x_t$, it may fail. This means that the final solution can largely depend on where the initial solution is, which is especially true for nonlinear multimodal functions.

This method can be modified to solve optimization problems. For example, for a single objective function $f(x)$, the minimal and maximal values should occur at stationary points $f'(x) = 0$, which becomes a root-finding problem. Thus, the maximum or minimum of $f(x)$ can be found by modifying the Newton's method as the following iterative formula:

$$x_{t+1} = x_t - \frac{f'(x_t)}{f''(x_t)}. \tag{2}$$

For a D-dimensional problem with an objective $f(\mathbf{x})$ with independent variables $\mathbf{x} = (x_1, x_2, \ldots, x_D)$, the above iteration formula can be generalized to a vector form

$$\mathbf{x}^{t+1} = \mathbf{x}^t - \frac{\nabla f(\mathbf{x}^t)}{\nabla^2 f(\mathbf{x}^t)}, \tag{3}$$

where we have used the notation convention $\mathbf{x}^t$ to denote the current solution vector at iteration t (not to be confused with an exponent).

In general, an algorithm A can be written as

$$\mathbf{x}^{t+1} = A(\mathbf{x}^t, \mathbf{x}_*, p_1, \ldots, p_K), \tag{4}$$

which represents that fact that the new solution vector is a function of the existing solution vector $\mathbf{x}^t$, some historical best solution $\mathbf{x}_*$ during the iteration history and a set of algorithm-dependent parameters $p_1, p_2, \ldots, p_K$. The exact function forms will depend on the algorithm, and different algorithms are only different in terms of the function form, number of parameters and the ways of using historical data.

2.2 Optimization

In general, an optimization problem can be formulated as

$$\text{minimize} \quad f(\mathbf{x}), \quad \mathbf{x} = (x_1, x_2, \ldots, x_D) \in \mathbb{R}^D, \tag{5}$$

subject to

$$h_i(\mathbf{x}) = 0, \quad (i = 1, 2, \ldots, I), \quad g_j(\mathbf{x}) \leq 0, \quad (j = 1, 2, \ldots, J), \tag{6}$$

where h_i and g_j are the equality constraints and inequality constraints, respectively. In most cases, the problem functions $f(\mathbf{x})$, $h_i(\mathbf{x})$ and $g_j(\mathbf{x})$ are all nonlinear, and such nonlinear optimization problems can be challenging to solve. There are a wide class of optimization techniques, including linear programming, quadratic programming, convex optimization, interior-point method, trust-region method, conjugate-gradient and many others [7, 26, 33].

2.3 What's Wrong with Traditional Algorithms

One may wonder what is wrong with traditional algorithms? A short answer is that there is nothing wrong. Traditional algorithms work well for the types of problems they can solve, but most traditional algorithms are local search.

- As traditional algorithms are mostly local search, there is no guarantee for global optimality for most optimization problems, except for linear programming and convex optimization. Consequently, the final solution will often depend on the initial starting points (except for linear programming and convex optimization).
- Traditional algorithms tend to be problem-specific because they usually use some information such as derivatives about the local objective landscape. Other methods such as k-opt and branch and bound can heavily depend on the type of problems in implementation.
- Traditional algorithms cannot solve highly nonlinear, multimodal problems effectively, and they struggle to cope with problems with discontinuity, especially when gradients are needed.
- Almost all traditional algorithms, except for hill-climbing with random restart, are deterministic algorithms. The final solutions will be identical if starting with the same initial points. No random numbers are used. Consequently, the diversity of the obtained solutions can be limited.

2.4 Heuristics and Metaheuristics

In order to remedy the above disadvantages, contemporary algorithms tend to be heuristic and metaheuristic. Heuristic algorithms use a trial-and-error approach in generating new solutions, while metaheuristic algorithms are a higher-level heuristics with the use of memory, solution history and other forms of 'learning' strategy. Nowadays, most metaheuristic algorithms are nature-inspired algorithms and most such algorithms are based swarm intelligence inspired by nature [13, 27, 37]. In contrast with traditional algorithms, metaheuristics are mainly designed for global search and tend to have the following advantages and characteristics:

- As they are global optimizers, it is more likely to find the true global optimality.
- They often treat problems as a black box without specific knowledge, thus they can solve a wider range of problems.
- Metaheuristic algorithms are usually gradient-free methods and they do not use any derivative information, and thus can deal with highly nonlinear problems and problems with discontinuity.
- Stochastic components in terms of random numbers and random walks are often used, and thus such algorithms are stochastic. Thus, no identical solutions can be obtained, even starting with the same initial points, but the final solutions can be sufficient close and they often enable the algorithm to escape any local modes (thus less likely to get stuck in local regions).

Despite these advantages, nature-inspired algorithms do have some disadvantages. In general, the computational efforts are higher than those for traditional algorithms because more iterations are needed, which can become too computationally expensive if the evaluation of a single objective requires a long time by a simulator (e.g., by finite element methods). In addition, the final solutions obtained by such algorithms cannot be repeated exactly, and multiple runs should be carried out to ensure consistency and some meaningful statistical analysis.

2.5 Deterministic or Stochastic

A key feature of traditional algorithms is that they are mainly deterministic and no randomness is used in generating new solutions. This can enhance the exploitation ability, but lacks exploration capabilities. On the other hand, nature-inspired metaheuristic algorithms use a certain degree of randomness, and these algorithms have stochastic components. A good degree of randomness will increase the exploration ability, but may reduce the exploitation abilities.

Some questions arise naturally: Which is better? How much randomness should an algorithm have? As we discussed earlier, both traditional deterministic algorithms and stochastic metaheuristic algorithms have some advantages and disadvantages. From the global optimization perspective, the advantages of stochastic algorithms far outweigh their disadvantages. Both empirical observations and simulation suggest that randomness can be largely beneficial to the overall performance of algorithms. As to the right degree of randomness, it is very difficult to say because such randomness can depend on the algorithmic structure, type of problems and the solution quality desired for a given type of problems. In fact, this is still an open problem.

3 Nature-Inspired Optimization Algorithms

There are many nature-inspired algorithms, it is estimated that there are over a hundred different algorithms and their variants [37]. Obviously, it is not possible to include even a good fraction of these algorithms. Therefore, our emphasis is on the algorithms that can be considered as representatives, especially those algorithms based on swarm intelligence. In addition, instead of giving detailed descriptions and background about each algorithm, our emphasis here is on the similarity and differences of different algorithms and the ways used for generating new solutions, selection of the best solutions and other major characteristics.

3.1 Genetic Algorithms

The genetic algorithm (GA), first developed by John Holland [18], is an evolutionary algorithm based on the Darwinian evolution of biological systems. Its main characteristics are the three genetic operators: crossover, mutation, and selection [16, 37]. A set of solutions from a population that are encoded as binary or real strings, called chromosomes. A new population of solutions are generated using such genetic operators.

Two child solutions can be generated from two parent solutions by crossover, which essentially swaps one segment or multiple segments of one parent solution with its counterparts. On the other hand, a new solution can be generated by mutating one bit or multiple bits of one solution. Mutation can be simply flipping between 0 and 1 for binary strings at one or more locations. The quality of a solution is determined by its fitness that is a normalized value associated with the function values of the objective. In case of maximization problems, the fitness can be proportional to the objective. Selection is done by choosing the most fittest solution according to their fitness.

In general, crossover occurs more often, typically with a probability of 0.6–0.95, while the rate of mutation is often lower, ranging from 0.001 to 0.05. Crossover can help exploit and enhance the key characteristics in the population, thus can enhance convergence. On the other hand, mutation can provide better diversity with a higher ability for exploration to allow the population to explore the search space more effectively. However, if the mutation rate is too high, it will generate solutions that may be far from existing solutions, leading to a slower convergence rate.

As an interesting note, genetic algorithms usually do not have any explicit equations in terms of generating new solutions. It is a detailed procedure, though some mathematical analysis can be done using binomial distributions and other tools [16].

3.2 Ant Colony Optimization

The ant colony optimization (ACO) was developed by Marco Dorigo in 1992 [6], and ACO attempts to mimic the foraging behaviour of social ants in a colony. All ants/agents use a chemical messenger, called pheromone, to communicate with other ants, their interactions are local, based on local information. Pheromone is deposited by each agent, and such chemical will also evaporate. The model for pheromone deposition and evaporation may vary slightly, depend on the variants of ACO. However, in most cases, incremental deposition and exponential decay are used in the literature.

From the implementation point of view, for example, a solution in a network optimization problem can be a path or route. Each agent will explore the network paths and deposit pheromone when it moves. The quality of this solution is related to the pheromone concentration on the path. At the same time, pheromone will evaporate as time. At a junction with multiple routes, the probability of choosing a particular route is determined by a decision criterion, depending on the normalized concentration of the route, the desirability of the route (for example, the distance of the overall path), and relative fitness of this route, comparing with all others.

It is worth pointing out that ACO is a mixed of procedure and some simple equations such as pheromone deposition and evaporation as well as the path selection probability.

3.3 Particle Swarm Optimization

Particle swarm optimization (PSO) was developed by Kennedy and Eberhart in 1995 [20], which uses equations to simulate the swarming characteristics of birds and fish. Both ACO and PSO are the primary examples of the so-called swarm intelligence (SI).

For the ease of discussions below, let us use $\mathbf{x}_i$ and $\mathbf{v}_i$ to denote the position (solution) and velocity, respectively, of a particle or agent i. In PSO, there are n particles as a population, thus $i = 1, 2, \ldots, n$. There are two equations for updating positions and velocities of particles, and they can be written as follows:

$$\mathbf{v}_i^{t+1} = \mathbf{v}_i^t + \alpha \epsilon_1 [\mathbf{g}^* - \mathbf{x}_i^t] + \beta \epsilon_2 [\mathbf{x}_i^* - \mathbf{x}_i^t], \tag{7}$$

$$\mathbf{x}_i^{t+1} = \mathbf{x}_i^t + \mathbf{v}_i^{t+1}, \tag{8}$$

where ϵ_1 and ϵ_2 are two uniformly distributed random numbers in [0, 1]. The learning parameters α and β are usually in the range of [0, 2]. In the above equations, $\mathbf{g}^*$ is the best solution found so far by all the particles in the population, and each particle has an individual best solution $\mathbf{x}_i^*$ by itself during the entire past iteration history.

It is clearly seen that the above algorithmic equations are linear in the sense that both equation only depends on $\mathbf{x}_i$ and $\mathbf{v}_i$ linearly. Selection is carried out by the attractor or converged state $\mathbf{g}^*$, which is also evolving. Randomization is done by two uniformly distributed random numbers.

PSO has been applied in many applications, and it has been extended to solve multiobjective optimization problems. However, there are some drawbacks because PSO can often have so-called premature convergence in which the population may get stuck locally with almost no diversity and thus lose its exploration ability. In addition, the use of velocities can also have some disadvantages. For example, high velocities (thus high energy) can destabilize the system, leading to slower convergence. As a result, various variants and remedies have been attempted in the literature with some degree of success. For example, some variants introduced a so-called inertia weight parameter, which is essentially equivalent to putting some mass on the particles as to as stabilize them.

3.4   Differential Evolution

Differential evolution (DE), developed by Storn and Price [25], uses a mutation operator in terms of the difference of two different solution vectors $\mathbf{x}_p$ and $\mathbf{x}_q$. This mutation vector is scaled by a parameter $F \in [0, 2]$ and is then used to perturb an existing solution $\mathbf{x}_r$ to generate a new solution

$$\mathbf{x}_k = \mathbf{x}_r + F(\mathbf{x}_p - \mathbf{x}_q). \tag{9}$$

In general, p, q, r should be different integers, corresponding to different solution vectors in the population.

On the other hand, the mutated solutions can also be applied by a crossover operator, which is either binomial or exponential. In addition, the selection mechanism for any new solution $\mathbf{u}^t$ is to ensure to record the solutions that are better than previous solutions in terms of fitness. For a minimization problem, we can write as

$$\mathbf{x}_i^t = \begin{cases} \mathbf{u}^t & \text{if } f(\mathbf{u}^t) \le f(\mathbf{x}_i^{t-1}) \\ \mathbf{x}_i^{t-1} & \text{otherwise.} \end{cases} \tag{10}$$

Since there are different ways of perturbing a solution in terms of the mutation operator, there are more than ten different variants [25].

3.5   Bat Algorithm

Bat algorithm (BA), developed by Xin-She Yang in 2010, uses some characteristics of frequency-tuning and echolocation of microbats [34, 35]. It also uses the variations of pulse emission rate r and loudness A to control exploration and exploitation. In the bat algorithm, main algorithmic equations are

$$f_i = f_{\min} + (f_{\max} - f_{\min})\beta, \tag{11}$$

$$\mathbf{v}_i^t = \mathbf{v}_i^{t-1} + (\mathbf{x}_i^{t-1} - \mathbf{x}_*)f_i, \tag{12}$$

$$\mathbf{x}_i^t = \mathbf{x}_i^{t-1} + \mathbf{v}_i^t, \tag{13}$$

where $\beta \in [0, 1]$ is a random vector drawn from a uniform distribution so that the frequency can vary from $f_{\min}$ to $f_{\max}$. In addition, these updating equations are also associated with r and loudness A via a uniformly distributed random number ε. Selection is done by the current best solution $\mathbf{x}_*$ found so far by all the virtual bats, which acts a similar role as the $\mathbf{g}^*$ in PSO.

From the above equations, we can see that both equations are linear in terms of $\mathbf{x}_i$ and $\mathbf{v}_i$. But, the control of exploration and exploitation is carried out by the variations of loudness $A(t)$ from a high value to a lower value and the emission rate r from a lower value to a higher value. That is

$$A_i^{t+1} = \alpha A_i^t, \quad r_i^{t+1} = r_i^0(1 - e^{-\gamma t}), \tag{14}$$

where $0 < \alpha < 1$ and $\gamma > 0$ are two parameters. As a result, the actual algorithm can have a weak nonlinearity. Consequently, BA can have a faster convergence rate in comparison with PSO. BA has been extended to multiobjective optimization and hybrid versions [35].

3.6 Firefly Algorithm

Firefly algorithm (FA), developed by Xin-She Yang in 2008, is an algorithm inspired by the swarming and light-flashing behaviour of tropical fireflies [32]. FA uses a nonlinear system by combing the exponential decay of light absorption and inverse-square law of light variation with distance. The main algorithmic equation in FA is

$$\mathbf{x}_i^{t+1} = \mathbf{x}_i^t + \beta_0 e^{-\gamma r_{ij}^2}(\mathbf{x}_j^t - \mathbf{x}_i^t) + \alpha \, \epsilon_i^t, \tag{15}$$

where α is a scaling factor controlling the step sizes of the random walks, while γ is a scale-dependent parameter controlling the visibility of the fireflies (and thus search modes). In addition, β_0 is the attractiveness constant when the distance between two fireflies is zero (i.e., $r_{ij} = 0$). This system is a nonlinear system, which may lead to rich characteristics in terms of algorithmic behaviour.

In fact, since FA is a nonlinear system, it has the ability to automatically subdivide the whole swarm into multiple subswarms. This is because short-distance attraction is stronger than long-distance attraction, and the division of swarm is related to the mean range of attractiveness variations. After division into multi-swarms, each sub-swarm can potentially swarm around a local mode. Consequently, FA is naturally suitable for multimodal optimization problems. Furthermore, there is no explicit use of the best solution $\mathbf{g}^*$, thus selection is through the comparison of relative brightness according to the rule of 'beauty is in the eye of the beholder'. Perturbation is done by a random walk with a scaling factor α.

It is worth pointing out that FA has some significant differences from PSO. Firstly, FA is nonlinear, while PSO is linear. Secondly, FA has an ability of multi-swarming, while PSO cannot. Thirdly, PSO uses velocities (and thus have some drawbacks), while FA does not use velocities. Finally, FA has some scaling control by using γ, while PSO has no scaling control. All these differences enable FA to search the design spaces more effectively for multimodal objective landscapes.

3.7   *Cuckoo Search*

As a very different approach, Cuckoo search (CS), developed by Yang and Deb, is
another nonlinear system by using a power-law, scale-free search mechanism [39].
CS was based on the intriguing brooding parasitism of some cuckoo species and their
co-evolution with host bird species such as warblers. CS uses a combination of both
local and global search capabilities, controlled by a switching probability p_a. There
are two algorithmic equations in CS, and one equation is

$$\mathbf{x}_i^{t+1} = \mathbf{x}_i^t + \alpha s \otimes H(p_a - \epsilon) \otimes (\mathbf{x}_j^t - \mathbf{x}_k^t), \tag{16}$$

where $\mathbf{x}_j^t$ and $\mathbf{x}_k^t$ are two different solutions selected randomly by random permutation, $H(u)$ is a Heaviside function, ϵ is a random number drawn from a uniform
distribution, and s is the step size. This step is primarily local, though it can become
global search if s is large enough. However, the main global search mechanism is
realized by the other equation with Lévy flights:

$$\mathbf{x}_i^{t+1} = \mathbf{x}_i^t + \alpha L(s, \lambda), \tag{17}$$

where the Lévy flights are simulated (or drawn random numbers) by

$$L(s, \lambda) \sim \frac{\lambda \Gamma(\lambda) \sin(\pi \lambda/2)}{\pi} \frac{1}{s^{1+\lambda}}, \quad (s \gg 0). \tag{18}$$

Here $\alpha > 0$ is the step size scaling factor.

By looking at the equations in CS carefully, we can clearly see that CS is a nonlinear system due to the Heaviside function, switch probability and Lévy flights. There
is no explicit use of global best $\mathbf{g}^*$, but selection is done by ranking and elitism
where the current best is passed onto the next generation. In addition, the use of
Lévy flights can enhance the search capability because a fraction of steps generated
by Lévy flights are larger than those used in Gaussian. Thus, the search steps in CS
are heavy-tailed [22, 23].

In addition, from the implementation point of view, Lévy flights can be approximated by a power-law type of distribution, the search steps are also scale-free. From
empirical observations and simulations, CS can have scale-free, self-similar structural characteristics in terms of its moves and search regions [39]. Consequently, CS
can be very effective for nonlinear optimization problems and multiobjective optimization [36, 41]. Cuckoo search has become powerful in solving many problems
such as software testing, scheduling, engineering optimization [40] and many others
[9, 10, 14, 21, 42, 46].

The above algorithms such as ACO, bat algorithm, PSO, cuckoo search and firefly
algorithms are all based on the swarming behaviour, and thus these algorithms are
often called swarm intelligence (SI) based algorithms. However, population-based
algorithms are not all SI-based. For example, both genetic algorithms and differential

evolution are not SI-based, but they are population-based. Another population-based algorithm is flower pollination algorithm that will be introduced next.

3.8 Flower Pollination Algorithm

Flower pollination algorithm (FPA) is a population-based algorithm, inspired by the pollination characteristics of flowering plants [37, 44]. FPA intends to mimic some key characteristics of biotic and abiotic pollination as well as the co-evolutionary flower constancy between certain flower species and pollinators such as insects and animals.

There are two main equations for this algorithm, and the global search is carried out by

$$\mathbf{x}_i^{t+1} = \mathbf{x}_i^t + \gamma L(\lambda)(\mathbf{g}_* - \mathbf{x}_i^t), \tag{19}$$

where γ is a scaling parameter, $L(\lambda)$ is the random number vector drawn from a Lévy distribution governed by the exponent λ. Here $\mathbf{g}_*$ is the best solution found so far, which acts as a selection mechanism. The current solution $\mathbf{x}_i^t$ is modified by varying step sizes because Lévy flights can have a fraction of large step sizes in addition to many small steps. The local search is carried out by

$$\mathbf{x}_i^{t+1} = \mathbf{x}_i^t + U(\mathbf{x}_j^t - \mathbf{x}_k^t), \tag{20}$$

which mimics local pollination and flower constancy. Here, U is a uniformly distributed random number. Furthermore, $\mathbf{x}_j^t$ and $\mathbf{x}_k^t$ are solutions representing pollen from different flower patches.

The equations are linear in terms of solutions $\mathbf{x}_i^t$, $\mathbf{x}_j^t$ and $\mathbf{x}_k^t$, but there is a switch probability p to activate which pollination activities (global or local). As a result, the system becomes somehow quasi-linear. The randomization is achieved by three components: Lévy flights, a uniform distribution and a switch probability. As a result, FPA can typically have a higher explorative ability. At the same time, the local branch provides a mechanism to remain a strong exploitation ability. Theoretical analysis using Markov chain theory has shown that FPA can have guaranteed global convergence under the right conditions [17]. FPA has been applied to solve many optimization problems such as solar photovoltaic parameter estimation, economic and emission dispatch, and EEG-based identification [1, 2, 24]. In addition, FPA has been extended to multiobjective optimization [44].

3.9 Other Algorithms

Obviously, there are many other metaheuristic algorithms such as artificial bee colony, gravitational search, artificial immune system and others. However, we will not discuss them due to page limits. Instead, our emphasis will be on the discussion and analyses of algorithms to gain more insight into the search mechanisms of various algorithms.

4   Why Nature-Inspired Algorithms Work

Though we know all the above algorithms we have discussed and other algorithms can work well in practice and they are able to solve a diverse range of problems, we rarely understand how they work exactly. To gain a truly comprehensive, in-depth understanding of all algorithms, it requires a multidisciplinary approach by combing mathematical analysis, numerical analysis, computational complexity, dynamical systems and other relevant tools. Therefore, we will not attempt such challenging tasks here. Instead, we will focus on analyzing the basic characteristics of algorithms, their components, search mechanisms and behaviour so as to gain a better insight into such algorithms. After such qualitative analysis in this section, we will try to provide some mathematical analyses in the next section.

4.1   Characteristics of Nature-Inspired Algorithms

First, let us look at nature-inspired algorithms by their basic steps, search characteristics and algorithm dynamics.

- All algorithms use a population of multiple agents (e.g., particles, ants, bats, cuckoos, fireflies, bees, etc.), each agent corresponds to a solution vector. Among the population, there is often the best solution $\mathbf{g}^*$ in terms of objective fitness. Different solutions in a population represent both diversity and different fitness.
- The evolution of the population is often achieved by some operators (e.g., mutation, crossover), often in terms of some algorithmic formulas or equations. Such evolution is typically iterative, leading to evolution of solutions with different properties. When all solutions become sufficiently similar, the system can be considered as converged.
- The moves of an agent represents a zigzag piecewise path in the search space, and such moves are quasi-deterministic. Thus, randomization techniques are often used to generate new solution vectors or moves. Such randomization provides a mechanism to perturb the states (or solutions) of the algorithm, which potentially allows it to escape any local optima (thus minimizing the probability of getting stuck locally).
- All algorithms try to carry out some sort of both local and global search. If the search is mainly local, it increases the probability of getting stuck locally. If the search focuses too much on global moves, it will slow down the convergence. Different algorithms may use different amount of randomization and different por-

Table 1 Characteristics of nature-inspired algorithms.

Components/characteristics	Role or properties
Population	Diversity and sampling
Randomization/perturbations	Escape local optima
Selection and elitism	Driving force for convergence
Algorithmic equations	Iterative evolution of solutions

tion of moves for local or global search. However, it is not clear yet what the right amount of randomness is and what the ratio of global search to local search should be.

- Selection of the better or best solutions is carried out by the 'survival of the fittest' or simply elitism so that the best solutions $\mathbf{g}^*$ are kept in the population in the next generation. Such selection essentially acts a driving force to drive the diverse population into a converged population with reduced diversity but with a more organized structure.

These basic components, characteristics and their properties can be summarized in Table 1, and this can form a basis for comparison with the characteristics of self-organization to be discussed in the next subsection.

4.2 Self-organization

Another way of looking at nature-inspired algorithms is from the perspective of self-organization. Loosely speaking, a complex system can self-organize when the size of the system is sufficiently large with a high number of degrees of freedom, perturbations and a driving mechanism, giving enough time for the system to evolve from noise and far from equilibrium states [3, 19]. Mathematically speaking, a system with multiple states S_i can evolve with time t towards the self-organized states S_*, driven by a driving mechanism M which can be written schematically as

$$S_i \overset{M}{\Longrightarrow} S_*. \tag{21}$$

Now let us look at an algorithm using self-organization, an algorithm can indeed be considered as a self-organization system, starting from a population of solutions $\mathbf{x}_i(i = 1, 2, \dots, n)$ (states), evolving towards some optimal solution/state $\mathbf{x}_*$. This is driven by the selection mechanism in an algorithm $A(p, t)$ with a set of parameter p, evolving with pseudo-time t. In essence, an algorithm for minimization can also written schematically as

$$f(\mathbf{x}_i) \overset{A(p,t)}{\Longrightarrow} f_{\min}(\mathbf{x}_*). \tag{22}$$

Table 2 Self-organization and algorithms

Self-organization	Characteristics	Algorithm	Properties
States	Complexity	Population	Diversity and sampling
Noise, perturbations	Far from equilibrium	Randomization	Escape local optima
Selection mechanism	Organization	Selection	Convergence
Re-organization	State changes	Iteration	Evolution of solutions

Furthermore, we can systematically compare the similarities and differences between self-organization and algorithms, which is summarized Table 2.

Despite these similarities, there are some crucial differences. First, for self-organization, the exact avenue to self-organization may not be clear, but for algorithms, the ways of solution generations are often clear. Second, for self-organization, time is not an important factor per se, but for algorithms, the number of iterations (pseudotime) is crucially important because an effective algorithm should be able to find the optimal solutions using as least amount of computational efforts as possible. Third, the structure in self-organization is important, while the converged solutions themselves (not necessarily the structure) are most relevant. Finally, exact conditions of self-organization may be physically maintained, but for algorithms, the conditions for convergence can often lead to undesired premature convergence, and it is still not clear yet how to avoid such premature convergence in algorithms.

4.3 Exploration and Exploitation

Another way of analyzing algorithms is to look at their exploration and exploitation abilities. Exploration provides diversification, which allows the algorithm to search different regions in the design space and thus increases the probability of finding the true global optimality.

Exploration is often achieved by randomization or random numbers in terms of some predefined probability distributions [5]. In most cases, random numbers drawn from a uniform distribution or a Gaussian distribution are used. Exploration can be consideration as a global, explorative mechanism. For example, cuckoo search has a strong ability of exploration due to the use of Lévy flights. PSO uses two uniformly distributed random numbers to enable its exploration.

On the other hand, exploitation uses local information such as gradients to search local regions more intensively, and such intensification can enhance the rate of convergence. Exploitation can make the population less diverse, and strong local guidance can even make the population relatively uniform in terms of solution variations. For example, in PSO and bat algorithm, the best solution g^* is used to exploit the current best solution and its locality in the design space.

Too much exploration and too little exploitation can slow down the convergence of an algorithm, while too much exploitation and too little exploration can sacrifice the possibility of finding true global solutions. Therefore, there is a fine balance between exploration and exploitation, which may depend on the algorithmic structure and type of problems.

4.4 Crossover, Mutation and Selection

Alternatively, we can also analyze the algorithm components in terms of their role. Borrowing the terminologies from genetic algorithms, we can look at mutation, crossover and selection.

Most algorithms use mutation. For example, differential evolution uses a vectorized mutation operator $(\mathbf{x}_j - \mathbf{x}_k)$, and firefly algorithm uses an isotropic random walk. All other algorithms such as PSO and bat algorithm use vectorized mutation in a similar way as that in differential evolution. However, cuckoo search and flower pollination algorithm use Lévy flights in terms of non-isotropic random walks, which makes the algorithms more efficient due to the power-law, scale-free search properties of Lévy flights.

Crossover is a mechanism that can enhance the mixing ability of the population, but not all algorithms use crossover. For example, differential evolution uses binomial and exponential crossover, but PSO, bat algorithm, cuckoo search and others do not use crossover explicitly. However, many variants of PSO, cuckoo search and flower pollination algorithm introduced some form of crossover, and they achieved enhanced performance.

Selection is a driving mechanism to ensure convergence among the populations. All algorithms have to have some good selection mechanisms. Genetic algorithms use elitism and survival of the fittest, while PSO uses both the best solution $\mathbf{g}^*$ and individual best $\mathbf{x}_i^*$ as selection. Firefly algorithm uses the brightest fireflies implicitly as an attraction mechanism. Other algorithms such as cuckoo search do not use $\mathbf{g}^*$, while flower pollination algorithm uses $\mathbf{g}^*$ explicitly. The use of $\mathbf{g}^*$ is something like a double-edged sword. If the selection mechanism is too strong, the diversity of the population can be limited. For example, in PSO, the use of both $\mathbf{g}^*$ and individual best solutions may be too strong for some problems, and the solutions can get stuck at some local regions, leading to potential premature convergence in this case. On the other hand, if the selection is weak, many solutions are not well-selected, and the convergence may be significantly slowed down. Again it needs a fine balance of selection strength as well as a good combination of crossover and mutation.

4.5 Biased Monto Carlo

From the sampling point of view, nature-inspired algorithms share some similarity with the well-known Monte Carlo method. In many algorithms, the initialization is

done by random sampling of the search space, often using some uniformly distributed random numbers, and the initial population generated by such randomization is
essentially the same as those by Monte Carlo. In addition, a solution vector can also
be considered as a sampling point in the design space, and in this sense, the set of
solutions during iterations form a sampling set.

However, there are some crucial differences. The samples generated by Monte
Carlo sampling and its many variants tend to be distributed relatively uniformly in
the design space and sometimes they are far away from each other in case of low dis-
crepancy random numbers, while samples generated by nature-inspired algorithms
will gradually aggregate towards some preferred regions based on the fitness of the
solutions. Thus, the overall sampling process in algorithms is biased towards some
promising regions where the optima and global optima may lie. The biased moves
are guided by the fitness and the local information from the objective landscape.
For example, the current best solution $\mathbf{g}^*$ in PSO acts a local guide to attract biased
moves. In this sense, we can consider all nature-inspired algorithms as information-
guided biased Monte Carlo.

4.6   Random Walks

From probability theories, we know that the moves to generate solutions can be considered as random walks, modifying an existing solution $\mathbf{x}_N$ at step N by a perturbation $\mathbf{w}_N$. Mathematically speaking, a random walk can be written as

$$\mathbf{x}_{N+1} = \mathbf{x}_N + \mathbf{w}_N, \tag{23}$$

where $\mathbf{w}_N$ is a vector of random numbers (steps) drawn from a known probability
solution. If $\mathbf{w}_N$ is drawn from a Gaussian distribution, then the random walks are
isotropic. The movements in this case are often referred to as normal diffusion or
Brownian motion. The expected distance moved (R) can be estimated by

$$R(N) \propto \sqrt{N}, \tag{24}$$

which has a square-root scaling property.

If the steps are drawn from a fat-tailed distribution such as Lévy distribution or
Cauchy distribution, the diffusion becomes anomalous. In general, the above scaling
property becomes

$$R(N) \propto N^q, \quad q > 0. \tag{25}$$

If $q \geq 1/2$, the diffusion is called super-diffusion [22]. Both Lévy distribution and
Cauchy distribution for step sizes can have a fraction of large steps, which will lead
to super-diffusion. This means that averaged distance increases faster than that for
normal diffusion, which can potentially lead to a higher search efficiency if used
properly in algorithms. For example, for Lévy flights, we have

$$q = \frac{3 - \lambda}{2}, \tag{26}$$

where $1 < \lambda \le 2$ is the exponent of the power-law approximation to Lévy distribution

$$L(w) \sim |w|^{-1-\lambda}, \tag{27}$$

where $\sim$ denotes to draw random numbers from a distribution on the right-hand side. In fact, cuckoo search and flower pollination algorithm have used such fat-tailed Lévy flights for global search.

4.7 No Free Lunch Theorems

Though there are many algorithms in the literature, different algorithms can have different advantages and disadvantages and thus some algorithms more suitable to solve certain types of problems than others. However, it is worth pointing out that there is no single algorithm that can be most efficient to solve all types of problems as dictated by the no-free-lunch (NFL) theorems [30]. Their rigorous proof requires some simplifications and assumptions. Two noticeable assumptions are (1) the set of points/solutions visited by an algorithm must be close under permutation and (2) the points found in the iteration history are non-revisiting in subsequent iterations. In addition, the performance measure is based on the averaged performance over all possible functions and problems.

An informal way of looking at the no-free-lunch theorem is as follows: For any univariate objective function $\phi(x)$ in a domain $[a, b]$, the mean of the function is

$$\mu = \frac{1}{b - a} \int_a^b \phi(x) dx. \tag{28}$$

Using the mean value theorem, we have

$$\frac{1}{b - a} \int_a^b \phi(x) dx = \phi(c), \quad c \in (a, b), \tag{29}$$

which suggests that the mean is a constant $\phi(c)$. If $\phi(x)$ can take any values and forms (including random values), we can treat $\phi(x)$ as a random variable. Then, $\mu = \phi(c)$ is the expectation. In a special case if all functions $\phi(x)$ are scaled to $[0, 1]$, then it can be expected that $\mu = 1/2$. Since μ is a constant, this means that the averaged objective landscape of all possible functions $\phi(x)$ becomes 'flat', which in turn means that there is no selection pressure for evolution of solutions. Consequently, it is no surprise that any algorithm (including a random search) can have equal efficiency. However, in practice, we do not need to solve all problems.

Even the no-free-lunch theorems hold under certain conditions, but these conditions may not be rigorously true for actual algorithms. For example, one condition for proving these theorems is the so-called no-revisiting condition. That is, the points during iterations form a path, and these points are distinct and will not be visited exactly again, though their nearby neighbourhood can be revisited. This condition is not strictly valid because almost all algorithms for continuous optimization will revisit some of their points in history. Such minor violation of assumptions can potentially leave room for free lunches. It has also been shown that under the right conditions such as co-evolution, certain algorithms can be more effective [31].

In addition, as we can see from Chap. 2 (of this book) by T. Joyce and J.M. Heremann on the review of no-free-lunch (NFL) theorems, free lunches may exist for a finite set of problems, especially those algorithms that can exploit the objective landscape structure and knowledge of optimization problems to be solved. If the performance is not averaged over all *possible* problems, then free lunches can exist. In fact, for a given finite set of problems and a finite set of algorithms, the comparison is essentially equivalent to a zero-sum ranking problem. In this case, some algorithms can perform better than others for solving a *certain type* of problems. In fact, almost all research papers published about comparison of algorithms use a few algorithms and a finite set (usually under 100 benchmarks), such comparisons are essentially ranking. However, it is worth pointing out that for a finite set of benchmarks, the conclusions (e.g., ranking) obtained can only apply for that set of benchmarks, they may not be valid for other sets of benchmarks and the conclusions can be significantly different. If interpreted in this sense, such comparison studies and their conclusions are consistent with NFL theorems.

5 Mathematical Analysis

The above analyses are mainly qualitative, but they can give some insight into the fundamental forms of search mechanisms and their role and properties. However, more insights can be gained by looking at algorithms mathematically. There are different ways of analyzing algorithms with mathematical rigour, but such analyses may have stringent assumptions that can be also unrealistic in some cases. Loosely speaking, mathematical framework can be dynamic systems, fixed-point theory, Markov chain theory, self-organization, filtering and others. To build a solid mathematical framework to analyze algorithms may require a long-term, multidisciplinary approach, thus we will not be too ambitious here. Instead, we would like to highlight a few approaches so as to inspire more research in this area.

5.1 Fixed-Point Theory

Numerical analysis often places emphasis on the iterative nature of an algorithm $A(\mathbf{x}_t)$ and tries to figure out how the solution $\mathbf{x}_t$ sequence may evolve as a pseudo-time iteration counter t. From the discussion in Sect. 2, we know that an algorithm can be written as

$$\mathbf{x}^{t+1} = A(\mathbf{x}^t, \mathbf{x}_*, p_1, \dots, p_K), \tag{30}$$

for $t \geq 0$. As the iteration continues, it is possible that

$$\lim_{t \to \infty} \mathbf{x}^{t+1} = \mathbf{x}_\infty, \tag{31}$$

where $\mathbf{x}_\infty$ is a fixed point. Obviously, if $\mathbf{x}_\infty$ does not exist, we can say the algorithm diverges. In a special case when $\mathbf{x}_\infty = \mathbf{x}_*$, we can safely say that the algorithm has found the true optimal solution. But if $\mathbf{x}_\infty \neq \mathbf{x}_*$, it may indicate that the iteration sequence becomes prematurely converged.

It is worth pointing out that the above solutions usually have some randomness and noise for metaheuristic algorithms, and, therefore, the above equation should be interpreted as the mean. That is

$$< \lim_{t \to \infty} \mathbf{x}^{t+1} > = < \mathbf{x}_\infty > . \tag{32}$$

The general fixed-point theory dictates how an iterative formula may evolve and lead to a fixed point in the search space [26]. It is worth pointing out that there may be multiple fixed points, and each iteration sequence may only find one fixed point at a time, though it is possible for some algorithms such as the firefly algorithm to find multiple fixed points simultaneously.

For a population of solutions in any nature-inspired algorithms, the population interact with each other and may lead to potentially multiple fixed points, depending on the algorithm dynamics of each algorithm. It can be expected that the ultimate $\mathbf{g}^*$ (not the best at each iteration) acts as a fixed point in PSO, while there are multiple fixed points in the firefly algorithm. Therefore, we can hypothesize that there is a single fixed point in BA, PSO, simulated annealing, FPA and bee algorithm, while multiple fixed points can exist in FA, CS, ACO and genetic algorithms if the conditions are right. However, it is not clear yet what these conditions can be and how to maintain these conditions in practice. In addition, these conditions may also be problem dependent. It is highly necessary to carry out more research in this area.

5.2 Dynamic System

The first analysis of PSO using a dynamic system theory was carried out by Clerc and Kennedy [8], and they linked the governing equations of PSO with the dynamic behaviour of particles under different parameter settings. Using matrix algebra, we can rewrite Eqs. (7) and (8) as the following dynamic system:

$$\begin{pmatrix} \mathbf{x}_i \\ \mathbf{v}_i \end{pmatrix}^{t+1} = \begin{pmatrix} 1 & 1 \\ -(\alpha\epsilon_1 + \beta\epsilon_2) & 1 \end{pmatrix} \begin{pmatrix} \mathbf{x}_i \\ \mathbf{v}_i \end{pmatrix}^{t} + \begin{pmatrix} 0 \\ \alpha\epsilon_1 \mathbf{g}^* + \beta\epsilon_2 \mathbf{x}_i^* \end{pmatrix}. \tag{33}$$

They did not use the eigenvalues of the above system because the matrices contain random numbers. Instead, they made additional assumptions and their analysis suggested that the PSO system is governed by the eigenvalues of a system matrix

$$\lambda_{1,2} = 1 - \frac{\gamma}{2} \pm \frac{\sqrt{\gamma^2 - 4\gamma}}{2}, \tag{34}$$

which leads a bifurcation at $\gamma = \alpha + \beta = 4$. This kind of analysis can indeed provide some insight into the working mechanism and main characteristics, but it may be difficult to provide a full picture of the system because of simplifications used in the analysis.

For the bat algorithm, we can rewrite the algorithmic equations as

$$\begin{pmatrix} \mathbf{x}_i \\ \mathbf{v}_i \end{pmatrix}^{t+1} = \begin{pmatrix} 1 & 1 \\ f_i & 1 \end{pmatrix} \begin{pmatrix} \mathbf{x}_i \\ \mathbf{v}_i \end{pmatrix}^{t} + \begin{pmatrix} 0 \\ f_i \end{pmatrix}, \tag{35}$$

where $f_i = f_{\min} + (f_{\max} - f_{\min})\beta$. A quick look seems to show that this system is very similar to (33), but we have not considered the variations of the pulse emission rate r and loudness A in the above equations. The similarity allows to do some similar analysis, but the incompleteness of this system to capture the full functionalities of the bat algorithm means that the analysis may not provide much useful information in practice. In principle, we can use the similar method to analyze other algorithms, however, it becomes difficult to extend to a generalized system. For example, in FA, CS and ACO, the nonlinearity makes it difficult to figure out the eigenvalues because the matrix will depend on the current solution, randomization and other factors. Furthermore, nonlinearity in algorithms such as FA also means that the characteristics can be much richer than simple linear dynamics such as PSO. Thus, this method may become intractable in practice, and some linearization and approximations may be needed.

5.3 Markov Chain Theory

As we mentioned earlier, algorithms can be considered as biased Monte Carlo since the solutions generated by an algorithm is a statistical sampling method such as Monte Carlo [12]. In general, Monte Carlo methods are closely associated with Markov chains. A Markov chain is a chain whose next state will depend only on the current state and the transition probability.

The solution set generated by an algorithm essentially form a system of Markov chains and thus it is natural that Markov chain theory can provide a generalized framework for analyzing nature-inspired algorithms. For example, Suzuki carried a simple analysis of genetic algorithms using Markov chain theory [28], while He et al. used a discrete-time Markov chain approach and have proved that the flower pollination algorithm can have guaranteed global convergence [17].

At an even higher level, we can view algorithm systems as systems of multiple, interacting Markov chains that evolve with time. For example, a generalized approach has been designed using a Markov chain Monte Carlo for global optimization [15]. In practice, this approach may converge slower than nature-inspired algorithms, and one of the reasons is that the selection mechanism is relatively weak in generalized Markov chain model. Despite this, this methodology can provide a quite general framework for optimization.

Mathematically speaking, Markov chain theory can provide some significant insight into the algorithms. The largest eigenvalue of a proper Markov chain is unity, while the second largest eigenvalue λ_2 of the transition probability matrix essentially controls the rate of convergence of the Markov chain. However, it is very challenging to find this eigenvalue in practice. Even some estimates can be difficult. Therefore, the information and insight we can obtain is limited in practice, which may also limit its practical use.

5.4 Computational Complexity

Computational complexity can be estimated for each algorithm, and this can help to under the computational efforts needed. For example, most algorithm such as PSO, FPA and bat algorithm, has a complexity of $O(nT)$ where n is the population size and T is the total number of iterations. Firefly algorithm has a computational complexity of $O(n^2T)$. Since n is relatively small compared with T, this usually does not increase the computation efforts substantially. In general, the computational complexity of nature-inspired algorithms is low.

On the other hand, the complexity of problems to be solved can be very high, even non-deterministic polynomial-time (NP) hard. For example, the well-known travelling salesman problems are NP-hard. Even nature-inspired algorithms are relatively simple, and studies have indicated that they can solve complex problems and even NP-hard problems. It still remains a bit mystery how such algorithms with low algorithmic complexity can solve highly complex problems and be able to find good solutions and even optimal solutions in practice.

5.5 Filter Theory

In telecommunications and signal processing, signals are processed and filtered so as to gain certain desired properties [37, 38, 45]. If we consider the solutions during iterations are signals, the action of an algorithm is to filter out undesired signals (solutions) and let desired signals (good solutions including the best solution) to pass through the system. As filtering occurs at multiple stages during iterations, the final well-filtered solutions are essentially the converged solution set.

In this sense, the design of algorithms is equivalent to the design of filters. Therefore, for linear algorithms such as PSO and bat algorithm, it is possible to use filter theory and signal processing techniques to analyze them. Though we have not seen such studies in the literature, it is no doubt future research will investigate this route further.

5.6   Self-organization

In the earlier discussion about self-organization, we have compared the similarity and differences between algorithms and self-organized systems, which was summarized in Table 2.

The self-organization theory is complex, which cannot directly transferred to analyze the characteristics of algorithms. Though comparison can provide some qualitative insight, it lacks crucial details about how the self-organized states in physical systems and/or algorithmic systems emerge, under what conditions and how quickly such converged states can be reached. Key information and properties may need to obtain by other means, unless new theory about self-organization emerges in the near future.

5.7   Statistical Analysis and Other Approaches

There are other approaches using statistical analysis and time series theory. However, it may not be easy to put some studies into a fixed category, though their results can be equally useful [29].

For example, Zaharie carried out a variance analysis of population and the effect of crossover in differential evolution [47]. The variance of the population $\text{var}(P_t)$ at time t is governed by

$$\text{var}(P_t) = Q(F, p_m, t) \, \text{var}(P_0), \tag{36}$$

where F is a constant and p_m is the effective mutation probability. This relationship links the variance of the current population P_t of n solutions with that of the initial population P_0 [47]. In addition, we have

$$Q(F, p_m, t) = \left[1 - 2F^2 p_m - \frac{p_m(2 - p_m)}{n} \right]^t, \tag{37}$$

which defines a critical value of F when $Q = 1$.

Indeed, variance analysis provides some information about the diversity of the population during the iterations. However, this kind of analysis requires that the system is linear, and thus it cannot directly be extended to analyze nonlinear systems such as the firefly algorithm.

5.8 Multidisciplinary Approach

The above analyses clearly indicate that a particular method can only look at the algorithms from one perspective. Different approaches and perspectives can provide different insights, potentially complementary to each other. Therefore, multidisciplinary approaches are needed to analyze algorithms from multiple angles so as to provide a fuller picture, including convergence analysis, stability analysis, sensitivity analysis and robustness analysis as well as parallelism in implementation.

It can be expected that a multidisciplinary framework can be formulated to analyze algorithms comprehensively, and such a framework requires all the above disciplines and methodologies to work together. It is hoped that this work can inspire more research in this area.

6 Conclusions

In this book chapter, we have attempted to analyze nature-inspired algorithms from both qualitative and quantitative perspectives. On the one hand, we have characterized algorithms using different components and the comparison of their role and properties with those of self-organized systems. On the other hand, we have tried to analyze algorithms using fixed point theory, dynamical system theory and Markov chain Monte Carlo framework. All these have provided some useful insights into the working mechanisms of algorithms.

Even with the above approaches, there are still many issues that need to be addressed in further research. Firstly, the mathematical framework need to be formulated in a more rigorous way so as to provide more detailed guides on how to analyze algorithms with mathematical rigour. Secondly, convergence analysis, stability analysis and sensitivity analysis should be closely linked with parameter tuning and parameter control of algorithm-dependent parameters of an algorithm [11]. Ideally, algorithms should be able to self-tune themselves to suit for a given type of problems [43]. Thirdly, robustness should also be studied so as to find suitable methods for solving optimization problems with noise and uncertainties in data and material properties. Fourthly, further studies should also focus on how to balance exploration and exploitation in algorithms so as to gain in-depth understanding. Finally, more large-scale case studies should be carried out to solve a diverse range of challenging problems in real-world applications.

References

1. Abdelaziz, A.Y., Ali, E.S., Abd Elazim, S.M.: Combined economic and emission dispatch solution using flower pollination algorithm. Int. J. Electr. Power Energy Syst. **80**(2), 264–274 (2016)
2. Alam, D.F., Yousri, D.A., Eteiba, M.B.: Flower pollination algorithm based solar PV parameter estimation. Energy Convers. Manage. **101**(2), 410–422 (2015)
3. Ashby, W.A.: Principles of the self-organizing system. In: Von Foerster, H., Zopf Jr., G.W. (eds.) Principles of Self-Organization: Transactions of the University of Illinois Symposium, pp. 255–278. Pergamon Press, London, UK (1962)
4. Bekdas, G., Nigdeli, S.M., Yang, X.S.: Sizing optimization of truss structures using flower pollination algorithm. Appl. Soft Comput. **37**(1), 322–331 (2015)
5. Blum, C., Roli, A.: Metaheuristics in combinatorial optimization: overview and conceptual comparision. ACM Comput. Survey **35**(2), 268–308 (2003)
6. Bonabeau, E., Dorigo, M., Theraulaz, G.: Swarm Intelligence: From Natural to Artificial Systems. Oxford University Press, Oxford (1999)
7. Boyd, S., Vandenberghe, L.: Convex Optimization. Cambridge Univeristy Press, Cambridge (2004)
8. Clerc, M., Kennedy, J.: The particle swarm–explosion, stability, and convergence in a multidimensional complex space. IEEE Trans. Evol. Comput. **6**(1), 58–73 (2002)
9. Dhivya, M., Sundarambal, M.: Cuckoo search for data gathering in wireless sensor networks. Int. J. Mobile Commun. **9**(4), 642–656 (2011)
10. Durgun, I., Yildiz, A.R.: Structural design optimization of vehicle components using cuckoo search algorithm. Mater. Test. **3**(3), 185–188 (2012)
11. Eiben, A.E., Smit, S.K.: Parameter tuning for configuring and analyzing evolutionary aglorithms. Swarm Evol. Comput. **1**(1), 19–31 (2011)
12. Fishman, G.S.: Monte Carlo: Concepts, Algorithms and Applications. Springer, New York (1995)
13. Fisher, L.: The Perfect Swarm: The Science of Complexity in Everyday Life. Basic Books (2009)
14. Gandomi, A.H., Yang, X.S., Alavi, A.H.: Cuckoo search algorithm: a metaheuristic approach to solve structural optimization problems. Eng. Comput. **29**(1), 17–35 (2013)
15. Ghate, A., Smith, R.: Adaptive search with stochastic acceptance probabilities for global optimization. Oper. Res. Lett. **36**(3), 285–290 (2008)
16. Goldberg, D.E.: Genetic Algorithms in Search, Optimisation and Machine Learning. Reading, Mass, Addison Wesley (1989)
17. He, X.S., Yang, X.S., Karamanoglu, M., Zhao, Y.X.: Global convergence analysis of the flower pollination algorithm: a discrete-time Markov chain approach. Procedia Comput. Sci. **108**(1), 1354–1363 (2017)
18. Holland, J.: Adaptation in Natural and Arficial Systems. University of Michigan Press, Ann Arbor, USA (1975)
19. Keller, E.F.: Organisms, machines, and thunderstorms: a history of self-organization, part two. Complexity, emergence, and stable attractors. Hist. Stud. Nat. Sci. **39**(1), 1–31 (2009)
20. Kennedy, J., Eberhart, R.C.: Particle swarm optimization. In: Proceedings of IEEE International Conference on Neural Networks, pp. 1942–1948. Piscataway, NJ (1995)
21. Moravej, Z., Akhlaghi, A.: A novel approach based on cuckoo search for DG allocation in distribution network. Electr. Power Energy Syst. **44**(1), 672–679 (2013)
22. Pavlyukevich, I.: Lévy flights, non-local search and simulated annealing. J. Comput. Phys. **226**(2), 1830–1844 (2007)
23. Reynolds, A.M., Rhodes, C.J.: The Lévy fligth paradigm: random search patterns and mechanisms. Ecology **90**(4), 877–887 (2009)
24. Rodrigues, D., Silva, G.F.A., Papa, J.P., Marana, A.N., Yang, X.S.: EEG-based person identificaiton through binary flower pollination algorithm. Expert Syst. Appl. **62**(1), 81–90 (2016)

25. Storn, R., Price, K.: Differential evolution: a simple and efficient heuristic for global optimization over continuous spaces. J. Global Optim. **11**(4), 341–59 (1997)
26. Süli, E., Mayer, D.: An Introduction to Numerical Analysis. Cambridge University Press, Cambridge (2003)
27. Surowiecki, J.: The Wisdom of Crowds. Anchor Books (2004)
28. Suzuki, J.A.: A Markov chain analysis on simple genetic algorithms. IEEE Trans. Sys. Man Cybern. **25**(4), 655–9 (1995)
29. Villalobos-Arias, M., Colleo, C.A.C., Hernández-Lerma, O.: Asypmotic convergence of metaheuristics for multiobjective optimization problems. Soft Comput. **10**(11), 1001–5 (2005)
30. Wolpert, D.H., Macready, W.G.: No free lunch theorem for optimization. IEEE Trans. Evol. Comput. **1**(1), 67–82 (1997)
31. Wolpert, D.H., Macready, W.G.: Coevolutionary free lunches. IEEE Trans. Evol. Comput. **9**(6), 721–735 (2005)
32. Yang, X.S.: Firefly algorithm, stochastic test functions and design optimisation. Int. J. Bio-Inspired Comput. **2**(2), 78–84 (2010)
33. Yang, X.S.: Engineering Optimization: An Introduction with Metaheuristic Applications. Wiley, Hoboken, NJ (2010)
34. Yang, X.S.: A new metaheuristic bat-inspired algorithm. In: Nature-Inspired Cooperative Strategies for Optimization (NICSO 2010), pp. 65–74, SCI vol. 284. Springer (2010)
35. Yang, X.S.: Bat algorithm for multi-objective optimisation. Int. J. Bio-Inspired Comput. **3**(5), 267–274 (2011)
36. Yang, X.S.: Cuckoo Search and Firefly Algorithm: Theory and Applications. Studies in Computational Intelligence, vol. 516. Springer (2014)
37. Yang, X.S.: Nature-Inspired Optimization Algorithms. Elsevier Insight, London (2014)
38. Yang, X.S., Chien, S.F., Ting, T.O.: Bio-Inspired Computation in Telecommunications. Morgan Kaufmann, Waltham (2015)
39. Yang, X.S., Deb, S.: Cuckoo search via Lévy flights. In: Proceedings of World Congress on Nature & Biologically Inspired Computing (NaBic 2009), Coimbatore, India, pp. 210–214. IEEE Publications, USA (2009)
40. Yang, X.S., Deb, S.: Engineering optimization by cuckoo search. Int. J. Math. Model. Num. Optim. **1**(4), 330–343 (2010)
41. Yang, X.S., Deb, S.: Multiobjective cuckoo search for design optimization. Comput. Oper. Res. **40**(6), 1616–1624 (2013)
42. Yang, X.S., Deb, S.: Cuckoo search: recent advances and applications. Neural Comput. Appl. **24**(1), 169–174 (2014)
43. Yang, X.S., Deb, S., Loomes, M., Karamanoglu, M.: A framework for self-tuning optimization algorithm. Neural Comput. Appl. **23**(7–8), 2051–2057 (2013)
44. Yang, X.S., Karamanoglu, M., He, X.S.: Flower pollination algorithm: a novel approach for multiobjective optimization. Eng. Optim. **46**(9), 1222–1237 (2014)
45. Yang, X.S., Papa, J.P.: Bio-Inspired Computation and Applications in Image Processing. Academic Press, London (2016)
46. Yildiz, A.R.: Cuckoo search algorithm for the selection of optimal machine parameters in milling operations. Int. J. Adv. Manuf. Technol. **64**(1), 55–61 (2013)
47. Zaharie, D.: Influence of crossover on the behaviour of the differential evolution algorithm. Appl. Soft Comput. **9**(3), 1126–38 (2009)

A Review of No Free Lunch Theorems, and Their Implications for Metaheuristic Optimisation

Thomas Joyce and J. Michael Herrmann

Abstract The No Free Lunch Theorem states that, averaged over all optimisation problems, all non-resampling optimisation algorithms perform equally well. In order to explain the relevance of these theorems for metaheuristic optimisation, we present a detailed discussion on the No Free Lunch Theorem, and various extensions including some which have not appeared in the literature so far. We then show that understanding the No Free Lunch theorems brings us to a position where we can ask about the specific dynamics of an optimisation algorithm, and how those dynamics relate to the properties of optimisation problems.

Keywords No Free Lunch (NFL) · Optimisation · Search · Metaheuristics

1 Introduction

In science, computing, and engineering it is common to encounter situations in which a function can be evaluated on any inputs, and one wants to find the inputs that produce the highest (or equivalently, lowest) output. When the number of possible inputs is too large for it to be feasible to simply try them all, then one must instead rely on some strategy for finding a satisfactory solution. There are many such strategies, and here we look at a subset of these strategies called metaheuristic optimisers. Metaheuristic optimisers can be though of as simple, broadly applicable strategies for finding inputs to functions that result in desirable outputs.

There is a large literature on metaheuristic optimisers, and much work goes in to developing more effective optimisation algorithms and refining existing approaches. However, a fundamental result in optimisation, the No Free Lunch Theorem, shows that, all non-resampling optimisation algorithms perform equally, averaged over all problems. Understanding this result is of central importance for anyone working in

T. Joyce (✉) · J.M. Herrmann
The University of Edinburgh, Edinburgh, Scotland
e-mail: t.joyce@ed.ac.uk

J.M. Herrmann
e-mail: michael.herrmann@ed.ac.uk

© Springer International Publishing AG 2018
X.-S. Yang (ed.), *Nature-Inspired Algorithms and Applied Optimization*,
Studies in Computational Intelligence 744, https://doi.org/10.1007/978-3-319-67669-2_2

27

optimisation, and in this chapter we give a detailed overview of the No Free Lunch (NFL) Theorems for optimisation, covering both the original theorems, and a number of extensions and refinements. We then examine how the results should influence research into meatheuristic optimisation.

2 Preliminaries

We start by introducing the central definitions and notation used throughout the chapter, most importantly functions and data. Functions are used as formal representations of the problems we want to solve, and data allows us to represent partial knowledge of these functions.

2.1 Functions

In this chapter we consider functions $f : X \to Y$, where X and Y are finite sets with $|X| = n$ and $|Y| = m$. We assume that neither X nor Y are empty. We denote the set of all such functions $\mathscr{F}$, with $|\mathscr{F}| = m^n$.

As we need to represent functions frequently throughout the chapter we introduce a concise representation scheme. Without loss of generality assume that $X = \{1, 2, \dots, n\}$, then for any $f \in \mathscr{F}$ we can write an ordered list of y values $y_1, y_2, \dots, y_n$ where $y_i = f(i)$. This ordered list of y values uniquely and fully describes f. When there is no ambiguity (e.g. in the case where $Y = \{0, 1\}$) we will sometimes omit the commas between elements.

Example 1 (Function) Let $X = \{1, 2, 3\}$ and $Y = \{0, 1\}$. We can then write out all $f \in \mathscr{F}$: $000, 001, 010, 011, 100, 101, 110, 111$. For example, 110 corresponds to the case where $f(1) = 1, f(2) = 1$ and $f(3) = 0$.

In the optimisation literature there are many terms used to refer to the function being optimised. It is called variously the "cost function", "fitness function", "target function", "objective function", "test function", "problem" and even simply "the function". Here we generally use (problem) function. We now introduce some extensions to the basic function.

Definition 1 *(Bijection)* A function $f : X \to Y$ is a bijection iff $\forall x_1, x_2 \in X$, $x_1 \neq x_2 \implies f(x_1) \neq f(x_2)$ and $\forall y \in Y$, $\exists x \in X$ such that $f(x) = y$.

Definition 2 *(Permutation)* A permutation is a bijection from a set onto itself. That is, $\phi : X \to X$ is a permutation iff ϕ is a bijection.

Definition 3 *(Function Permutation)* Let ϕ be a permutation $\phi : X \to X$, and let $f : X \to Y$ be an arbitrary function. We call f_ϕ a function permutation where we

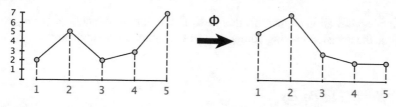

Fig. 1 Function permutation example, $f = 2, 5, 2, 3, 7$ (left) is permuted by $\phi = 2, 5, 4, 1, 3$, resulting in $f_\phi = 5, 7, 3, 2, 2$ (right)

define $f_\phi(x) = f(\phi(x))$. A function permutation can be thought of as re-ordering the output values. The resulting function has the same outputs, but they now correspond to different inputs.

Example 2 (Function Permutation) Let $f : \{1, 2, 3, 4, 5\} \rightarrow \{1, 2, 3, 4, 5, 6, 7\}$ with $f = 2, 5, 2, 3, 7$ and let ϕ be a permutation $\phi = 2, 5, 4, 1, 3$, then $f_\phi = 5, 7, 3, 2, 2$. See Fig. 1 for a graphical representation of this example.

2.2 Data

When optimising a function $f : X \rightarrow Y$ we assume we know X and Y, but only know some (or none) of the values $y = f(x)$. We represent such partial knowledge as a function $d : X \rightarrow Y \cup \{?\}$ which we call data. The question mark "?" represents an x value for which the $f(x)$ value is unknown. For all x values where we know $y = f(x)$ let $d(x) = f(x)$, for all other values of x we let $d(x) = ?$. In other words, if for some $x \in X$ we have $d(x) = ?$ then we currently do not know $f(x)$. Let $\mathcal{D}$ be the set of all possible data for functions in $\mathcal{F}$. It follows that $|\mathcal{D}| = (m + 1)^n$. We refer to the data where $\forall x \in X, \; d(x) = ?$ as "no data", as this represents the situation in which we know nothing about the problem function.

Example 3 (Data) Let $X = \{1, 2, 3, 4\}$ and $Y = \{0, 1\}$. Assume we have partial knowledge of some $f \in \mathcal{F}$. In particular, we know that $f(2) = 1$ and $f(4) = 1$. We can represent this data as a function d where $d(1) = ?, \; d(2) = 1, \; d(3) = ?$ and $d(4) = 1$. As discussed in Sect. 2.1, this can be represented more succinctly as: ?1?1. Note that, assuming the data is accurate, there are four possible functions consistent with the data: 0101, 0111, 1101 and 1111.

3 Optimisation Algorithms

We have introduced functions, and data, to represent the problems we want to solve, and partial knowledge of these problems, respectively. We now introduce a representation of optimisation algorithms themselves. We first introduce a *sampling policy*, which is the decision making component of the optimisation algorithm.

3.1 Sampling Policy

The situation discussed throughout this chapter is the following: there is a function $f : X \rightarrow Y$ which can be evaluate at any $x \in X$, and the current knowledge of the f is represented by data d. A sampling policy is a rule for deciding where to sample next, based on the current data. In other words, given the current data it specifies which of the potentially numerous x we should evaluate f at next.

Definition 4 (*Sampling Policy*) A sampling policy s is a function $s : \mathcal{D} \rightarrow X$, where $\mathcal{D}$ is the set of all possible data, as defined in Sect. 2.2.

Example 4 (*Sampling Policy*) Let $X = \{1, 2\}$ and $Y = \{0, 1\}$. An example sampling policy s is: $s(??) = 1$, $s(0?) = 2$, $s(1?) = 1$, $s(?0) = 1$, $s(?1) = 1$. We do not need to specify $s(00)$, $s(01)$, $s(10)$, or $s(11)$, as we are only concerned with the behaviour of the sampling policy when some points remain unsampled.

Definition 5 (*Non-Repeating Sampling Policy*) A non-repeating sampling policy s is a function $s : \mathcal{D} \rightarrow X$ s.t. if $s(d) = x$ then either $d(x) = ?$ or $|d| = m$.

Example 5 (*Non-Repeating Sampling Policy*) Let $X = \{1, 2\}$ and $Y = \{0, 1\}$. Then the set of possible data $\mathcal{D} = \{??, ?0, ?1, 0?, 1?, 10, 01, 00, 11\}$ and $s(??) = 1$, $s(?0) = 1$, $s(?1) = 1$, $s(0?) = 2$, $s(1?) = 2$, is a non-repeating sampling policy.

Here, we will consider deterministic non-repeating sampling policies, unless otherwise stated, and when there is no ambiguity we will just call them sampling policies.

3.2 Optimisation Algorithms

So far we have defined sampling policies, which decide where to sample given data. However, we can add to our data by evaluating the problem function f at some x and updating d by setting $d(x) = f(x)$. Essentially, an optimisation algorithm is the repeated use of a sampling policy, adding to our data each time we make a sample, until a termination condition is reached. For the time being we fix our termination condition to: terminate iff we have sampled f at every $x \in X$. Given this fixed termination condition an optimisation algorithm is fully specified by a choice of sampling policy.

Definition 6 (*Optimisation Algorithm*) An optimisation algorithm A based on sampling policy s is iterated use of that sampling policy and data updating:

1. Set d to be the initial data (generally "no data", but see Sect. 3.3).
2. If there are no unsampled $x \in X$, then terminate.
3. Sample at $x = s(d)$ and add the result to the data d, i.e. set $d(x) = f(x)$.
4. Go to step 2.

3.3 On-Policy and Off-Policy Behaviour

One useful observation on the behaviour of deterministic optimisation algorithms is the existence of what can be thought of as on-policy and off-policy behaviour. This distinction is important for later proofs, and so we make it clear here.

The on-policy behaviour of an algorithm is the the behaviour on inputs potentially seen when the algorithm is run until termination *starting with no information about* f. The on-policy behaviour is generally not the full behaviour, as there are data inputs that will never be seen in the normal execution of the algorithm, regardless off the function f being sampled. This is clarified in the following example:

Example 6 (On-Policy Behaviour) Consider the sampling policy described in Example 5. $s(??) = 1$, and so an optimisation algorithm using this sampling policy will initially sample $f(1)$, and thus the data ?0 and ?1 will never be seen, as those data are the possible results from evaluating $f(2)$ first. It is not that $f(2)$ will never be evaluated, rather that if we follow the policy, $f(2)$ will never be evaluated *first*.

Off-policy behaviour is the sampling policy restricted to exactly those data inputs that do not appear in the on-policy behaviour. Thus together the off-policy and on-policy behaviours describe the full behaviour. Broadly speaking, the on-policy behaviour is all the behaviour that is possible in the normal running of the optimiser, and off-policy behaviour is that that results from starting the optimisation algorithm with some initial (off-policy) knowledge of the problem function. In this chapter we restrict attention to on-policy behaviour unless explicitly stated.

3.4 Representing Optimisation Algorithm Behaviour

When discussing optimisation algorithms it is often helpful to consider their behaviour represented as a behaviour graph, which we now define:

Definition 7 (*Behaviour Graph*) An optimisation algorithm's behaviour can be represented as a directed graph, in which nodes represent data and the edges show all potential transitions between data resulting from sampling the problem function in accordance with the optimiser's sampling policy. We call such a graph the behaviour graph of the optimiser. See Fig. 2 for an example of a behaviour graph.

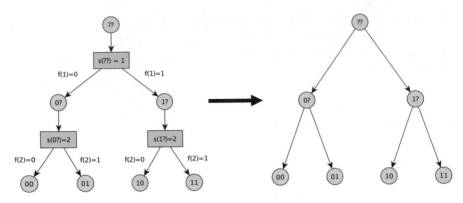

Fig. 2 Left: The on-policy behaviour of the optimisation algorithm using the sampling policy from Example 5. The yellow circular nodes represent data, the blue rectangular nodes show the sampling policy's decision based on data. Right: The same tree shown more compactly by removing the explanatory details. The tree still contains the same information and is the format we will generally use, for compactness

We now show that if we restrict attention to on-policy behaviour of deterministic, non-resampling optimisation algorithms, then the behaviour graph is in fact a tree. This is a key observation and is used variously throughout the rest of the chapter.

Theorem 1 (Optimisation Algorithm Tree Representation) *A deterministic non-resampling optimisation algorithm's on-policy behaviour can be uniquely represented as a directed graph. In fact this directed graph is a balanced tree.*

Proof From the definition of on-policy behaviour we start the optimisation with no data about the function, which we represent as the root node of the tree. The algorithm's sampling policy determines which x should be sampled given no data, say x_1. When the algorithm performs the sample at x_1 there are $|Y| = m$ possible results, and each of these results necessarily leads to different data (as the data is just a description of the result). Thus, we can potentially transition to any of m new data nodes by appending the result to the data, setting $d(x_1) = f(x_1)$.

At this point, either the node we are at represents data with no unknown values, in which case the algorithm halts and we are at a leaf, or the node contains at least one value for which $d(x) = ?$. In this second case the sampling policy will select one of these x to be sampled (as it is non-resampling), and again m possible results exist, we follow one of the possible edges (dependent on the results of the sample) and then repeat the process until we do eventually reach a leaf.

We now show that all paths do eventually lead to a leaf. Whenever an x is sampled an unknown $f(x)$ becomes known. As the algorithm terminates exactly when all $f(x)$ are known, and there are only $n = |X|$ unknown $f(x)$ at the start, and as, because the sampling policy is non-resampling, the algorithm only ever samples x for which $f(x)$ is unknown, then the algorithm necessarily terminates after exactly n samples, and we reach a leaf.

We now show that no two paths arrive at the same node. First we note that all paths through the tree start at the root node. Assume we have two paths that at some point separate. If they were to rejoin they must both eventually arrive at the same data set. However, the very fact that they separated means that for some x they produced different $f(x)$ values. Thus, their data can never be the same, and they will never rejoin at a node. It follows that the graph is a tree.

Finally, as every path necessarily terminates after exactly n steps, and every non-leaf node leads to exactly the same number m of immediate children, it follows that the tree is balanced.

Theorem 2 (Tree Representation Details) *The tree representing the on-policy behaviour of an optimiser for functions mapping $X \to Y$ where $|X| = n$ and $|Y| = m$ has $n + 1$ layers. If we label these layers $0, 1, \ldots, n$ starting from the root then layer i contains m^i nodes, and the tree contains $1 + m + m^2 + \cdots + m^n$ nodes in total. The final layer consists of m^n leaves with exactly one representing each of the m^n functions $f \in \mathscr{F} = Y^X$.*

Proof Each node represents particular data $d \in \mathscr{D}$. The root node is always d s.t. $\forall\, x \in X$, $d(x) = ?$. Call this root node layer 0. Each node on the i-th layer, for $i \in \{0, \ldots, n\}$, will represent data with exactly $n - i$ unsampled x values. Thus, the n-th layer will consist of leaves, and will be the final layer.

Every non-leaf node leads to exactly $m = |Y|$ immediate children. We have also seen in the proof of Theorem 1 that no node is the child of more that one node. Thus, as in layer zero there is 1 node in layer 1 there will be, m, in layer 2 there will be m^2 nodes, and in layer i there will be m^i nodes for i ranging from 0 to n.

On the n-th layer there are m^n leaves, each containing data describing a different $f \in \mathscr{F}$. As $|\mathscr{F}| = m^n$ every $f \in \mathscr{F}$ must correspond to a leaf.

In Fig. 3 we show example on-policy trees for three different domains. It can be seen that even for $|X| = 4$ and $|Y| = 2$ the tree becomes fairly large. Although trees are possible for any finite X and Y, we will generally not be able to show them explicitly.

3.5 Paths down Trees

We have seen that the on-policy behaviour of an optimisation algorithm can be represented as a balanced tree (Theorem 1). We now show that running an optimiser on a particular function corresponds to taking a particular path down the optimiser's tree. Figure 4 provides a graphical example of these paths down trees.

Theorem 3 (Paths Down Trees) *When the on-policy behaviour of an optimisation algorithm is represented as a rooted tree, then paths down that tree biject with the functions $f \in \mathscr{F}$, and the path shows the choices that the algorithm will make when optimising the corresponding f.*

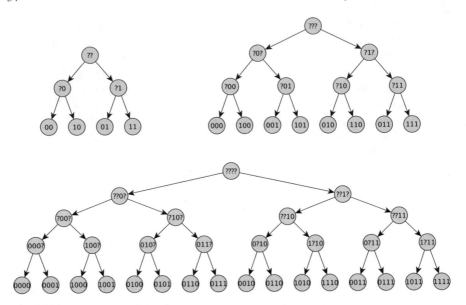

Fig. 3 The example policies, represented as trees. Top left: A policy for functions from $X = \{1, 2\}$ to $Y = \{0, 1\}$. Top right: A policy for functions from $X = \{1, 2, 3\}$ to $Y = \{0, 1\}$. Bottom: A policy for functions from $X = \{1, 2, 3, 4\}$ to $Y = \{0, 1\}$

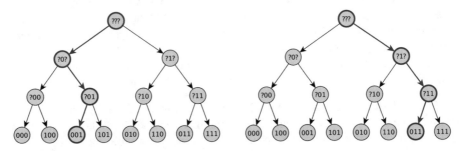

Fig. 4 Two example paths taken by the optimisation algorithm down its behaviour tree. The left path shows the route when the algorithm is run on $f = 001$ and the right path shows the behaviour on $f = 011$

Proof Recall that every leaf is the result of optimising one $f \in \mathscr{F}$. The existence of the bijection follows directly from the fact that the graph structure is a tree, and there is thus only a single path from the route to each leaf.

4 No Free Lunch

The original No Free Lunch (NFL) theorem for optimisation [1, 2] states that no optimisation algorithm can outperform any other under any metric over all problems. The theorem first appears in Wolpert and Macready's 1995 paper "No Free Lunch Theorems for Search" [1], but became more widely known through the same authors 1997 paper, "No Free Lunch Theorems for Optimization" [2]. There are many formulations of the result in the literature, with different emphasis. Here we present a representative selection, starting with Wolpert and Macready's own characterisation:

1. The average performance of any pair of algorithms across all possible problems is identical [2].
2. For all possible metrics, no search algorithm is better than another when its performance is averaged over all possible discrete functions [3].
3. On average, no algorithm is better than random enumeration in locating the global optimum [4].
4. The histogram of values seen, and thus any measure of performance based on it is independent of the algorithm if all functions are considered equally likely [5].
5. No algorithm performs better than any other when their performance is averaged over all possible problems of a particular type [6].
6. With no prior knowledge about the function $f : X \to Y$, in a situation where any functional form is uniformly admissible, the information provided by the value of the function in some points in the domain will not say anything about the value of the function in other regions of its domain [7].

As this selection shows, NFL allows a broad range of assertions. We will now formally state and prove the NFL theorem, after two preliminary definitions.

Definition 8 (*Traces*) The trace of an optimisation algorithm A running on a function f is the ordered list of (x, y) pairs sampled (where $y = f(x)$). We write $T_A(f)$ for the trace of algorithm A running on function f. We also define $T_A^{(k)}(f)$ to be the trace after k function evaluations. As we restrict attention to non-resampling optimiser, it follows that if n is the size of the domain of f then $T_A^{(n)}(f) = T_A(f)$. We call $T_A(f)$ the (full) trace and $T_A^{(k)}(f)$ a partial trace for any $k < n$. Let $\mathcal{T}_A$ be the set of all the traces that algorithm A produces on functions mapping X (where $|X| = n$) to Y, that is:

$$\mathcal{T}_A = \cup_{f \in \mathcal{F}} \cup_{k=0}^{n} T_A^{(m)}(f)$$

We also define a trace of just the inputs $T_A(f)_X$ and a trace of just the outputs $T_A(f)_Y$ as ordered lists of just the x and y values respectively from the full trace. We call these the input trace and the output trace. The output trace is sometimes called the performance vector or range trace in the literature. Similarly, we define $\mathcal{T}_{A,X}$ and $\mathcal{T}_{A,Y}$ all possible input traces and all possible output traces, respectively.

Example 7 (*Traces*) Let $X = \{1, 2, 3\}$ and $Y = \{0, 1\}$ and let A be an optimisation algorithm using sampling policy s, with $s(???) = 2$, $s(?0?) = 3$ and $s(?1?) = 3$

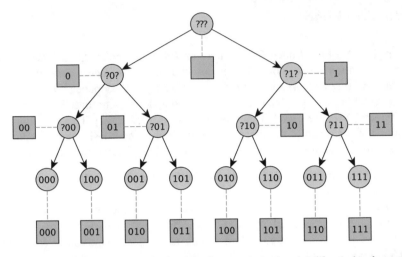

Fig. 5 The yellow nodes show the behaviour tree for an optimisation algorithm (using the sampling policy defined in Example 7) on functions between $\{1, 2, 3\}$ and $\{0, 1\}$. The green squares attached to each yellow node show the corresponding output trace at that point. If we let A be the algorithm represented by the tree, then it can see that, for example, $T_{A(010),Y} = 100$ by reading the trace at the 010 leaf

(A is shown graphically in Fig. 5). Let $f : X \to Y$ with $f = 010$. Then $T_A(f) = \{(2, 1), (3, 0), (1, 0)\}$, $T_A^{(2)}(f) = \{(2, 1), (3, 0)\}$, $T_A(f)_X = \{2, 3, 1\}$ and $T_A(f)_Y = \{1, 0, 0\}$, $\mathscr{T}_{A,Y} = \{\{0, 0, 0\}, \{0, 0, 1\}, \{0, 1, 0\}, \{0, 1, 1\}, \{1, 0, 0\}, \{1, 0, 1\}, \{1, 1, 0\}, \{1, 1, 1\}, \{0, 0\}, \{0, 1\}, \{1, 0\}, \{1, 1\}, \{0\}, \{1\}, \{\}\}$.

As noted in [8] optimisation algorithms have no "intrinsic purpose" and their behaviour needs to be qualified by some external metric. A metric provides a way to evaluate an algorithm's performance. We now make this idea of a metric precise.

Definition 9 (*Optimisation Metric*) An optimisation metric M is a function that maps output traces to $\mathbb{R}$. A metric can be thought of as assigning a value, or score, to an output trace, $M : \mathscr{T}_Y \to \mathbb{R}$.

An alternative but equivalent way to think of a metric is as a function M^* from an algorithm, a function, and a number of samples $t \in \mathbb{N}$, to a rating of how well the algorithm performs on that function after that many samples, that is, $M^* : \mathscr{A} \times \mathscr{F} \times \mathbb{N} \to \mathbb{R}$ where $M^*(A, f, t) = M(T_{A(f),Y}^{(t)})$.

Example 8 (*Optimisation Metrics*) A simple metric for measuring minimisation performance is a function that returns the minimum Y value in the trace (i.e. the smallest y value sampled so far). In fact, although simple, this metric is often used to compare optimisers, especially in cases where the true global minimum of the function is unknown.

We are now in a position to formally state and prove NFL. We follow the style of proof used by English [8–10], as this is in our opinion clearest, and naturally yields various generalisations to the theorem. It involves explicit reference to the representation of optimisers as trees. Another particularly clear approach to the proof is using the "fundamental matrix" method from [11]. We have chosen a tree based method because, in our opinion, it makes the sequential decision aspect of the algorithm producing a trace (see Definition 8) more explicit. However, we strongly recommend [11] to any readers wanting an alternative approach.

Theorem 4 (NFL) *All optimisation algorithms produce the same set of traces when run over all possible functions between two finite sets. More formally, let* $Y^X = \{f \mid f : X \to Y\}$, *where X and Y are arbitrary finite sets. For all optimisation algorithms A, B,* $\{T_{A(f),Y} \mid f \in Y^X\} = \{T_{B(f),Y} \mid f \in Y^X\}$.

Proof As we have shown in Sect. 3.4, the behaviour of an optimisation algorithm is uniquely representable as a tree. This tree has $|Y|^{|X|}$ leaves and the trace at each leaf is unique. However, by a counting argument there are only $|Y|^{|X|}$ possible distinct traces of length $|X| = n$. Therefore, every optimisation algorithm produces the same set of traces, namely every possible trace exactly once.

It is worth noting that the proof above is very succinct, this is in part due to the reference to the proof that the behaviour of an optimisation algorithm is uniquely representable as a tree, but also partly a result of the decision tree proof style.

5 Basic No Free Lunch Extensions

We now present some additional results and observations that are of interest in themselves, and will also be used in refinements of NFL below. Theorem 4 showed that every algorithm is equivalent when full traces are considered. We first generalise to the case where we stop the optimisation after k steps, then we generalise further to the case of arbitrary stopping conditions. After this we give a proof that the NFL result still holds if we allow stochastic sampling policies.

5.1 k-Step No Free Lunch

In real applications, optimisation algorithms are not usually run until the entire domain has been sampled. In the original no free lunch papers by Wolpert et al. [1, 2] they show that the no free lunch theorem still applies if algorithms are run for some fixed number of steps. We restate this theorem here and prove it using tree representations. We first define a multi-set, which is used in the theorem.

Definition 10 (*Multi-set*) A multi set is a set in which elements can occur multiple times. Another way to think of a multi-set is as a set in which each element has an associated count.

Theorem 5 (*k*-step NFL) *All optimisation algorithms produce the same set of traces when run over all possible functions between two finite sets for k steps. More formally, let* $Y^X = \{f \mid f : X \to Y\}$*, where X and Y are arbitrary finite sets. For any optimisation algorithms A, B,* $\{T_A^{(k)}(f)_Y \mid f \in Y^X\} = \{T_B^{(k)}(f)_Y \mid f \in Y^X\}$*. In fact they produce the same multi-set, in that every possible trace appears the same number of times, with the exact value depending on k and* $|X|$*.*

Proof If we prune the full behaviour tree after *k* steps the resulting tree will have $|Y|^k$ leaves. Each leaf has corresponds to a unique output trace and these traces are of length *k*. Thus, as there are only $|Y|^k$ possible output partial traces of length *k*, each partial trace must be present exactly once in the leaves. This is the case for all optimisation algorithms.

Because we have pruned the tree after *k* steps, each leaf in the pruned tree (and thus each partial output trace) will result when optimising multiple problem functions. However, as the full behaviour tree was a balanced tree, it follows that each partial trace will be obtain the same number of times when all possible functions are considered.

We defined the behaviour tree for an optimiser in Sect. 3.4. We now define a similar but distinct tree representation, the trace tree. Essentially the behaviour tree show both the *x* and *y* values of the algorithms samples, the trace tree in contrast only shows the *y* values.

Definition 11 (*Trace Tree*) Let *A* be an optimisation algorithm for functions $f : X \to Y$. The trace tree for *A* is the behaviour tree with the nodes labelled with the trace, rather than the data.

An example trace tree is given in Fig. 6. It is the trace tree of the optimisation algorithm in Example 5, the behaviour of which is shown in Fig. 5. An example of a trace tree for a restricted number of steps is shown in Fig. 7.

We now show that trace tree only depends on the range and the size of the domain of the functions being optimised. The detail that we lose when we switch from behaviour trees to trace trees is exactly the detailed that differentiated the optimisers.

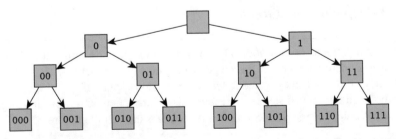

Fig. 6 The behaviour tree showing only the trace values. When considering just the traces, all algorithms have the same behaviour tree (see Theorem 6). What differentiates algorithms is that for different algorithms a given function will produce different *paths* in this tree

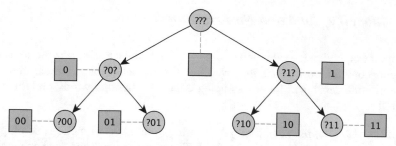

Fig. 7 The tree in Fig. 6 after $k = 2$ steps. As $|Y| = |\{0,1\}| = 2$ there are $|Y|^k = 2^2 = 4$ leaves, each having one of the four possible two bit traces

Theorem 6 (Identical Trace Trees) *All optimisation algorithms for functions mapping X to Y produce the same trace tree. In fact, all optimisation algorithms for functions mapping Z to Y also produce the same trace tree, as long as $|Z| = |X|$.*

Proof Consider an optimisation algorithm A for functions $f : X \to Y$. We will show that A's trace tree does not depend on any of the specific details of A, and thus would be the same for any optimisation algorithm for functions $f : X \to Y$.

From the definition of a trace tree we know that the algorithm's trace tree has the same structure as the algorithm's behaviour tree. In particular, as we are only considering non-resampling optimisers we know that the trace tree will have $|X| + 1$ layers, and that each node in the tree will be either a leaf or the parent of exactly $|Y|$ other nodes.

Now we simply observe that if we consider a non-leaf node in the trace tree then we can detail the node's children without knowledge of A. Suppose the node we are considering has trace $y_1, y_2, \ldots, y_k$, then its $|Y|$ children will have the traces $y_1, y_2, \ldots, y_k, y$ for each $y \in Y$.

Next we note that the only influence the domain X has on the trace tree is in setting the number of layers to be $|X| + 1$, thus it is only the size of the domain that is important and the trace tree will be the same for any domain of that size.

Corollary 1 *The leaves of the trace tree are all possible problem functions.*

This was shown in Theorem 2 but is restated as a corollary above, as it is important in the proofs to follow.

Continuing our generalisations of NFL to early stopping scenarios, we now consider the more general situation in which the optimisation process can terminate based on the results so far, rather than simply after a fixed number of steps. As we will see, a no free lunch result still pertains.

5.2 Stopping Condition No Free Lunch

A stopping condition is a rule for when to stop the optimiser. We make the concept of stopping condition formal and then we show that the set of traces over all functions is algorithm independent for any stopping condition. In other words, an NFL result still holds.

Definition 12 (*Stopping condition*) A stopping condition is a function mapping output traces to either 0 or 1, $S : \mathscr{T}_Y \to \{0, 1\}$. The stopping condition can be thought of as looking at the results of the optimisation algorithm so far and deciding whether to continue searching. After each function evaluation an optimisation algorithm using a stopping condition S evaluates S on its current output trace T_Y and then stop iff $S(T_Y) = 1$.

Using the above definition we now state and prove a no free lunch result for arbitrary stopping conditions.

Theorem 7 (stopping-condition NFL) *For any stopping condition S, all optimisation algorithms produce the same set of traces when run over all possible functions between two finite sets. More formally, let $Y^X = \{f \mid f : X \to Y\}$, where X and Y are arbitrary finite sets. For any optimisation algorithms A, B, $\{T_{A|S}(f)_Y \mid f \in Y^X\} = \{T_{B|S}(f)_Y \mid f \in Y^X\}$, where $T_{A|S}(f)_Y$ is the output trace generated by algorithm A using stopping condition S running on function f.*

Proof From Theorem 1, we know that the behaviour of an algorithm can be represented by a tree. A stopping condition can be seen as a pruning of this tree. Whereas in the k-step NFL proof we cut each branch after the same number of steps, a stopping condition can potentially prune branches after differing numbers of steps.

As we have seen above in Theorem 6, all algorithms produce the same trace tree (see Fig. 6). The stopping condition leads to a pruning of this trace tree, specifically we prune all the children from any leaf with a trace T such that $S(T) = 1$.

A particular stopping condition, then, leads to a particular pruning. As the trace tree and the pruning are both algorithm independent the resulting pruned trace tree will be the same for all algorithms, and thus its leaves (the final traces produced) will be the same. It follows that, for any optimisation algorithms A, B, $\{T_{A|S}(f)_Y \mid f \in Y^X\} = \{T_{B|S}(f)_Y \mid f \in Y^X\}$.

5.3 Stochastic No Free Lunch

We now state and prove the basic no free lunch result for stochastic optimisation algorithms. The stochastic case was also considered in Wolpert and Macready's original papers [1, 2].

Definition 13 (*Stochastic Optimiser*) A stochastic optimisation algorithm is an optimiser that uses a stochastic sampling policy to choose where it samples. Previously a sampling policy was a function mapping $s : \mathscr{D} \to X$. In the case of a stochastic optimiser the sampling policy is instead a function $s : \mathscr{D} \to P_X$ where P_X is a probability distribution over X. To select the next x to sample given data d a sample is drawn from the distribution $s(x)$.

Theorem 8 (stochastic NFL) *Let* $Y^X = \{f \mid f : X \to Y\}$, *where X and Y are arbitrary finite sets. For any stochastic optimisation algorithms A, B, if we sample f uniformly at random from Y^X then $P(T_A(f) = t) = P(T_B(f) = t) = \frac{1}{|Y|^{|X|}}$ for all full length traces* $t \in \mathscr{T}$.

Proof Let A be a stochastic optimisation algorithm. This stochastic behaviour is equivalent to sampling a deterministic algorithm from some probability distribution over algorithms, then running that. However, we know from the original no free lunch result that regardless of the deterministic algorithm chosen, each trace is equally probable when each function is equally probable, thus each trace has probability $\frac{1}{|Y|^{|X|}}$ regardless of the stochastic optimiser used.

6 Refined and Generalised No Free Lunches

Since their original publication the NFL theorems have been augmented and specialised in various ways. In this section we survey these extensions, providing intuitive explanations of the results, as well as proofs and examples. We start with two definitions that are used in the extensions.

Definition 14 (*CUP*) Let G be a set of functions mapping X to Y. We say G is closed under permutation, or CUP, iff for any permutation $\phi : X \to X, f \in G \implies f_\phi \in G$.

Definition 15 (*Permutation Closure*) Let G be a set of function mapping X to Y. We define G_{cup} as the smallest set containing G that is closed under permutation.

6.1 Optimisation Algorithms Are Bijections

In this section we show that an optimisation algorithm can be seen as defining a function mapping $\mathscr{F}$ to itself, and that this function is a bijection, and thus a permutation. We also look at the behaviour of an optimiser when run on a particular function f, and we will see that the output trace produced can be interpreted as a function permutation of the input f. Results relating optimisation algorithms to permutations are worked through in detail in [12]. We start with a definition:

Definition 16 (*Shuffle Permutation*) $\pi : Y^X \to Y^X$ is a shuffle permutation if it is a bijection and, if $\pi(f) = g$ then there exists a permutation $\phi : X \to X$ such that $\forall x \in X, f(x) = g(\phi(x))$. That is, g is a function permutation (see Definition 3) of f. Intuitively, a shuffle permutation maps a function to new function with the same output values, but rearranged to correspond to different inputs.

Now we will show the sense in which an optimisation algorithm can be thought of as a map from $\mathscr{F}$ to itself. Recall that without loss of generality we assume that $X = \{1, 2, \ldots, n\}$. Recall also that $T_A(f)_Y$ is the full output trace of optimiser A running on function f. Thus $T_A(f)_Y$ is an ordered list of n output (i.e. $y \in Y$) values. Given this trace a new function, g, can be defined as $g(i) = T_A(f)_Y(i)$, where $T_A(f)_Y(i)$ is the i-th value of the output trace (i.e. the i-th y value encountered during the optimisation).

We now show that any optimisation algorithm naturally implies a map from Y^X to itself, and in fact this map is a shuffle permutation.

Theorem 9 (Optimisation Algorithms Imply Shuffle Permutations) *Let A be an arbitrary optimisation algorithm, then $T_A(.)_Y$ is a bijection $T_A(.)_Y : Y^X \to Y^X$. Further this bijections is a shuffle permutation.*

Proof That the implied map is a bijection follows from Theorem 1. Next we note that from their definition the optimisers are non-resampling and eventually sample every point. It follows that for any input function the output trace is just a reordering of the function's y values. Thus, the bijection is a shuffle permutation. ∎

6.2 Representation Invariance

The representation of a problem is generally considered important for optimisation. However, an interesting corollary of NFL is that the representation doesn't matter when considering the ensemble of all possible problems. In other words, there is a representational no free lunch: No representation scheme is better than any other under any metric for any optimisation algorithm when average performance over all problems is considered. This representation invariance was made explicit in [13].

Theorem 10 (Representation Invariance *[13]*) *Given a function $h : X \to Y$ we can re-represent the problem by introducing a set C and a surjective map $g : C \to X$ and then considering a new function $f : C \to Y$ where $f(c) = h(g(c))$. As C is surjective then we know $|C| \geq |X|$. Then for any optimisers A, B,*

$$|C| = |X| \implies \{T_A(h)_Y | h \in Y^X\} = \{T_B(h)_Y | h \in Y^X\} = \{T_A(f)_Y | f \in Y^C\}$$

$$|C| > |X| \implies \{T_A(f)_Y | f \in Y^C\} = \{T_B(f)_Y | f \in Y^C\}$$

A full proof can be found in [13], however, it can be seen that the case when $|C| = |X|$ follows directly from Theorem 6.

6.3   Sharpened No Free Lunch

The Sharpened No Free Lunch Theorem (SNFL) was first presented in [14]. Whereas the original no free lunch theorem is a statement about algorithms on the set of all functions, SNFL shows that the result still holds even when we restrict consideration to certain subsets of function. In particular, SNFL states that all optimisation algorithms are equivalent over any subset of functions closed under permutation (see Definition 14).

Theorem 11 (SNFL [14]) *Let $G \subseteq Y^X$ be closed under permutation, then for any optimisation algorithms A, B, $\{T_A(f)_Y \mid f \in G\} = \{T_B(f)_Y \mid f \in G\}$.*

Proof We have seen in Theorem 9 that the output trace produced when an optimiser is run on a function f is always some permutation of f. In the same theorem we also saw that for any optimisation algorithm A, for any two functions $f, g \in \mathcal{F}$ $f \neq g \implies T_A(f)_Y \neq T_A(g)_Y$. From these two facts it follows that if the input set is permutation closed, then, regardless of the optimisation algorithm used, the set of output traces will be this same set of functions. In other words, any optimisation algorithm is a permutation on any permutation closed set of functions. $\{T_A(f)_Y \mid f \in G\} = G$ for any optimisation algorithm A. It follows that $\{T_A(f)_Y \mid f \in G\} = \{T_B(f)_Y \mid f \in G\}$.

Many researchers have examined the realism of the closed under permutation condition for real problems. In particular, Igel and Toussaint [15] show that the proportion of subsets of functions that are closed under permutation tends to zero double-exponentially as the size of the domain of the functions increases.

6.4   Focused No Free Lunch

The Focused No Free Lunch Theorem (FNFL) is an extension of SNFL (see Sect. 6.3). Essentially it shows that, when only considering a restricted set of optimisation algorithms, a (potentially very small) subset of the permutation closure of a test function is enough for NFL to hold. This result was first presented in [3]. Intuitively, because of the restriction to a subset of algorithms a more focused result is possible as there are fewer requirements to satisfy.

Theorem 12 (FNFL [3]) *Let β be a set of test functions, $\beta = \{f_1, f_2, \ldots, f_m\}$, and let $\mathscr{A}$ be a set of optimisation algorithms, $\mathscr{A} = \{A_1, A_2, \ldots, A_n\}$. Then there exists a "focused set" $F_{\mathscr{A}}(\beta)$, with $\beta \subseteq F_{\mathscr{A}}(\beta) \subseteq \beta_{CUP}$ such that all algorithms in $\mathscr{A}$ produce the same set of traces over $F_{\mathscr{A}}(\beta)$. Moreover, this focused set $F_{\mathscr{A}}$ can potentially be* *much smaller than* β_{CUP} *(note* β_{CUP} *is just the permutation closure of* β, *see Definition 15).*

The proof can be found in [3], here we omit the proof and instead include some simple illustrative examples.

Example 9 (Simple FNFL Example) Consider optimising functions mapping between $\{1, 2, 3, 4, 5\}$ and $\{0, 1\}$. Let $\beta = \{01010\}$, call this function f_1 (i.e. $f_1 = 01010$). Consider two optimisation algorithms: A, which inspects the function from left to right, and B, which inspects the function from right to left. $T_A(f_1)_Y = 01010 = T_B(f_1)_Y$, and thus $F_{\mathscr{A}}(\beta)$ for $\mathscr{A} = \{A, B\}$ is just $\{01010\}$.

Example 10 (Second FNFL Example) Again consider optimising functions mapping between $\{1, 2, 3, 4, 5\}$ and $\{0, 1\}$. Let $\beta = \{00110\}$, let $f_1 = 00110$. Consider two optimisation algorithms, A, which inspects the function from left to right, and B, which inspects the function from right to left. $T_A(f_1)_Y = \{00110, 01100\} = T_B(f_1)_Y$, and thus $F_{\mathscr{A}}(\beta)$ for $\mathscr{A} = \{A, B\}$ is $\{00110, 01100\}$.

6.5 Almost No Free Lunch

Another important extension to the no free lunch theorem is the so called Almost No Free Lunch Theorem. The Almost No Free Lunch Theorem (ANFL) shows that if a stochastic optimiser (Definition 13) performs well on a given function then there is a function of similar complexity on which it performs badly [16].

Theorem 13 (ANFL [16]) *Let H be a randomised optimisation strategy, $X = \{1, \ldots, 2^k\}$, $Y = \{1, \ldots, m\}$ and $f : X \to Y$. Define $c = 2^{k/3}$. Then there exist at least m^{c-1} functions $g : X \to Y$ that differ from f on at most c inputs, such that the probability that H finds the optimum of g within c steps is less than or equal to c^{-1}.*

A proof is given in [16]. Functions of similar complexity here means any of evaluation time, circuit size representation, and Kolmogorov complexity.

Example 11 Let $k = 6$ and $m = 2$. Then $X = \{1, 2, 3, \ldots, 64\}$ and $Y = \{0, 1\}$. In this case ANFL asserts that there are at least $2^3 = 8$ functions $g : X \to Y$ that agree with f on all but at most 4 inputs such that H finds the optimum of g within 4 steps with probability bounded above by $\frac{1}{4}$.

6.6 Restricted Metric No Free Lunch

In this section we introduce a Restricted Metric No Free Lunch Theorem (RNFL) as an extension of the FNFL. Towards the end of [8] English notes the need for an NFL theory for the case of restricted metrics. There has been some work towards

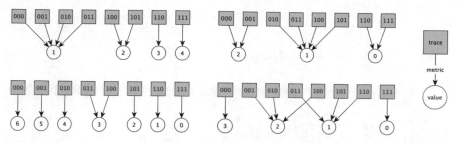

Fig. 8 Four example metrics showing mappings of traces to values. The values are determined as follows (clockwise from top left): (**1**) number of samples until a zero is found. (**2**) number of zeros in first two samples. (**3**) number of zeros in the whole trace. (**4**) value based on how many zeros are found, and how quickly they are found

this goal, for example in [17] they restrict attention to maximisation and show that the correspondence between NFL holding and the set of functions considered being closed under permutation breaks down when only considering optimisers running for k-steps. Specifically, they show that under a maximisation metric, the set of functions being closed under permutation is not necessary for NFL type results. This is an example of a restricted metric free lunch in an k-step optimisation setting.

The original NFL theorem and its successors consider arbitrary performance metrics, or equivalently all performance metrics. Here we show how a restriction on the set of metrics considered—a choice of a single metric for instance—yields NFL results on subsets of functions. This is similar to FNFL, except as well as considering restricted sets of algorithms and functions it also considers a restricted set of metrics. The key idea is that we compare the multi-sets of scores assigned to traces rather than the multi-sets of traces themselves. The sets of scores often has fewer unique elements (and can never have more) an d thus there are more situations in which no free lunch results hold. Figure 8 shows how metrics can reduce the set of traces to a smaller set of scores.

Theorem 14 (RNFL) *Let β be a set of test functions, $\beta = \{f_1, f_2, \ldots, f_m\}$, let $\mathscr{A}$ be a set of optimisation algorithms, $\mathscr{A} = \{A_1, A_2, \ldots, A_n\}$, and let $\mathscr{M}$ be a set of optimisation metrics, $\mathscr{M} = \{m_1, m_2, \ldots, m_k\}$. Then there exists a "restricted set" $R_{\mathscr{A},\mathscr{M}}(\beta)$, with $\beta \subseteq R_{\mathscr{A},\mathscr{M}}(\beta) \subseteq F_{\mathscr{A}}(\beta) \subseteq \beta_{CUP}$ such that all algorithms in $\mathscr{A}$ have the same average performance over $R_{\mathscr{A},\mathscr{M}}(\beta)$. A restricted set $R_{\mathscr{A},\mathscr{M}}(\beta)$ always exists, regardless of the choice of β, $\mathscr{A}$ and $\mathscr{M}$. In some cases it is identical to the focus set $F_{\mathscr{A}}(\beta)$ from the FNFL Theorem, but it can also be strictly smaller than the focus set.*

Proof We give a proof by providing an example in which the restricted set is smaller than the focused set. The fact that a restricted set always exists follows from the fact a focused set always exists, thus we need only to show that the restricted set is potentially a subset of the focused set. Let $X = \{1, 2, 3, 4, 5, 6\}$, $Y = \{0, 1\}$, $\beta = \{101001, 001011, 001111\}$, $\mathscr{A} = \{A_1, A_2\}$ where A_1 deterministically samples from left to right and A_2 deterministically samples $f(2), f(4), f(6), f(1), f(3), f(5)$ in that

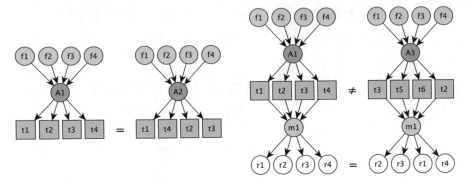

Fig. 9 On the left is a visualisation of a FNFL result for $\beta = \{f_1, f_2, f_3, f_4\}$ and $\mathscr{A} = \{A_1, A_2\}$. Over this set of four functions both optimisers produce the same set of traces, namely $\{t_1, t_2, t_3, t_4\}$. On the right we show that a RNFL can hold when a FNFL does not. For $\beta = \{f_1, f_2, f_3, f_4\}$ and $\mathscr{A} = \{A_1, A_3\}$ FNFL does not hold as A_1 produces the traces $\{t_1, t_2, t_3, t_4\}$ where as A_3 produces $\{t_2, t_3, t_5, t_6\}$. However, if we set $\mathscr{M} = \{m_1\}$ then a RNFL result holds, as both sets of traces lead to the same set of scores, $\{r_1, r_2, r_3, r_4\}$

order. Finally, let $\mathscr{M} = \{m_1\}$ where m_1 just returns the number of samples until the first 1 is found.

If follows from the definitions of the algorithms that, when run on the functions in β, A_1 produces the traces $\{101001, 001011, 001111\}$ and A_2 produces the traces $\{001110, 001011, 011011\}$. These sets of traces are not equal, and when we put these sets of traces through the metric m_1 they result in the multi-set of values $\{1, 3, 3\}$ and $\{3, 3, 2\}$, respectively, which are also not equal.

If we define $R_{\mathscr{A}, \mathscr{M}}(\beta) = \{101001, 001011, 001111, 010010\}$ then running A_1 on the functions in $R_{\mathscr{A}, \mathscr{M}}(\beta)$ produces the traces $\{101001, 001011, 001111, 010010\}$ and similarly running A_2 produces the traces $\{001110, 001011, 011011, 100001\}$. These sets of traces are still not equal, thus $R_{\mathscr{A}, \mathscr{M}}(\beta)$ is not the "focus set" from the FNFL. However, when we put these sets of traces through the metric m_1 they result in the same multi-set of values $\{1, 2, 3, 3\}$. Thus, $R_{\mathscr{A}, \mathscr{M}}(\beta)$ is a "restricted set" and we are done. See Fig. 9 for a visualisation of the proof.

6.7 Multi-objective No Free Lunch

So far we have considered metrics that map to a scalar. However, within the optimisation literature there is much work on so-called multi-objective optimisation problems, where the metric assigns a vector rather than a scalar. A natural question to ask is whether no free lunch results generalise to these situations. The answer is yes [18, 19]. The proof used in [18] works by defining a bijection between multi-objective problems and scalar problems. A further multi-objective result in [18] is that a no free lunch holds over the set of all multi-objective functions with any particular shape of Pareto front.

However, as they show in [20], in the case of multi-objective optimisation real world constraints (such as the algorithm only having finite memory) can readily result in algorithms with differing performances. The key idea is that, in scalar optimisation keeping track of the current best solution is straight forward, whereas in multi-objective optimisation problems, where one is searching from the Pareto front, it becomes more practically difficult to store the current best (in this case a set of points making up the Pareto front) efficiently, and so even fairly weak restrictions on the algorithm can mean that practically it is necessary to instead store some sort of approximation of the Pareto front.

Thus, theoretically NFL holds for multi-objective functions, but when it comes to implementation multi-objective optimisers more often need to violate the NFL assumptions for reasons of pragmatism.

6.8 Block Uniform Distributions

In [10] English presents an intuitive necessary and sufficient condition for NFL. He defined *block uniform* distributions, and proved that NFL holds if and only if functions are sampled from a block uniform probability distribution. We state the theorem below, and a proof can be found in the paper.

Definition 17 (*Block-uniform distribution*) A probability distribution over the set of functions $\{f : X \rightarrow Y\}$ is block uniform iff $\forall f$, $\forall \phi$, $P(f) = P(f_\phi)$, where ϕ is a function permutation (see Definition 3) (Fig. 10).

Theorem 15 (NFL iff Block Uniform) *For any metric, all optimisation algorithms have the same expected performance if and only if there is a block uniform probability distribution over functions.*

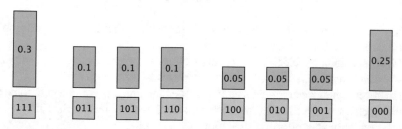

Fig. 10 An example of a block uniform probability distribution over functions $f : \{1, 2, 3\} \rightarrow \{0, 1\}$. There are four "blocks", corresponding to the four constituent permutation closed subsets of functions. The figure is not to scale

6.8.1 ε-Block Uniform

In [21] Everitt investigates what happens when a distribution is *almost* block uniform. He proves that the amount of free lunch available increases at most linearly with increasing ε, where ε characterises the distance of a distribution from block-uniform.

Thus, beyond the statement of Theorem 15, the amount of free lunch available is bounded by the distance of the probability distribution over functions from a block uniform distribution.

6.9   Infinite and Continuous Lunches

It should be noted that in this exploration of the NFL we have not considered infinite domains and continuous extensions. However, work to this end can be found in the literature [22–24]. This decision is in part due to the fact that the algorithms in which we are interested will be running on finite sets of possible inputs, and in part because optimisation algorithms in general use run on finite procession machines.

7   Comparing Optimisers After No Free Lunch

In the above sections we have covered in detail the original NFL results, and various extensions. It is evident that, as emphasised in [25], the existence and importance of free lunches is far from a straight forward question, and the opinions of researchers vary. This said, we now try to briefly summarise what the results, when considered as a whole, seem to really mean for algorithm comparison and evaluation:

1. The no free lunch results preclude meaningful comparison of optimisation algorithm's *exploration behaviour* without reference to specific problems.
2. However, almost all restrictions on the set of problem functions result in possible free lunches.
3. Similarly, but more generally, almost all probability distributions over problem functions result in possible free lunches.
4. More specifically, block-uniform distributions capture exactly the scenarios where no free lunch results hold for any metric.
5. However, when we are interested in no free lunch results with respect to particular metrics, and for limited numbers of samples, then free lunches are possible even under block-uniform distributions.
6. When free lunches are possible, their prominence tends to depend crucially on the optimisation metric used.
7. When free lunches are possible, the algorithms that achieve them are *aligned* with the probability distribution over problem functions.

8. When considering more than just the exploration behaviour of an algorithm, algorithms can be ranked. For example, some optimisers are simpler, some are faster and some tend to resample less than others.

Pragmatically the results mean that benchmarking alone cannot be used to evaluate an algorithm, but must be used in combination with clear underlying assumptions on the distribution of problem functions. The benchmark functions must be representative of the problems and there must be some smoothness, in the sense that being good at a problem means that the algorithm is likely to be good at similar problems.

8 Metaheuristic Optimisation After NFL

In this chapter we have explored in detail NFL results, and seen how they preclude superiority of any particular optimisation algorithm in many settings. However, there are many real problems that need solving through optimisation, and many popular, successful metaheuristic optimisers in common use, such as [26, 27]. The questions then, are firstly how can the NFL results be reconciled with the existence of effective optimisation methods, and secondly, how can we use a thorough understanding of NFL to improve research into and development of metaheursitc optimisers?

Optimisation algorithms are able to perform better than others in situations where free lunches exist, which can most generally be understood as situations in which block uniform distributions over functions, as discussed in Sect. 6.8, do not pertain. Just as a compression algorithm cannot universally compress, but must exploit expected input structure to compress well on average, so an optimisation algorithm cannot work well on all possible inputs, but must exploit expected input structure to optimise well on average over the inputs it receives. Luckily, in reality, problems tend to exhibit certain structure, such as local smoothness, or symmetries, and optimisers that exploit this structure *can* do better on average, for that class of problems, than those that do not. Successful metaheuristic optimisers then, are those that effectively exploit common problem structure.

However, despite a general awareness of the existence of the NFL theorems, it is still sometimes the case that new optimisers are presented as a panacea. Of course, the NFL results tell us specifically that this can never be true.

Instead, we must try to characterise the dynamics of optimisation algorithms, to understand their search behaviour, so that we can better understand which algorithms should be used for which problems, and how the algorithms hyper-parameters influence the search dynamics. This sort of investigation has been undertaken for PSO for example [28, 29], where the effects of the hyper-parameters on the search behaviour are considered in detail.

Pursuing a Bayesian understanding of metaheuristic optimisation is another potential approach to the problem of characterising optimisers and understanding which sorts of functions they work best for. Recently, Serafino has been emphasising the important close relationship between Bayesian optimisation and no free lunch results [7, 30]. Roughly speaking, in a Bayesian framing of optimisation the standard NFL result becomes the claim that informative induction can not be done without prior assumptions. In fact, this is a widely known maxim, the necessity of an inductive bias voiced by various researchers: "A basic insight of machine learning is that prior knowledge is a necessary requirement for successful learning" [31], in other words, "[Y]ou can't do inference ... without making assumptions" [32]. Making this relationship more precise, Streeter has shown in [33] that NFL applies only when a certain form of Bayesian learning is impossible.

Metaheursitic optimisation algorithms essentially make implicit assumptions about the kinds of problem function they will be used on, and we should aim to make these implicit assumptions as explicit as possible when either developing an optimiser, or investigating one's behaviour. However, uncovering the particular biases and affinities of an optimisation algorithm has proven to be very difficult, and it is not clear how one can represent these alignments in a general way. As metaheuristic optimisation continues to develop, it will be necessary for understanding of these problems to develop, too.

References

1. Wolpert, D.H., Macready, W.G.: No free lunch theorems for search. Technical report. SFI-TR-95-02-010, Santa Fe Institute (1995)
2. Wolpert, D.H., Macready, W.G.: No free lunch theorems for optimization. In: IEEE Trans. Evol. Comput. **1.1**, 67–82 (1997)
3. Whitley, D., Rowe, J.: Focused no free lunch theorems. In: Proceedings of the 10th Annual Conference on Genetic and Evolutionary Computation, pp. 811–818. ACM (2008)
4. Whitley, D.: Functions as permutations: regarding no free lunch, walsh analysis and summary statistics. In: Parallel Problem Solving from Nature PPSN VI, pp. 169–178. Springer (2000)
5. Culberson, J.C.: On the futility of blind search: an algorithmic view of 'no free lunch. Evol. Comput. **6**(2), 109–127 (1998)
6. Lattimore, T., Hutter, M.: No free lunch versus Occam's razor in supervised learning. In: Algorithmic Probability and Friends. Bayesian Prediction and Artificial Intelligence, pp. 223–235. Springer (2013)
7. Serafino, L.: No Free Lunch Theorem and Bayesian probability theory: two sides of the same coin. Some implications for black-box optimization and metaheuristics. In: (2013). arXiv:1311.6041
8. English, T.: No more lunch: analysis of sequential search. In: Proceedings of the 2004 Congress on Evolutionary Computation CEC2004, Vol. 1, pp. 227–234. IEEE (2004)
9. English, T.: Optimization is easy and learning is hard in the typical function. In: Proceedings of the 2000 Congress on Evolutionary Computation, vol. 2, pp. 924–931. IEEE (2000)
10. English, T.: On the structure of sequential search: beyond 'no free lunch'. In: Evolutionary Computation in Combinatorial Optimization, pp. 95–103. Springer (2004)
11. Ho, Yu-Chi, Pepyne, D.L.: Simple explanation of the no-free-lunch theorem and its implications. J. Optim. Theory Appl. **115**(3), 549–570 (2002)
12. Duéñez-Guzmán, E.A., Vose, M.D.: No free lunch and benchmarks. Evol. Comput. **21**(2), 293–312 (2013)

13. Radcliffe, N.J., Surry, P.D.: Fundamental limitations on search algorithms: evolutionary computing in perspective. In: Computer Science Today, pp. 275–291. Springer (1995)
14. Schumacher, C., Vose, M.D., Whitley, L.D.: The no free lunch and problem description length. In: Proceedings of the Genetic and Evolutionary Computation Conference (GECCO-2001), pp. 565–570 (2001)
15. Igel, C., Toussaint, M.: A no-free-lunch theorem for non-uniform distributions of target functions. J. Math. Model. Algorithm. 3(4), 313–322 (2005)
16. Droste, S., Thomas, J., Ingo, W.: Optimization with randomized search heuristics—the (A) NFL theorem, realistic scenarios, and difficult functions. Theor. Comput. Sci. 287(1), 131–144 (2002)
17. Griffiths, E.J., Orponen, P.: Optimization, block designs and no free lunch theorems. Inf. Process. Lett. 94(2), 55–61 (2005)
18. Corne, D.W., Knowles, J.D.: No free lunch and free leftovers theorems for multiobjective optimisation problems. In: Evolutionary Multi- Criterion Optimization. Springer (2003)
19. Service, T.C.: A no free Lunch theorem for multi-objective optimization. Inf. Process. Lett. 110(21), 917–923 (2010)
20. Corne, D., Knowles, J.: Some multiobjective optimizers are better than others. In: The 2003 Congress on Evolutionary Computation, Vol. 4, pp. 2506–2512. IEEE (2003)
21. Everitt, T.: Universal indution and optimisation: no free lunch? In: Thesis (2013)
22. Auger, A., Teytaud, O.: Continuous lunches are free plus the design of optimal optimization algorithms. Algorithmica 57(1), 121–146 (2010)
23. Rowe, J., Vose, M., Wright, A.: Reinterpreting no free lunch. Evol. Comput. 17(1), 117–129 (2009)
24. Alabert, A. et al.: No-free-lunch theorems in the continuum. In: (2014). arXiv:1409.2175
25. Yang, X.-S.: Free lunch or no free lunch: that is not just a question? Int. J. Artif. Intell. Tools 21(03), 1240010 (2012)
26. Yang, X.-S., Deb, S.: Cuckoo search via Lévy flights. In: World Congress on Nature & Biologically Inspired Computing, pp. 210–214. IEEE (2009)
27. Kennedy, J.: Particle swarm optimization. In: Encyclopedia of Machine Learning, pp. 760–766. Springer (2011)
28. Poli, R.: Mean and variance of the sampling distribution of particle swarm optimizers during stagnation. IEEE Trans. Evol. Comput. 13(4), 712–721 (2009)
29. Erskine, A., Joyce, T., Herrmann, J.M.: Parameter selection in particle Swarm optimisation from stochastic stability analysis. In: International Conference on Swarm Intelligence, pp. 161–172. Springer (2016)
30. Serafino, L.: Optimizing without derivatives: what does the no free lunch theorem actually say? In: Notices of the AMS 61.7 (2014)
31. Ben-David, S., Srebro, N., Urner, R.: Universal learning versus no free lunch results. In: Philosophy and Machine Learning Workshop NIPS (2011)
32. David, J.C.M.: Information theory, inference and learning algorithms. Cambridge University Press (2003)
33. Streeter, M.J.: Two broad classes of functions for which a no free lunch result does not hold. In: Genetic and Evolutionary Computation-GECCO, pp. 1418–1430 Springer (2003)

Global Convergence Analysis of Cuckoo Search Using Markov Theory

Xing-Shi He, Fan Wang, Yan Wang and Xin-She Yang

Abstract The cuckoo search (CS) algorithm is a powerful metaheuristic algorithm for solving nonlinear global optimization problems. In this book chapter, we prove the global convergence of this algorithm using a Markov chain framework. By analyzing the state transition process of a population of cuckoos and the homogeneity of the constructed Markov chains, we can show that the constructed stochastic sequences can converge to the optimal state set. We also show that the algorithm structure of cuckoo search satisfies two convergence conditions and thus its global convergence is guaranteed. We then use numerical experiments to demonstrate that cuckoo search can indeed achieve global optimality efficiently.

Keywords Cuckoo search · Convergence rate · Global convergence · Markov chain theory · Optimization · Swarm intelligence

1 Introduction

Nature-inspired algorithms have become widely used for optimization and computational intelligence [11, 12, 26–28, 30]. Many new optimization algorithms are based on the so-called swarm intelligence with diverse characteristics in mimicking natural systems. However, there is a significant gap between theory and practice. Most metaheuristic algorithms have very successful applications in practice, but their mathematical analysis lags far behind. In fact, apart from a few limited results about the convergence and stability concerning particle swarm optimization, genetic algorithms, simulated annealing and others [4, 10, 16], many algorithms do not have any

X.-S. He · F. Wang · Y. Wang
College of Science, Xi'an Polytechnic University, Xi'an, People's Republic of China

X.-S. Yang (✉)
School of Science and Technology, Middlesex University, London NW4 4BT, UK
e-mail: x.yang@mdx.ac.uk; xy227@cam.ac.uk

© Springer International Publishing AG 2018
X.-S. Yang (ed.), *Nature-Inspired Algorithms and Applied Optimization*,
Studies in Computational Intelligence 744, https://doi.org/10.1007/978-3-319-67669-2_3

53

theoretical analysis. Therefore, we may know they can work well in practice, but we rarely understand why they work and how to improve them with a good understanding of their working mechanisms.

In this work, we will try to prove the convergence of the cuckoo search (CS) so as to gain insight into its search mechanisms. The rest of this paper is organized as follows: we will introduce the details of the cuckoo search algorithm in Sect. 2, followed by the introduction of the convergence criteria in Sect. 3 and the detailed convergence analysis in Sect. 4. Then, we validate the cuckoo search algorithm by numerical experiments and observe its convergence behaviour in Sect. 5. Finally, we conclude by summarizing the main results in Sect. 6.

2 Cuckoo Search

Cuckoo search (CS) is one of the recent nature-inspired metaheuristic algorithms, developed in 2009 by Xin-She Yang and Suash Deb [23]. CS is based on the brood parasitism of some cuckoo species. In addition, this algorithm is enhanced by the so-called Lévy flights [15], rather than by simple isotropic random walks. Recent studies show that CS is potentially far more efficient than PSO and genetic algorithms [8, 24]. A relatively comprehensive review of the studies up to 2014 was carried out by Yang and Deb [25].

2.1 Standard Cuckoo Search

Cuckoo behaviour is intriguing because of the so-called brood parasitism reproduction strategy. Some species such as the *ani* and *Guira* cuckoos lay their eggs in communal nests, though they may remove others' eggs to increase the hatching probability of their own eggs. Quite a number of species engage obligate brood parasitism by laying their eggs in the nests of other host birds (often other species such as warblers). In addition, the eggs laid by cuckoos may be discovered and thus abandoned with a probability, around 1/4 to 1/3, depending on species and the average number of eggs in a nest.

For simplicity in describing the cuckoo search, we now use the following three idealized rules [23]:

1. Each cuckoo lays one egg at a time, and dumps it in a randomly chosen host nest.
2. The best nests with high-quality eggs will be carried over to the next generations.
3. The number of available host nests is fixed, and the egg laid by a cuckoo is discovered by the host bird with a probability $p_a \in [0, 1]$. In this case, the host bird can either get rid of the egg, or simply abandon the nest and build a completely new nest at a new location.

As a further approximation, this last assumption can be approximated by a fraction p_a of the n host nests are replaced by new nests (with new random solutions). For a maximization problem, the quality or fitness of a solution can simply be proportional to the value of the objective function.

From the implementation point of view, we can use the following simple representation, that each egg in a nest represents a solution, and each cuckoo can lay only one egg (thus representing one solution), the aim is to use the new and potentially better solutions (cuckoos) to replace a not-so-good solution in the nests. Obviously, this algorithm can be extended to the more complicated case where each nest has multiple eggs representing a set of solutions. For this present work, we will use the simplest approach where each nest has only a single egg. In this case, there is no distinction between an egg, a nest, or a cuckoo, as each nest corresponds to one egg which also represents one cuckoo, corresponding to a single solution vector.

This algorithm uses a balanced combination of a local random walk and the global explorative random walk, controlled by a switching parameter p_a. The local random walk can be written as

$$\mathbf{x}_i^{t+1} = \mathbf{x}_i^t + \beta s \otimes H(p_a - \epsilon) \otimes (\mathbf{x}_j^t - \mathbf{x}_k^t), \tag{1}$$

where $\mathbf{x}_j^t$ and $\mathbf{x}_k^t$ are two different solutions selected randomly by random permutation, $H(u)$ is a Heaviside function, ϵ is a random number drawn from a uniform distribution, and s is the step size. Here β is the small scaling factor. On the other hand, the global random walk is carried out by using Lévy flights

$$\mathbf{x}_i^{t+1} = \mathbf{x}_i^t + \alpha \otimes L(s, \lambda), \tag{2}$$

where

$$L(s, \lambda) \sim \frac{\lambda \Gamma(\lambda) \sin(\pi \lambda / 2)}{\pi} \frac{1}{s^{1+\lambda}}, \quad (s \gg 0). \tag{3}$$

Here $\alpha > 0$ is the step size scaling factor, which should be related to the scales of the problem of interest. Here '$\sim$' denotes that the fact that the random numbers $L(s, \lambda)$ should be drawn from the Lévy distribution on the right-hand side, which is approximated by a power-law distribution with an exponent λ. In addition, $\otimes$ is an entry-wise operation.

The above equation is essentially the stochastic equation for a random walk. In general, a random walk is a Markov chain whose next status/location only depends on the current location (the first term in the above equation) and the transition probability (the second term). However, a substantial fraction of the new solutions should be generated by far field randomization and their locations should be far enough from the current best solution; this will make sure that the system will not be trapped in a local optimum [23, 25].

2.2   Cuckoo Search in Applications

Cuckoo search has been applied in many areas of optimization, engineering design, data ming and computational intelligence with promising efficiency. For example, in the engineering design applications, cuckoo search has superior performance over other algorithms for a range of continuous optimization problems such as spring design and welded beam design problems [8, 24, 25].

In addition, a modified cuckoo search by Walton et al. [21] has demonstrated to be very efficient for solving nonlinear problems such as mesh generation. Vazquez [20] used cuckoo search to train spiking neural network models. Yildiz [32] has used cuckoo search to select optimal machine parameters in milling operation with enhanced results. Then Durgun and Yildiz [7] used CS for the optimization of vehicle components, while Zheng and Zhou [33] provided a variant of cuckoo search using Gaussian process. In the context of data fusion and wireless sensor network, cuckoo search has been shown to be very efficient [5, 6].

Among the diverse applications, an interesting performance enhancement has been obtained by using cuckoo search to train neural networks as shown by Valian et al. [18] and reliability optimization problems [19].

For complex phase equilibrium applications, Bhargava et al. [2] have shown that cuckoo search offers a reliable method for solving thermodynamic calculations. Furthermore, Moravej and Akhlaghi [13] have solved DG allocation problem in distribution networks with good convergence rate and performance. Taweewat and Wutiwiwatchi have combined cuckoo search and supervised neural network to estimate musical pitch with reduced size and higher accuracy [17].

As a further extension, Yang and Deb [31] developed a multiobjective cuckoo search (MOCS) for design engineering applications. For multiobjective scheduling problems, another progress was made by Chandrasekaran and Simon [3] using cuckoo search algorithm, which demonstrated the superiority of their proposed methodology. Recent studies have demonstrated that cuckoo search can performance significantly better than other algorithms in many applications [8, 14, 29, 32, 33].

2.3   Simplified Cuckoo Search

In the cuckoo search algorithm, a set of two updating equations are used. One equation is mainly for global moves, while the other is mainly for local exploitation. Whether it is global or local is largely determined by the step sizes of the moves of new solutions from the existing solutions in the population. However, since Lévy flights can have both small steps and occasionally large steps, it can carry out both local and global moves simultaneously. Thus, it is difficult to put into a fixed category. However, in order to simplify the analysis and also to emphasize the global search capability, we now use a simplified version of cuckoo search. That is, we use only the global branch with a random number $r \in [0, 1]$, compared with a discovery/switching probability p_a. Now we have

$$\begin{cases} \mathbf{x}_i^{(t+1)} \leftarrow \mathbf{x}_i^{(t)} & \text{if } r < p_a, \\ \mathbf{x}_i^{(t+1)} \leftarrow \mathbf{x}_i^{(t)} + \alpha \otimes L(s, \lambda) & \text{if } r > p_a. \end{cases} \quad (4)$$

Obviously, due to the stochastic and iterative nature of the cuckoo search algorithm, we have to focus on the key steps. Therefore, we use the following steps to represent the simplified cuckoo search [22]:

- Step 1. Generate randomly an initial population of n nests at the positions, $\mathbf{X} = \{\mathbf{x}_1^0, \mathbf{x}_2^0, ..., \mathbf{x}_n^0\}$, then evaluate their objective values and record the initial best $\mathbf{g}_t^0$.
- Step 2. Generate new solutions/moves by

$$\mathbf{x}_i^{(t+1)} = \mathbf{x}_i^{(t)} + \alpha \otimes L(s, \lambda). \quad (5)$$

- Step 3. Draw a uniformly distributed random number r from $[0, 1]$. Update $\mathbf{x}_i^{(t+1)}$ if $r > p_a$. Then, evaluate the new solutions and update the new global best $\mathbf{g}_t^*$ at iteration t.
- Step 4. Stop if the stopping criterion is satisfied and output the global best $\mathbf{g}_t^*$. Otherwise, go to step (2).

Though this is a simplified version of cuckoo search, it captures all the main characteristics of the standard cuckoo search. Thus, the proof of its global convergence will be equivalent to the proof of the global convergence of the original algorithm.

3 Markov Chains and Convergence Criteria

For the ease of analysis and notations, let us first use $<\Omega_s, f>$ to denote the optimization problem with an objective f in the search space Ω_s. This problem is to be solved by a stochastic search algorithm A. The solution obtained at the t-th iteration can be written as

$$\mathbf{x}_{t+1} = A(\mathbf{x}_t, \xi), \quad (6)$$

where Ω_s is the feasible solution space. ξ denotes the set of the visited solutions of algorithm A during the iterative process.

Loosely speaking, the infimum of the search in the Lebesgue measure space can be defined as

$$\phi = \inf \left(t : v(x \in \Omega_s | f(x) < t] > 0 \right), \quad (7)$$

where $v[X]$ denotes the Lebesque measure on the set X. In essence, Eq. (7) represents the non-empty set in the search space, and the region or regions for optimal solutions can be defined as

$$R_{\epsilon,M} = \begin{cases} \{x \in \Omega_s | f(x) < \phi + \epsilon\} & \text{if } \phi \text{ is finite,} \\ \{x \in \Omega_s | f(x) < -C\} & \text{if } \phi = -\infty, \end{cases} \qquad (8)$$

where $\epsilon > 0$ and $C \gg 1$ is a sufficiently large positive number. Loosely speaking, the set $R_{\epsilon,M}$ is a set that can be belong to different regions in the search space, depending on the objective landscapes. As long as this set is accessible, for any solution or a point falling into $R_{\epsilon,M}$ during the iteration, we can say that algorithm A has reached the optimal set and thus found the globally optimal solution or its best approximation.

The two conditions for convergence are as follows [9, 10]:

- 1 If $f(A(x,\xi)) \leq f(x)$ and $\xi \in \Omega_s$, we have

$$f(A(x,\xi)) \leq f(\xi). \qquad (9)$$

Here we focus on minimization problems. For maximization problems, the inequality is reversed, but the rest are the same.

- 2 For any set $S \in \Omega_s$ with $v(S) > 0$, we have

$$\prod_{k=0}^{\infty}(1 - u_k(S)) = 0, \qquad (10)$$

where $u_k(S)$ corresponds to the probability measure on S at the kth iteration of the algorithm A.

Before we proceed, let us use the results about the global convergence of an algorithm, based on existing studies without repeating the proofs [9, 10]:

Theorem 1 *If the objective f is measurable and its feasible solution space Ω_s forms a measurable subset in $\mathfrak{R}^n$, then algorithm A can indeed satisfy the above two conditions with the search sequence $\{x_k\}_{k=0}^{\infty}$, which will lead to*

$$\lim_{k\to\infty} P(x_k \in R_{\epsilon,M}) = 1. \qquad (11)$$

That means that algorithm A will converge globally with a probability one. Here $P(x_k \in R_{\epsilon,M})$ is the probability measure of the kth solution on $R_{\epsilon,M}$ at the kth iteration.

This same methodology has been used by He et al. to prove the global convergence of the flower pollination algorithm [9]. In this book chapter, we will use essentially the same procedure to prove the global convergence of cuckoo search by first proving the constructed Markov chains are proper and the conditions of convergence are satisfied.

4 Global Convergence Analysis

In order to simplify the presentations and analysis, let us first introduce some formal definitions and some preliminary results.

4.1 Preliminaries

Now we start to define the state and state space to be used later for proving the global convergence of the cuckoo search. For simplicity of notations, we use the standard non-bold case symbols for vectors and variables in the rest of this chapter.

Definition 1 The positions of a cuckoo/nest and its corresponding global best solution g in the search history forms the states of cuckoos: $y = (x, g)$ where $x, g \in \Omega_s$ and $f(g) \leq f(x)$ (for minimization). The set of all the possible states forms the state space, denoted by

$$Y = \{y = (x, g) | x, g \in \Omega_s, f(g) \leq f(x)\}. \tag{12}$$

The state and state space of the cuckoo population or group can be defined as follows:

Definition 2 The states of all n cuckoos/nests form the states of the group, denoted by $q = (y_1, y_2, ..., y_n)$. All the states of all the cuckoos form a state space for the group, denoted by

$$Q = \{q = (y_1, ..., y_i, ..., y_n), y_i \in Y, 1 \leq i \leq n\}. \tag{13}$$

As Q contains all the states found during the iterations, it also contains the historical global best solution g^* for the whole population as well as all individual best solutions $g_i (1 \leq i \leq n)$ in history. Obviously, the global best solution of the whole population is the best among all g_i, so that $f(g^*) = \min(f(g_i))$, $1 \leq i \leq n$.

Furthermore, the state transition for the positions of cuckoos representing solutions can be defined as follows. For $\forall y_1 = (x_1, g_1) \in Y$ and $\forall y_2 = (x_2, g_2) \in Y$, the state transition from y_1 to y_2 can be denoted by

$$T_y(y_1) = y_2. \tag{14}$$

4.2 Markov Chain Model for Cuckoo Search

One of the main tasks here is that we have to build a Markov chain model for cuckoo search algorithm, and the first step is to prove a theorem to be used later.

Theorem 2 *The transition probability from state y_1 to y_2 in the cuckoo search is*

$$P(T_y(y_1) = y_2) = P(x_1 \to x_1')P(g_1 \to g_1')P(x_1' \to x_2)P(g_1' \to g_2), \quad (15)$$

where $P(x_1 \to x_1')$ is the transition probability at Step 2 in cuckoo search, and $P(g_1 \to g_1')$ is the transition probability for the historical global best at this step. $P(x_1' \to x_2)$ is the transition probability at Step 3, while $P(g_1' \to g_2)$ is the transition probability of the historical global best.

Proof In the simplified cuckoo search, the state transition from y_1 to y_2 only has one middle transition state (x_1', g_1'), which means that $x_1 \to x_1', g_1 \to g_1', x_1' \to x_2$ and $g_1' \to g_2$ are valid simultaneously. Then, the probability for $P(T_y(y_1) = y_2)$ is

$$P(T_y(y_1) = y_2) = P(x_1 \to x_1')P(g_1 \to g_1')P(x_1' \to x_2)P(g_1' \to g_2). \quad (16)$$

From Eq. (5), the transition probability for $x_1 \to x_1'$ is

$$P(x_1 \to x_1') = \begin{cases} \frac{1}{|g-x_1|} & \text{if } x_1' \in [x_1, x_1 + (x_1 - g)], \\ 0 & \text{if } x_1' \notin [x_1, x_1 + (x_1 - g)]. \end{cases} \quad (17)$$

Since x and g are higher-dimensional vectors, the mathematical operations here should be interpreted as vector operations, while the $|\cdot|$ means the volume of the hypercube.

The transition probability of the historical best solution is

$$P(g_1 \to g_1') = \begin{cases} 1 & f(x_1') \leq f(g_1), \\ 0 & f(x_1') > f(g_1). \end{cases} \quad (18)$$

From Step 3 in the simplified cuckoo search algorithm, we know that a random number $r \in [0, 1]$ is compared with the discovery probability $p_a = 0.25 = 1/4$. If $r > p_a$, then the position/solution of a cuckoo can be changed randomly; otherwise, it remains unchanged. Therefore, the transition probability for $x_1' \to x_2$ is

$$P(x_1' \to x_2) = \begin{cases} 1 - p_a & \text{if } r > p_a, \\ p_a & \text{if } r \leq p_a \end{cases} = \begin{cases} \frac{3}{4} & \text{if } r > p_a, \\ \frac{1}{4} & \text{if } r \leq p_a. \end{cases} \quad (19)$$

The transition probability for the historical best solution is

$$P(g_1' \to g_2) = \begin{cases} 1 & f(x_2) \leq f(g_1), \\ 0 & f(x_2) > f(g_1). \end{cases} \quad (20)$$

Furthermore, the group transition probability in the cuckoo search can be defined as $T_q(q_i) = q_j$ for $\forall q_i = (y_{i1}, y_{i2}, ..., y_{in}) \in \Omega_s$ and $\forall q_j = (y_{j1}, y_{j2}, ..., y_{jn}) \in \Omega_s$.

Theorem 3 *In the simplified cuckoo search, the group transition probability from q_i to q_j in one step is*

$$P(T_q(q_i) = q_j) = \prod_{k=1}^{n} P(T_y(y_{ik}) = y_{jk}). \tag{21}$$

Proof If the group states can be transferred from q_i to q_j in one step, then all the states will be transferred simultaneously. That is, $T_y(y_{i1} = y_{j1}, T_y(y_{i2}) = y_{j2}, ..., T_y(y_{in}) = y_{jn}$, and the group transition probability can be written as the joint probability

$$P(T_q(q_i) = q_j) = P(T_y(y_{i1}) = y_{j1})P(T_y(y_{i2}) = y_{j2}) \cdots P(T_y(y_{in}) = y_{jn})$$

$$= \prod_{k=1}^{n} P(T_y(y_{ik}) = y_{jk}). \tag{22}$$

Theorem 4 *The state sequence $\{q(t); t \geq 0\}$ in the cuckoo search is a finite homogeneous Markov chain.*

Proof First, let us assume that all search spaces for a stochastic algorithm are finite. Then, x and g in any cuckoo/nest state $y = (x, g)$ are also finite, so that the state space for cuckoos/nests are finite. Since the group state $q = (y_1, y_2, ..., y_n)$ consists of n positions of the n cuckoos/nests where n is positive and finite, so group states q are also finite.

From the previous theorems, we know that the group transition probability

$$P(T_q(q(t-1)) = q(t), \tag{23}$$

for $\forall q(t-1) \in Q$ and $\forall q(t) \in Q$ is the group transition probability $P(T_y(y_i(t-1)) = y_i(t))$ for $1 \leq i \leq n$. From Eq. (16), we have the transition probability for any cuckoo is

$$P(T_y(y(t-1)) = y(t)) = P(x(t-1) \rightarrow x'(t-1))P(g(t-1) \rightarrow g'(t-1))$$

$$\times P(x'(t-1) \rightarrow x(t))P(g'(t-1) \rightarrow g(t)), \tag{24}$$

where $P(x(t-1) \rightarrow x'(t-1))$, $P(g(t-1) \rightarrow g'(t-1))$, $P(x'(t-1) \rightarrow x(t))$ and $P(g'(t-1) \rightarrow g(t))$ are all only depend on x and g at $t-1$. Therefore, $P(T_q(q(t-1)) = q(t))$ also only depends on the states $y_i(t-1), 1 \leq i \leq n$ at time $t-1$. Consequently, the group state sequence $\{q(t); t \geq 0\}$ has the property of a Markov chain.

Finally, $P(x(t-1) \rightarrow x'(t-1))$, $P(g(t-1) \rightarrow g'(t-1))$, $P(x'(t-1) \rightarrow x(t))$ and $P(g'(t-1) \rightarrow g(t))$ are all independent of t, so is $P(T_y(y(t-1)) = y(t))$. Thus, $P(T_q(q(t-1)) = q(t)$ is also independent of t, which implies that this state sequence is also homogeneous.

In summary, the group state sequence $\{q(t); t \geq 0\}$ is a finite, homogeneous Markov chain.

4.3   Global Convergence of Cuckoo Search

For the globally optimal solution g_b for an optimization problem $<\Omega_s, f>$, the optimal state set is defined as $R = \{y = (x, g)|f(g) = f(g_b), y \in Y\}$. In addition, for the globally optimal solution g_b to an optimization problem $<\Omega_s, f>$, the optimal group state set can be defined as

$$H = \{q = (y_1, y_2, ..., y_n)|\exists y_i \in R, 1 \leq i \leq n\}. \tag{25}$$

With the above results and definitions, we are now ready to prove the following theorems:

Theorem 5 *Given the position state sequence $\{y(t); t \geq 0\}$ in cuckoo search, the state set R of the optimal solutions corresponding to optimal nests/cuckoos form a closed set on Y.*

Proof For $\forall y_i \in R, \forall y_j \notin R$, the probability for $T_y(y_j) = y_i$ is $P(T_y(y_j) = y_i) = P(x_j \rightarrow x_j')P(g_j \rightarrow g_j')P(x_j' \rightarrow x_i)P(g_j' \rightarrow g_i)$. Since for $\forall y_i \in R$ and $\forall y_j \notin R$, it holds that $f(g_i) \geq f(g_j') = f(g_b) = \inf(f(a)), a \in \Omega_s$.

From Eqs. (18–20), we have $P(g_j \rightarrow g_j')P(g_j' \rightarrow g_i) = 0$, which leads to $P(T_y(y_j) = y_i) = 0$. This condition implies that R is closed on Y.

Theorem 6 *Given the group state sequence $\{q(t); t \geq 0\}$ in cuckoo search, the optimal group state set H is closed on the group state space Q.*

Proof From Eq. (21), the probability

$$P(T_q(q_j) = q_i) = \prod_{k=1}^{n} P(T_y(y_{jk}) = y_{ik}), \tag{26}$$

for $\forall q_i \in H, \forall q_j \in H$ and $T_q(q_j) = q_i$. Since $\forall q_i \in H$ and $\forall q_j \notin H$, in order to satisfy $T_q(q_j) = q_i$, there exists at least one cuckoo whose position will transfer from the inside of R to the outside of R. That is, $\exists T_y(y_{jk}) = y_{ik}, y_{jk} \in R, y_{ik} \notin R, 1 \leq k \leq n$. From the previous theorem, we know that R is closed on Y, which means that $P(T_y(y_{jk}) = y_{ik}) = 0$. Therefore,

$$P(T_q(q_j) = q_i) = \prod_{k=1}^{n} P(T_y(y_{jk}) = y_{ik}) = 0.$$

From the definition of a closed set, we can conclude that the optimal set H is also closed on Q.

Theorem 7 *In the group state space Q for cuckoos/nests, there does not exist a non-empty closed set B so that $B \cap H = \emptyset$.*

Proof Reductio ad absurdum. Assuming that there exists a close set B so that $B \cap H = \emptyset$ and that $f(g_j) > f(g_b)$ for $q_i = (g_b, g_b, ..., g_b) \in H$ and $\forall q_j = (y_{j1}, y_{j2}, ..., y_{jn}) \in B$, then Eq. (21) implies that

$$P(T_q(q_j) = q_i) = \prod_{k=1}^{n} P(T_y(y_{jk}) = y_{ik}).\tag{27}$$

For each $P(T_y(y_j) = y_i)$, it holds that $P(T_y(y_j) = y_i) = P(x_j \to x_j')P(g_j \to g_j')P(x_j' \to x_i)P(g_j' \to g_i)$. Since $P(g_j' \to g_i) = 1, P(g_j \to g_j'), P(x_j \to x_j')P(x_j' \to x_i) > 0$, then $P(T_y(y_j) = y_i) \neq 0$, implying that B is not closed, which contradicts with the assumption. Therefore, there exists no non-empty closed set outside H in Q.

Using the above definitions and results, it is straightforward to arrive another theorem:

Theorem 8 *Assuming that a Markov chain has a non-empty set C and there does not exist a non-empty closed set D so that $C \cap D = \emptyset$, then*

$$\lim_{n \to \infty} P(x_n = j) = \pi_j,$$

only if $j \in C$, and $\lim_{n \to \infty} P(x_n = j) = 0$ only if $j \notin C$.

Now using the above three theorems, it is straightforward to show

Theorem 9 *When the number of iteration approaches infinity, the group state sequence will converge to the optimal state/solution set H.*

This is the foundation for proving the global convergence theorem, which states

Theorem 10 *The cuckoo search with the Markov chain model outlined earlier has guaranteed global convergence.*

Proof Since the iteration process in cuckoo search always keeps/updates the current global best solution for the whole population, which ensures that it satisfies the first convergence condition. In addition, the previous theorem means that the group state sequence will converge towards the optimal set after a sufficiently large number of iterations or infinity. Thus, the probability of not finding the globally optimal solution is asymptotically 0, which satisfies the second convergence condition. Consequently, from Theorem 1, we can conclude that cuckoo search has guaranteed global convergence towards its global optimality.

5 Validation by Numerical Experiments

All new algorithms should be validated using various benchmarks to test their basic performance, rate of convergence and other properties. However, since the cuckoo search has been tested in the literature with a diverse range of benchmarks and design case studies, the numerical experiments we have done here are mainly to see if the global convergence can be reached easily and the rate of convergence. For this purpose, we have selected five benchmark functions with different modalities and objective landscapes:

The first function is the Ackley function [1]

$$f(\mathbf{x}) = -20 \exp\left[-\frac{1}{5}\sqrt{\frac{1}{d}\sum_{i=1}^{d} x_i^2} \right] - \exp\left[\frac{1}{d}\sum_{i=1}^{d} \cos(2\pi x_i) \right] + 20 + e, \qquad (28)$$

which has a global minimum $f_* = 0$ at $(0, 0, ..., 0)$. This function is highly nonlinear and multimodal.

De Jong's functions is unimodal and convex, which can be written as

$$f(\mathbf{x}) = \sum_{i=1}^{n} x_i^2, \quad -5.12 \le x_i \le 5.12, \qquad (29)$$

whose global minimum is obviously $f_* = 0$ at $(0, 0, ..., 0)$. It is also commonly referred to as the sphere function.

Rosenbrock's function

$$f(\mathbf{x}) = \sum_{i=1}^{d-1} \left[(x_i - 1)^2 + 100(x_{i+1} - x_i^2)^2 \right], \qquad (30)$$

has a narrow valley where lies its global minimum $f_* = 0$ at $\mathbf{x}_* = (1, 1, ..., 1)$ in the domain $-5 \le x_i \le 5$ where $i = 1, 2, ..., d$.

Xin-She Yang's forest-like function

$$f(\mathbf{x}) = \left(\sum_{i=1}^{d} |x_i| \right) \exp\left[-\sum_{i=1}^{d} \sin(x_i^2) \right], \quad -2\pi \le x_i \le 2\pi, \qquad (31)$$

has a global minimum $f_* = 0$ at $(0, 0, ..., 0)$. This function is highly nonlinear and multimodal, and its first derivatives do not exist at the optimal point due to the modulus $|.|$ factor.

Zakharov's function

$$f(\mathbf{x}) = \sum_{i=1}^{d} x_i^2 + \left(\sum_{i=1}^{d} \frac{ix_i}{2} \right)^2 + \left(\sum_{i=1}^{d} \frac{ix_i}{2} \right)^4, \qquad (32)$$

Fig. 1 Convergence of 5 test functions using cuckoo search

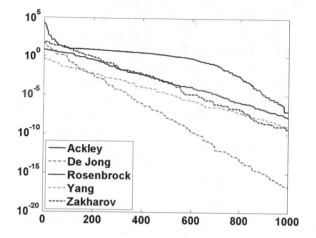

is nonlinear and has its global minimum $f(\mathbf{x}_*) = 0$ at $\mathbf{x}_* = (0, 0, ..., 0)$ in the domain $-5 \leq x_i \leq 5$.

All these functions have the global minimum $f_{\min} = 0$, and such simplicity allows to test the accuracy of an algorithm with various dimensions. For this reason, we set $d = 8$ for all these five functions.

For the implementation of cuckoo search algorithm, we have used $n = 25$, $\lambda = 1.5$, $p_a = 0.25$ and a fixed number of iterations $t = 1000$. The convergence graphs for all these functions are summarized and shown in Fig. 1 where the vertical axis is plotted using the logarithm scale. From the figure, it is clearly seen that the cuckoo search can converge quickly and the best objective values decrease in an almost exponential manner, except for Rosenbrock's function which has a narrow valley. However, as the search has gone through some part of the valley during iterations, its objective values once again decrease almost exponentially with a higher slope.

6 Conclusions

Cuckoo search is an efficient optimization algorithm with a wide range of applications. We have used the Markov chain theory and proved the global convergence of the simplified version of cuckoo search. Then, we have used a few benchmark functions with diverse properties to show that CS can indeed converge very quickly. In fact, cuckoo search has been used in many applications and the rate of convergence is usually very good in practice.

The current results are mainly for a simplified variant, derived from the standard cuckoo search. It can be expected that this methodology can be used to prove both standard cuckoo search algorithm and its variants. Therefore, it will be useful if further research can focus on the extension of the proposed methodology to analyze the convergence of other variants of the cuckoo search algorithm and other metaheuristic algorithms.

In addition, though we can show the cuckoo search will converge in the probabilistic sense, there is no information about how quickly it can convergence. Therefore, further research can also try to figure out the rate of convergence and its link to the algorithmic structure, parameter setting and even the modal shapes of the objective landscapes. After all, the rate of convergence is crucially important from the implementation point of view.

Furthermore, as the setting of parameters in an algorithm can affect the performance of the algorithm significantly, and consequently affect the rate of convergence. It would be useful to find the relationship between parameter values and the convergence rate, and then to control the rate of convergence by fine tuning the algorithm-dependent parameters.

Acknowledgements The authors would like to thank the financial support by Shaanxi Provincial Education Grant (12JK0744) and Shaanxi Provincial Soft Science Foundation (2012KRM58).

References

1. Ackley, D.H.: A Connectionist Machine For Genetic Hillclimbing. Kluwer Academic Publishers (1987)
2. Bhargava, V., Fateen, S.E.K., Bonilla-Petriciolet, A.: Cuckoo search: a new nature-inspired optimization method for phase equilibrium calculations. Fluid Phase. Equilib. **337**, 191–200 (2013)
3. Chandrasekaran, K., Simon, S.P.: Multi-objective scheduling problem: hybrid appraoch using fuzzy assisted cuckoo search algorithm. Swarm Evol. Comput. **5**(1), 1–16 (2012)
4. Clerc, M., Kennedy, J.: The particle swarm–explosion stability and convergence in a multidimensional complex space. IEEE Trans. Evol. Comput. **6**(1), 58–73 (2002)
5. Dhivya, M., Sundarambal, M., Anand, L.N.: Energy efficient computation of data fusion in wireless sensor networks using cuckoo based particle approach (CBPA). Int. J. Commun. Netw. Syst. Sci. **4**(4), 249–255 (2011)
6. Dhivya, M., Sundarambal, M.: Cuckoo search for data gathering in wireless sensor networks. Int. J. Mob. Commun. **9**(4), 642–656 (2011)
7. Durgun, I., Yildiz, A.R.: Structural design optimization of vehicle components using cuckoo search algorithm. Mater. Test. **3**(3), 185–188 (2012)
8. Gandomi, A.H., Yang, X.S., Alavi, A.H.: Cuckoo search algorithm: a metaheuristic approach to solve structural optimization problems. Eng. Comput. **29**(1), 17–35 (2013)
9. He, X.S., Yang, X.S., Karamanoglu, M., Zhao, Y.X.: Global convergence analysis of the flower pollination algorithm: a discrete-time Markov chain approach. Proced. Comput. Sci. **108**(1), 1354–1363 (2017)
10. Jiang, M., Luo, Y.P., Yang, S.Y.: Stochastic convergence analysis and parameter selection of the standard particle swarm optimization algorithm. Inf. Process. Lett. **102**(1), 8–16 (2007)
11. Kennedy, J., Eberhart, R.C.: Particle swarm optimization. In: Proceedings of IEEE International Conference on Neural Networks. Piscataway, NJ, pp. 1942–1948 (1995)
12. Koziel, S., Yang, X.S.: Computational Optimization Methods And Algorithms. Springer, Berlin (2011)
13. Moravej, Z., Akhlaghi, A.: A novel approach based on cuckoo search for DG allocation in distribution network. Electr. Power Energy Syst. **44**(1), 672–679 (2013)
14. Noghrehabadi, A., Ghalambaz, M., Vosough, A.: A hybrid power series–cuckoo search optimization algorithm to electrostatic deflection of micro fixed-fixed actuators. Int. J. Multidiscip. Sci. Eng. **2**(4), 22–26 (2011)

15. Pavlyukevich, I.: Lévy flights, non-local search and simulated annealing. J. Comput. Phys. **226**(2), 1830–1844 (2007)
16. Ren, Z.H., Wang, J., Gao, Y.L.: The global convergence analysis of particle swarm optimization algorithm based on Markov chain. Control Theory Appl. **28**(4), 462–466 (2011). (in Chinese)
17. Taweewat, P., Wutiwiwatchai, C.: Musical pitch estimation using a supervised single hidden layer feed-forward neural network. Expert Syst. Appl. **40**(2), 575–589 (2013)
18. Valian, E., Mohanna, S., Tavakoli, S.: Improved cuckoo search algorithm for feedforward neural network training. Int. J. Artif. Intell. Appl. **2**(3), 36–43 (2011)
19. Valian, E., Tavakoli, S., Mohanna, S., Haghi, A.: Improved cuckoo search for reliability optimization problems. Comput. Ind. Eng. **64**(1), 459–468 (2013)
20. Vazquez, R.A.: Training spiking neural models using cuckoo search algorithm. In: IEEE Congress on Eovlutionary Computation (CEC'11), pp. 679–686 (2011)
21. Walton, S., Hassan, O., Morgan, K., Brown, M.R.: Modified cuckoo search: a new gradient free optimization algorithm. Chaos Solitons Fractals **44**(9), 710–718 (2011)
22. Wang, F., He, X.S., Wang, Y., Yang, S.M.: Markov model and convergence analysis of cuckoo search algorithm. Comput. Eng. **38**(11), 180–185 (2012) (in Chinese)
23. Yang, X.S., Deb, S.: Cuckoo search via Lévy flights. In: Proceeding of World Congress on Nature and Biologically Inspired Computing (NaBic), pp. 210–214. IEEE Publications, Coimbatore, India, USA (2009)
24. Yang, X.S., Deb, S.: Engineering optimization by cuckoo search. Int. J. Math. Model. Numer. Optim. **1**(4), 330–343 (2010)
25. Yang, X.S., Deb, S.: Cuckoo search: recent advances and applications. Neural Comput. Appl. **24**(1), 169–174 (2014)
26. Yang, X.S.: Engineering Optimization: An Introduction With Metaheuristic Applications. Wiley (2010)
27. Yang, X.S.: A new metaheuristic bat-inspired algorithm. In: Nature-Inspired Cooperative Strategies for Optimization (NICSO), vol. 284, pp. 65–74. SCI, Springer (2010)
28. Yang, X.S.: Bat algorithm for multi-objective optimisation. Int. J. Bio-Inspir. Comput. **3**(5), 267–274 (2011)
29. Yang, X.S.: CUCKOO search and firefly algorithm: theory and applications. In: Studies in Computational Intelligence, vol. 516 Springer (2014)
30. Yang, X.S.: Nature-Inspired Optimization Algorithms. Elsevier Insight, London (2014)
31. Yang, X.S., Deb, S.: Multiobjective cuckoo search for design optimization. Comput. Oper. Res. **40**(6), 1616–1624 (2013)
32. Yildiz, A.R.: Cuckoo search algorithm for the selection of optimal machine parameters in milling operations. Int. J. Adv. Manuf. Technol. **64**(1), 55–61 (2013)
33. Zheng, H.Q., Zhou, Y.: A novel cuckoo search optimization algorithm based on Gauss distribution. J. Comput. Inf. Syst. **8**(10), 4193–4200 (2012)

On Efficiently Solving the Vehicle Routing Problem with Time Windows Using the Bat Algorithm with Random Reinsertion Operators

Eneko Osaba, Roberto Carballedo, Xin-She Yang, Iztok Fister Jr., Pedro Lopez-Garcia and Javier Del Ser

Abstract An evolutionary and discrete variant of the Bat Algorithm (EDBA) is proposed for solving the Vehicle Routing Problem with Time Windows, or VRPTW. The EDBA developed not only presents an improved movement strategy, but it also combines with diverse heuristic operators to deal with this type of complex problems. One of the main new concepts is to unify the search process and the minimization of the routes and total distance in the same operators. This hybridization is achieved by using selective node extractions and subsequent reinsertions. In addition, the new approach analyzes all the routes that compose a solution with the intention of enhancing the diversification ability of the search process. In this study, several variants of the EDBA are shown and tested in order to measure the quality of both metaheuristic

E. Osaba (✉) · R. Carballedo · P. Lopez-Garcia
Deusto Institute of Technology (DeustoTech), University of Deusto,
Av. Universidades 24, 48007 Bilbao, Spain
e-mail: e.osaba@deusto.es

R. Carballedo
e-mail: roberto.carballedo@deusto.es

P. Lopez-Garcia
e-mail: p.lopez@deusto.es

X.-S. Yang
School of Science and Technology, Middlesex University,
Hendon Campus, London NW4 4BT, UK
e-mail: x.yang@mdx.ac.uk

I. Fister Jr.
Faculty of Electrical Engineering and Computer Science, University of Maribor,
Smetanova 17, 2000 Maribor, Slovenia
e-mail: iztok.fister1@um.si

J. Del Ser
TECNALIA, 48160 Derio, Spain

J. Del Ser
University of the Basque Country (UPV/EHU), 48013 Bilbao, Spain

J. Del Ser
Basque Center for Applied Mathematics (BCAM), 48009 Bilbao, Spain
e-mail: javier.delser@tecnalia.com

© Springer International Publishing AG 2018
X.-S. Yang (ed.), *Nature-Inspired Algorithms and Applied Optimization*,
Studies in Computational Intelligence 744, https://doi.org/10.1007/978-3-319-67669-2_4

algorithms and their operators. The benchmark experiments have been carried out
by using the 56 instances that compose the 100 customers Solomon's benchmark.
Two statistical tests have also been carried out so as to analyze the results and draw
proper conclusions.

Keywords Bat algorithm · Discrete bat algorithm · Vehicle routing problem with
time windows · VRPTW · Combinatorial optimization · Traveling salesman
problem

1 Introduction

The rapid advance of technology has made the logistic management increasingly
important in ever-increasingly connected societies, which has led transport networks
to be very demanding. To meet such demands, companies have to be innovative
in designing their logistic networks, and a competitive logistic network can make
the difference between some companies and others. Consequently, the development
of efficient methods for proper logistics and routing planning is a hot topic in the
research community.

   To model and optimize a logistic network, all relevant issues have to be addressed
in an appropriate way using appropriate techniques. In this case, we focus our attention here on one of these areas: artificial intelligence. In fact, route planning problems
and their resolution is one of the most recurrent topics related to artificial intelligence. More specifically, the problems arisen in this field are normally named as
routing problems, and they fall into the combinatorial optimization category. The
most studied problems in this field are the Vehicle Routing Problem (VRP) and the
Traveling Salesman Problem (TSP). Besides the basic TSP and VRP, many variations of these problems can be found in the literature. In this chapter, the attention is
focused on one of these variants: the Vehicle Routing Problem with Time Windows,
or VRPTW. Briefly speaking, in the VRPTW, each client imposes a time window for
the start and the end of the service. This problem will be explained in greater detail
later.

   A few solution methods can be found in the literature to deal with this kind of
problems properly. The most well-known approaches for this purpose are probably
the exact methods [1], heuristics and metaheuristics. Here, we focus our attention
on metaheuristic methods. For example, some classical examples of local search-
based methods are Simulated Annealing [2] and Tabu Search [3]. On the other hand,
population-based techniques such as the Ant Colony Optimization [4], Genetic Algorithms (GA) [5, 6], and Particle Swarm Optimization [7] are some of the most used
alternatives.

   Although classical techniques can somehow manage to solve certain class of such
problems, they are not sufficiently effective, and thus the development of novel metaheuristics for tackling optimization problems, especially for routing problems, is a
hot topic in this area of research. Consequently, many different methods have been

proposed in recent years. Some examples of these methods are the Imperialist Competitive Algorithm, proposed by Atashpaz-Gargari and Lucas in 2007 [8], the Artificial Bee Colony, presented by Karaboga and Basturk in 2007 [9], and the bat algorithm developed by Yang in 2010 [11]. To the interested readers, some additional successful methods will be described in following sections.

For the current study, the method that we have selected for addressing the above mentioned VRPTW is the Bat Algorithm (BA). This metaheuristic is a nature-inspired algorithm, based on the echolocation behavior of micro-bats, which was proposed by Yang in 2010 [10]. From the review of some of the recent literature [11, 12], the BA has been successfully applied to wide variety of optimization fields and problems since its proposal. Furthermore, recent works such as [13, 14] confirm that BA still attracts a lot of interest from the scientific community. In this sense, despite the fact that the BA has been applied to many different optimization problems up to date, it has not been applied yet to the well-known VRPTW. Thus, this motivates us to carry out this current work. The detailed explanation of BA will be given in following sections.

It is worth highlighting that we have used some novel route optimization operators for enhancing the performance of the developed algorithm. These operators, which will be described in following sections, perform selective extractions of nodes in an attempt of minimizing the number of routes of the current solution. At this moment, these operators have only been used once in the literature, inside a Firefly Algorithm [15]. For this reason, this is the first time in the literature that such heuristic functions are used in the BA for routing problems.

For the purpose of proving that the implemented Evolutionary Discrete Bat Algorithm (EDBA) is a promising approach to solve the VRPTW, an experiment composed by 56 different instances has been conducted in this work. The results obtained by some variants of the EDBA are compared. In addition, two different statistical tests have been conducted with the results obtained: the non-parametric Friedmans test for multiple comparisons, and the post-hoc Holm's test.

Therefore, the rest of the paper is organized as follows. Section 2 presents the related background with an emphasis on routing problems and nature-inspired meta-heuristics for their resolution. After that, in Sect. 3, the philosophy of the basic BA is detailed. Then, in Sect. 4, a brief description of the VRPTW can be found. Then, the proposed EDBA and our route optimization operators are described in Sect. 5. Furthermore, in Sect. 6, the experimentation performed for the validation of the study is detailed. Finally, the paper concludes with with suggestions for further work in Sect. 7.

2 Background

Nowadays, route planning is one of the most studied fields. Problems arisen in this field are usually known as vehicle routing problems, which are a particular case of problems of combinatorial optimization. Probably, the most used and well-known routing problems are the Traveling Salesman Problem [16] and the Vehicle Routing Problem [17], which are the focus of a huge amount of studies in the literature [18, 19]. In addition, the VRPTW is the main problem of our attention here and is also one of the most cited and used, as can be seen in different works such as [20] and [21].

The reasons for the popularity and importance of these problems are two folds: the scientific aspect, and the social one. On the one hand, being NP-Hard, most of the problems arising in this field have an extraordinary complexity, and thus their solutions pose a major challenge for the scientific community. On the other hand, routing problems are usually built to address a real-world situation related to logistics or transportation, which is directly linked to the profit of a business service.

Even the problems are challenging to solve, several approaches can be found in the literature to tackle this kind of problems. The exact methods [1, 22], heuristics and metaheuristics have all been attempted. For example, as can be seen in the work by Braysy and Gendreau in 2005 [23], metaheuristics are a good approach for solving the VRPTW.

To be more specific within the category of metaheuristics, nature-inspired methods are among the most used approaches for tackling this sort of problems in the current literature [24]. In this sense, some of these recently proposed approaches that can be classified in this category are the Bat Algorithm (BA), Firefly Algorithm (FA), and Cuckoo Search (CS). The first one, and the one that is used in this work, is the BA. This metaheuristic was proposed by Yang in 2010 [10], and it is based on the echolocation behavior of microbats, which can find their prey and discriminate different kinds of insects even in complete darkness. Recent literature reviews and surveys [11, 12] show that BA has been successfully applied to different optimization fields and problems since its proposal. Focusing in routing problems, several recently published papers have shown that the BA is a promising technique also in this field. For example, in [25], which was published in 2015, an adapted variant of this algorithms for solving the well-known Capacitated VRP. The Adapted BA developed in that study allows a large diversity of the population and a balance between global and local search.

A more recent work is proposed in [26] by Zhou et al. in which the same Capacitated VRP is faced. In their paper, a hybrid BA with path relinking is described. This approach is constructed based on the framework of the continuous BA, in which the greedy randomized adaptive search procedure and path relinking are effectively integrated. Additionally, with the aim of improving the performance of the technique, the random subsequences and single-point local search are operated with a certain probability.

Regarding the second of above mentioned methods, that is FA, proposed by Yang in 2008 [27]. This a nature-inspired algorithm is based on the flashing behavior of fireflies, which acts as a signaling system to attract other fireflies. This metaheuristic algorithm has been also applied to a wide range of optimization fields and problems since its proposal [28, 29]. Like the BA, this method has also shown a promising performance for routing problems. In [30], for example, the first application of the FA was presented for solving the TSP. In order to do that, the authors adapted the FA, which was firstly proposed for tackling continuous problems, enhancing it with an evolutionary and discrete behavior.

Another interesting example of application is the one presented in [31], in which a hybrid variant of the FA is proposed to solve a time-dependent VRP with multi-alternative graph, in order to reduce the fuel consumption. The developed variant of FA is a Gaussian Firefly Algorithm. The most interesting part of that paper is the real-world case study, focused on a distribution company, established in Esfahan, Iran. More recently, FA has been compared with other nature-inspired heuristics for a bi-objective variant of the classical VRP problem with pickup and delivery deadlines, multiple concurrent vehicles and selectivity of nodes. Interestingly, in their work, the quality of routes is determined by the Pareto trade-off between the profit gained by the delivery of goods along the routes and a measure of fairness in the share of the revenues of the transport company [32].

The third of the algorithms previously mentioned is the CS, developed by Yang and Deb in 2009 [33]. It was inspired by the obligate brood parasitism of some cuckoo species by laying their eggs in the nests of other host birds (of other species such as warblers). The CS has also been modified to solve routing problems, as can be seen, for example, in the work published in 2014 by Ouaarab et al. [34]. In that paper, the authors presented the first adaptation of the CS to the well-known TSP, creating a discrete variant of the CS with promising results. The authors also tested their proposed discrete CS against a set of benchmarks of symmetric TSP from the well-known TSPLIB library.

More examples of the CS applied to the VRP can be found in the literature. In [35], for example, a discrete CS algorithm for the capacitated VRP is presented. The main novelty of this method is not only its application itself, but also the Taguchi-based Parameter Setting developed for the parameter optimization. Besides that, in 2016, the reputable Information Sciences journal published a paper in which four different soft computing methods were applied for solving also the Capacitated VRP [36]. One of these approaches was an advanced CS, which introduced new adjustments and features for improving its efficiency. Another example is the paper presented by Chen and Wang in 2016 [37], in which a hybrid CS was proposed for the solving the VRP in logistics distribution systems. This novel algorithm was based on the combination of Optical Optimization, Particle Swarm Optimization and CS. Specifically, in their method, optical optimization was introduced to initialize population for obtaining a group of initial values with high quality, which were then optimized according to PSO. After each iterative operation for keeping the optimal individual, CS was used to optimize the rest of the individuals.

Another metaheuristic is a music-inspired Harmony Search (HS). This technique was firstly proposed by Geem et al. in 2001 as a phenomenon-mimicking metaheuristic [38], inspired by the improvisation process of jazz musicians. There are a wide range of applications of HS in the literature [39–41]. The HS has also been applied to routing problems several times, showing also a promising performance. The paper presented by Geem et al. in 2005 collected some of the most interesting works up to that date on this topic [42]. Another research related to the HS was the one that can be found in [43], which presented a discrete variant of the HS in order to solve the challenging the Selective Pick-Up and Delivery VRP with Delayed Drop-Off. Additionally, the recent work by Bounzidi and Riffi in 2014 described the adaptation of the HS for solving the TSP.

Another meta-heuristic mentioned in this background section is the Gravitational Search algorithm (GS), proposed by Rashedi et al. in 2009 [44], and it was based on the metaphor of gravitational interaction between masses. GS has also been used in many applications [45–47]. Concerning routing problems, Nodehi et al. in 2016 [48] presented a randomized GS algorithm for the solving of the TSP. The GS implemented in this work was based on randomized search concepts using two of the four main parameters of velocity and gravitational force in physics. The performance of the developed method was compared with some additional well-known methods, such as the Genetic Algorithm, showing a promising performance.

Regarding VRP problems, the work [49] explored the application of a discrete variant of the GS to the Open VRP. Being firstly proposed to solve continuous problems, the main challenge of the authors of that paper was to adapt all the characteristics of the basic variant of the GS to the discrete optimization. As has been mentioned, the problem to solve in this case is the Open VRP, which is a variant in which vehicles are not required to return to the depot. Finally, the paper by Hosseinabad et al. in 2017 [50] presented another approach of the GS to solve the Capacitated VRP with enhanced performance.

There are many challenging issues related to VRPTW, and the number of publications related to this problem is increasing. In [51], for example, Desaulniers et al. presented a set of exact algorithms to tackle the electric VRPTW. On the other hand, Belhaiza et al. proposed in their work [52] a hybrid variable neighborhood tabu search approach for solving the VRPTW. A multiple ant colony system was developed for the VRPTW with uncertain travel times by Toklu et al. [53]. Finally, an a hybrid generational algorithm for the periodic VRPTW can be found in [54]. In relation to the above mentioned nature-inspired methods and the VRPTW, in [15], an evolutionary discrete firefly algorithm was proposed for the resolution of this problem, using the same operators in the experimentation. An additional paper is the one presented by [55], in which a hybrid variant of the HS was presented to deal with the VRPTW.

Since the literature in this area is expanding, it is not possible to review all the relevant work. Interested readers can refer to literature reviews in [11] about the BA, [29] about FA, and [56] about the CS. On the other hand, for additional information about the VRPTW and its solution methods, the work presented in [57, 58] is highly recommended. As mentioned in the introduction, this present work is the first time in the literature that the BA is applied to the VRPTW. In the rest of this chapter, we will describe our proposed approach in greater detail.

Algorithm 1: Pseudo code of the basic BA
1 Define the objective function $f(x)$;
2 Initialize the bat population $X = x_1, x_2, ..., x_n$;
3 **for** *each bat x_i in the population* **do**
4 $\quad$ Initialize the pulse rate r_i, velocity v_i and loudness A_i;
5 $\quad$ Define the pulse frequency f_i at x_i;
6 **end**
7 **repeat**
8 $\quad$ **for** *each bat x_i in the population* **do**
9 $\quad\quad$ Generate new solutions through Equations (1), (2) and (3);
10 $\quad\quad$ **if** *rand>r_i* **then**
11 $\quad\quad\quad$ Select one solution among the best ones;
12 $\quad\quad\quad$ Generate a local solution around the best one;
13 $\quad\quad$ **end**
14 $\quad\quad$ **if** *rand<A_i and $f(x_i)<f(x_*)$* **then**
15 $\quad\quad\quad$ Accept the new solution;
16 $\quad\quad\quad$ Increase r_i and reduce A_i;
17 $\quad\quad$ **end**
18 $\quad$ **end**
19 **until** *termination criterion not reached*;
20 Rank the bats and return the current best bat of the population;

3 Bat Algorithm

In this section, the basic variant of the BA is fully described before we proceed to introduce further modifications and enhancements. As we have briefly mentioned in previous sections, the BA is a nature-inspired metaheuristic, whose main idea is to imitate the echolocation features of microbats with some idealized rules outlined as follows [10]:

- All bats use echolocation to detect the distance and can differentiate between an obstacle and a prey (bad or good solutions, respectively).
- All bats fly randomly with a velocity v_i at position x_i with a varying frequency from f_{min} to f_{max}, loudness A_i and pulse emission rate r.
- In the real-world, the loudness and emission rates of bats can vary in many different ways. Here, we assume that the loudness varies monotonically from A_0 to a lower (quieter) value A_{min}, while r varies from a lower value to a higher value.

The main steps of this BA are summarized as the pseudocode as shown in Algorithm 1. Taking a quick look at this pseudo-code, it can be seen that the first six lines correspond to the initialization process. First, the objective function is defined, and the initial population is initialized. Each bat of the population represents a possible solution to the addressed problem, in this case, the VRPTW. After that, velocity v_i, frequency f_i, pulse rate r_i and loudness A_i parameters are initialized and defined.

After this initialization phase, the main evolution of solutions in the algorithm are executed. At each generation, each bat of the swarm moves through the search space by updating its velocity and position. More specifically, the following equations are used for this movement:

$$f_i = f_{min} + (f_{max} - f_{min})\beta \tag{1}$$

$$v_i^t = v_i^{t-1} + [x_i^{t-1} - x_*]f_i \tag{2}$$

$$x_i^t = x_i^{t-1} + v_i^t \tag{3}$$

where β is a uniformly distributed random number in $[0,1]$, and x_* represents the current best solution of the whole population. In addition, v_i^t and x_i^t denote the velocity and position, respectively, of a bat i at time step t. Furthermore, the results of Eq. (1) is used to control the pace and range of bats movement.

If a solution is selected among the best ones, a new solution for each bat is generated using a random walk

$$x_{new} = x_{old} + \varepsilon A^t \tag{4}$$

where ε is a randomly generated number within the interval $[-1, 1]$, and A^t is the average loudness of the swarm at time step t. Finally, the rate r_i and the loudness A_i of each bat are updated, only if the conditions shown in the line 14 of Algorithm 1 are met. This update is performed as follows:

$$r_i^{t+1} = r_i^0[1 - \exp(-\gamma t)] \tag{5}$$

$$A_i^{t+1} = \alpha A_i^t \tag{6}$$

where α and γ are constants. Thereby, for any $0 < \alpha < 1$ and $\gamma > 0$ we have

$$A_i^t \to 0, r_i^t \to r_i^0, \text{ as } t \to \infty \tag{7}$$

In most cases in the literature, $\alpha = \gamma$ is used in order to simplify the implementation of the method. In the present study, $\alpha = \gamma = 0.98$ is used. We have selected this value after an empirical experiment using a range of values from 0.90 to 0.99.

4   Vehicle Routing Problem with Time Windows

As we have pointed out in Sect. 2, the VRPTW is an extension of the classic and widely studied VRP. In addition to the basic constraints inherent from the VRP, each client that composes a VRPTW instance has an associated time window $[e_i,l_i]$. More specifically, this time window has a lower limit e_i and an upper limit l_i which must be respected by the vehicle that will attend the demand of the client. This means that the service in every customer must be performed after e_i and before l_i.

Obviously, a route is not feasible if a vehicle tries to serve any customer after the upper limit of this range. On the other hand, a route would be feasible if the vehicle reaches a client before its lower limit. In this last special situation, the client cannot be served before this limit, so that the vehicle should be waiting until e_i to start the delivery.

Besides that, the central depot, which is the starting and ending point of all the routes and vehicles, has also a time window, which restricts the period of the whole activity. Apart from this temporal window, the problem can also take into account the customer's service time. This parameter is the time that the vehicle needs to spend on the client in order to perform the delivery properly. This is a factor to be taken into account to calculate if the vehicle arrives on time to the next customer. Furthermore, the variant that we are using in this paper is the VRPTW with hard time windows. In this sense, there is also another variant that enables noncompliance with some time window (with a penalization in the objective function).

Being one of the most famous variant of the VRPs, this problem has been widely studied both in the past [20, 21], and nowadays [59, 60]. One reason why the VRPTW is so interesting is its dual nature, since it is considered as a two phase problem. The first of these phases concerns the vehicle routing, while the second one regards the planning phase or customer scheduling.

An additional reason for its popularity is its easy adaptation to the real-world applications. The great majority of distribution chains, customers have strong temporal constraints that have to be fulfilled, and the VRPTW perfectly fits with this kind of real-world situations.

Regarding the mathematical formulation of VRPTW, it can take several forms, using a different amount of variables [61, 62]. One of the most interesting formulations can be found in [63].

5 Our Proposed Approach for Solving the VRPTW

In this section, the description of our EDBA for the VRPTW is provided (Sect. 5.1). A more detailed description of the proposed novel route optimization operator will be given in Sect. 5.2.

5.1 An Evolutionary Discrete Bat Algorithm

Before starting with the description of our proposed method, it is worth mentioning that the original BA was firstly developed for solving continuous optimization problems, and thus the standard BA cannot be directly applied to solve any discrete problem such as the VRPTW. Hence, some modifications in the structure of the basic BA should be performed in order to prepare it to solve the VRPTW.

First, in the EDBA, each bat of the swarm represents a possible and feasible solution for the VRPTW. Since the VRPTW is a minimization problem, the most attractive bats are those with a lower objective function value. Regarding the philosophy of both r_i and A_i parameters, it has remained exactly in the same form as that in the standard BA. Furthermore, with the intention of simplifying the complexity of the algorithm, the parameter f_i has not been considered.

Furthermore, the "velocity", v_i, has been modified. In the continuous variant of the BA, this parameter is calculated as has been shown in Eq. (2). However, this formula cannot be used in the same way for solving a discrete problem such as the VRPTW. Thus, we have related v_i to a distance measure between the bat i and the best bat of the swarm. It is worth pointing out that all the quantities are treated as unitless, and thus there is no need to worry about the unit of velocity. Obviously, the true physical quantities have units and the solutions will be given the right units when the final solutions are interpreted. Thus, all the quantities in BA are considered as mathematical values without units. For this purpose, we have adapted v_i using the well-known Hamming Distance in the following way:

$$v_i^t = \text{Random}[1, \text{HammingDistance}(x_i^t, x_*)] \qquad (8)$$

This means that the v_i of a bat i at time step t is a random number, which follows a discrete uniform distribution between 1 and the difference between this i and the best bat of the swarm. This difference is represented by the Hamming Distance, which is the number of non-corresponding elements in the sequence. A detailed example of this application can be found in [15].

Additionally, regarding the new bats generation, in the classic variant of the Bat Algorithm the movement of the bats is performed using the Eq. (3). Similar with the v_i parameters, this equation cannot be applied directly to a discrete problem such as the VRPTW. Thus, a modification has been proposed, and the movement of a bat i is determined by the following equation:

$$x_i^t \leftarrow MovementFunction(x_i^{t-1}, v_i^t) \qquad (9)$$

In other words, every bat examines at every generation a v_i number of its neighbors, and it chooses the best one as its current movement. Explained in other way, the bat i conducts a v_i number of movements, and it chooses the best one. In the proposed EDBA, a single operator to simulate the movement of bats is used. This operator is described in the next section.

Furthermore, regarding the local search procedure represented in Lines 10–12 of Algorithm 1, whether $rand > r_i$, one solution is randomly chosen among the best ones (in our performed experiments, one bat among the 10 best ones; or less, if v_i is lower than 10), and a local solution is generated around this one, using the well-known 2-opt* operator. After that, if the new solution is accepted, it replaces the current bat.

Algorithm 2: Pseudocode of the route minimization operator.

input : $Solution_{current}$, optimizeRoutes, proximityReinsertion
1 $ejectionPool = initEjectionPool(Solution_{current})$;
2 $Solution_{new} = removeEmptyRoutes(Solution_{current})$;
3 **if** optimizeRoutes **then**
4 | optimizeRoutes($Solution_{new}$) ;
5 **end**
6 **if** proximityReinsertion **then**
7 | reinsert(ejectionPool,$Solution_{new}$) ;
8 **end**
9 **if** $ejectionPool \neq \oslash$ **then**
10 | $Solution_{new} = parallelReconstruction(ejectionPool,Solution_{new})$;
11 **end**
12 **if** $Solution_{new}$ better than $Solution_{current}$ **then**
13 | $Solution_{current} = Solution_{new}$;
14 **end**
output: $Solution_{current}$

Finally, regarding the termination criterion, each technique finishes its execution when it reaches the generation (iteration) 101, or when there are 20 generations without any improvement in the best solution found.

5.2 Description of the Bat Movement Operator

In this section the operator used to simulate the movement of the bats is described. This operator is responsible for creating the neighbor solutions generated when a bat is performed its movement (Line 9 of the Algorithm 1).

Using the inspiration by the concept of "ejection chains" [64], a family of operators (whose objective is the reduction of the number of routes) have been presented in a previous work related to the Firefly algorithm [65]. These operators combine the "ejection chains" technique with other simple measures (such as the size of a route and the proximity of the customers with respect to the "center of gravity of a route"). The proposed operators were designed to increase the diversification ability of the traditional node and arc interchange based operators.

Using the results obtained in our previous work focused on Firefly Algorithm [65], in the present work we center our attention only on one operator: the "Random Route Elimination Operator—RrE-opt". As the name suggests, the operator is based on the removal of a route at random and the subsequent reinsertion of the clients of that route in the remaining routes. The main objective is to reduce the number of routes. This is the first criterion of the classical evaluation function for VRPTW.

Figure 1 illustrates a simple worked example of the RrE-opt operator. Furthermore, Algorithm 2 shows the description of this operator:

E. Osaba et al.

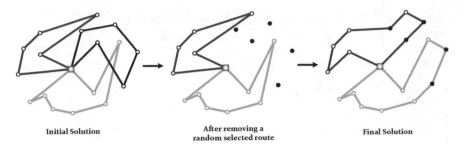

Initial Solution   After removing a   Final Solution
random selected route

Fig. 1   Example of the RrE-opt operator

- In the first step, a route is selected at random and it is removed from the current solution. The clients that were part of the removed route configure the *ejectionPool*. In the next steps the aim is to reinsert the customers in the remaining routes.
- After the route removal, two optional processes can be performed:

  - A local route optimization using the well-known Or-opt operator. The objective of this process is the reordering of the remaining routes to facilitate the reinsertion of the customers of the *ejectionPool*. Other optimization operators could be used but the Or-opt operator has been chosen for its speed and efficiency.
  - The reinsertion of the customers by proximity in the closest route. This process checks all clients that are in the *ejectionPool* and tries to insert them into the geographically most surrounded route. In this way, the total distance traveled tries to be reduced. This is the second criterion of the VRPTW evaluation function. To perform this reinsertion in an efficient way, the use of neighbor lists is recommended [66].

- The last step is to use a parallel initialization heuristic to reinsert clients that are still in the *ejectionPool*. In this step the heuristic of Campbell and Savelsbergh [67] is used for its speed and simplicity of implementation.

This new operator performs a more complex process than traditional VRPTW operators, but in spite of being more expensive in runtime, this operator has a great ability to reduce the number of routes during the search process. Reducing the number of routes in the context of VRPTW is often done as an independent process. With the proposed new operator, this process is implicitly integrated into the search process.

In the experimentation section below, four variants of the proposed EDBA will be compared. These variants will allow the evaluation of the two optional processes of the operator for the reduction of the number of routes. Its nomenclature will be: OR (only Optimize Routes process), PR (only Proximity Reinsertion process), FULL (both optional processes) and NONE (no optional process).

6 Experimentation

In this section the details of the experimentation conducted are described. The experimentation has two clear objectives: first, to show the use of the proposed EDBA algorithm; and second, to analyze the behavior of the new operator to reduce the number of routes for the VRPTW.

For the experimentation, Solomon's VRPTW benchmark has been used [68]. This set of problems consists of 56 instances of 100 customers classified into 6 categories (C1, C2, R1, R2, RC1 y RC2). The categories differ in the geographical distribution of the customers, the capacity of the vehicles and the compatibility of the time windows.

There are other VRPTW benchmarks with larger problems instances (such as Gehring & Homberger's[1]), but the objective of the work presented focuses on the use of the EDBA and the analysis of the new optimization operator for the VRPTW. For this reason, Solomon's benchmark is adequate and representative.

All the tests conducted in this work have been performed on an Intel Core i5-6200U CPU @ 2.40 GHz with 8 GB of RAM. The algorithms have been programed in Java and double precision is used for all numeric variables and parameters. The used operating systems has been Windows 7.

The evaluation function used is the classic hierarchical one that prioritizes first the number of routes (the minimum the best) and then the total travel distance (again the lower the best).

The experimentation has been performed with 4 variants of the proposed EDBA. Such variants differ in the use (or not) of the optional processes included in the optimization operator presented in Sect. 5.2. They are identified as: EDBA-OR (only Optimize Routes process), EDBA-PR (only Proximity Reinsertion process), EDBA-FULL (both optional processes) and EDBA-NONE (no optional process).

The parameterization for the EDBA used in the experimentation is the following:

- The swarm of bats (population) is composed of 25 individuals.
- The initial population is initialized at random.
- The termination criterion is: a maximum of 100 iterations or 20 iterations without improvement.
- New solutions are generated with the new operator described in Sect. 5.2.
- The local solution around the best new solution is generated using the well-known 2-opt* operator.
- α and β have been initialized to with 0.98.
- r_i^0 for each bat of the population has been initialized with a random value between 0.0 and 0.40.
- A_0 has been set with a random value between 0.70 and 1.0 for each bat.
- v_i has been initialized with a random value between 0.0 and the Hamming Distance between a bat and the best solution found.

---

[1]https://www.sintef.no/projectweb/top/vrptw/homberger-benchmark/.

Table 1 Results obtained by EDBA-OR

Class	T	AVG_V	SD_V	AVG_D	SD_D
C1	7921	10.978	0.093	1512.744	35.109
C2	14625	3.200	0.209	779.528	43.174
R1	9350	14.367	0.162	1529.738	6.871
R2	18043	3.164	0.041	1211.082	12.836
RC1	4525	14.925	0.112	1915.900	15.860
RC2	12862	3.750	0.153	1467.878	15.434

Table 2 Results obtained by EDBA-PR

Class	T	AVG_V	SD_V	AVG_D	SD_D
C1	266	12.889	0.091	2270.032	102.794
C2	1054	4.563	0.217	1817.001	148.536
R1	192	18.167	0.152	2193.207	41.194
R2	1421	4.704	0.087	1965.860	79.367
RC1	112	19.219	0.157	2645.416	32.932
RC2	728	5.594	0.120	2206.801	10.821

Table 3 Results obtained by EDBA-FULL

Class	T	AVG_V	SD_V	AVG_D	SD_D
C1	1375	10.967	0.105	1531.749	39.450
C2	4737	3.725	0.079	889.098	22.066
R1	1212	14.533	0.137	1634.549	30.242
R2	7040	3.237	0.064	1297.255	20.965
RC1	696	15.100	0.184	1960.489	35.536
RC2	4123	3.825	0.121	1566.401	39.975

Finally, in order to calculate proper statistics, each variant of the EDBA has been executed 10 times.

The results of the experimentation are shown in Tables 1, 2, 3 and 4. All the tables have the same structure: one row for each class of the Solomon's benchmark (summarizing the results of all the instances of a class) and five columns. Each column corresponds to the average runtime for all the instances of each class (T, in seconds), and average (AVG) and standard deviation (SD) for the number of vehicles (V) and the total cumulative travel distance (D).

Table 1 presents the results obtained by EDBA-OR. This variant of the algorithm is characterized by using only the route optimization process. This means that once the *ejectionPool* is generated, the routes that remain in the solution are optimized (using the Or-opt operator) to facilitate the reinsertion of the customers of the

Table 4 Results obtained by EDBA-NONE

Class	T	AVG_V	SD_V	AVG_D	SD_D
C1	1338	12.867	0.093	2183.340	64.583
C2	3385	4.575	0.112	1766.683	131.954
R1	1162	18.217	0.173	2206.293	26.755
R2	4461	4.854	0.138	1878.603	32.055
RC1	624	19.175	0.190	2702.481	109.543
RC2	2680	5.500	0.088	2293.172	66.782

removed route. According to the experimentation conducted, this variant obtained the best results (both in vehicles and traveled distance) for all the classes except C1. For the Class C1, this variant obtained the best results in terms of distance and the number of vehicles is only about 0.1% worse than the best one. The results obtained are consistent since the standard deviation for both vehicles and for distance does not exceed 6.5%. The results obtained confirm that the local optimization of the routes before reinserting the clients of the ejectionPool allows to obtain better solutions. However, the runtime time is significantly higher than the other variants.

In Table 2 the results of EDBA-PR are presented. In this case only nearest reinsertion process is performed. After the removal of the random selected route and before the final parallel initialization, the customers in the *ejectionPool* try to be reinserted in the geographically closest path. This variant is the fastest. However, together with the EDBA-NONE variant, it reports the worst results being 35.5% and 62% worse (than the best results) in terms of number of vehicles and total distance traveled.

EDBA-FULL results are shown in Table 3. In this case both processes are performed (route optimization and proximity reinsertion processes are carried out). This has obtained the second best results. The average percentage differences in number of vehicles and total distance traveled (for all classes) are 3.85% and 5.7%, respectively. In addition, it is the one that obtains the best result in number of vehicles for the class C1. Furthermore, analyzing standard deviations, it can be seen that the values obtained are the lowest. This implies that this method is more robust. One last important fact is the runtime. This variant obtains values significantly better than those obtained by the EDBA-OR variant.

Finally, Table 4 shows the results of EDBA-NONE. In this variant the customers of the removed route are reinserted directly using the parallel construction heuristic without any extra process. This variant, like EDBA-PR, gets poor results that are (on average for all classes) 36% worse in number of vehicles and 56% worse in distance traveled. On the other hand, the execution times are slightly higher than the EDBA-PR variant, but smaller than any of the two variants that get the best results. Finally, analyzing the standard deviations of the obtained results can be said that the algorithm is consistent (like the rest of variants).

To summarize, Table 5 shows the comparison of all variants and the difference with respect to the EDBA-OR (which reported the best results).

Table 5 Summary of the results and comparison between all the methods

	EDBA-OR				EDBA-FULL				EDBA-PR				EDBA-NONE			
	AVG_V	$\%_V$	AVG_D	$\%_D$	AVG_V	$\%_V$	AVG_D	$\%_D$	AVG_V	$\%_V$	AVG_D	$\%_D$	AVG_V	$\%_V$	AVG_D	$\%_D$
C1	**10.978**	0.1	**1512.744**	0.0	**10.967**	0.0	1531.749	1.3	12.889	17.5	2270.032	50.1	12.867	17.3	2270.032	50.1
C2	**3.200**	0.0	**779.528**	0.0	3.725	16.4	889.098	14.1	4.563	42.6	1817.001	133.1	4.575	43.0	1817.001	133.1
R1	**14.367**	0.0	**1592.738**	0.0	14.533	1.2	1634.549	2.6	18.167	26.4	2193.207	37.7	18.217	26.8	2193.207	37.7
R2	**3.164**	0.0	**1211.082**	0.0	3.237	2.3	1297.255	7.1	4.704	48.7	1965.860	62.3	4.854	53.4	1965.860	62.3
RC1	**14.925**	0.0	**1915.000**	0.0	15.100	1.2	1960.489	2.3	19.219	28.8	2645.416	38.1	28.5	0.088	2645.416	38.1
RC2	**3.750**	0.0	**1467.878**	0.0	3.825	2.0	1566.401	6.7	5.594	49.2	2206.801	50.3	5.500	46.7	2206.801	50.3

Table 6   Average ranking obtained by the Friedman's test

Algorithm	AVG_V	AVG_D
EDBA-OR	1.1667	1
EDBA-FULL	1.8333	2
EDBA-PR	3.5	3.5
EDBA-NONE	3.5	3.5

Once the results of the experimentation have been presented, two statistical tests (using the number of vehicles and traveled distance) have been made. These tests are based on the guidelines suggested by Derrac et al. [69]. The objective of this task is to ensure that comparisons between the different variants of the EDBA are fair and objective. First, the non-parametric Friedman's test for multiple comparison was conducted. This test aims to check for significant differences between the four variants of the EDBA.

Table 6 shows the average ranking obtained for each variant (the lower the value, the better the performance of the variant). The test has been conducted for both criteria of the objective function: number of vehicles and total traveled distance. Regarding the number of vehicles, the resulting Friedman statistic has been 15.2. Taking into account that the confidence interval has been stated at the 99.5% confidence level, the critical point in a χ^2 distribution with 3 degrees of freedom is 12.838. Because $15.2 > 12.838$, it can be concluded that there are significant differences among the results reported by the four compared algorithms, being EDBA-OR the one with the lowest rank. Finally, for this Friedman's test, the computed p-value has been 0.001653. On the other hand, in relation to the distance, the resulting Friedman statistic has been 16.2. In this case, taking the same confidence interval, the differences are again significant; and the EDBA-OR variant is the one that reports the best results. In this case, the computed p-value is 0.001032. These results confirm the superiority of the EDBA-OR variant.

Once discovered significant differences in the number of vehicles, it is appropriate to compare technique by technique. For this reason, a post-hoc Holm's test, using EDBA-OR as reference (which ranks first in number of vehicles), has been made. The results of this test are shown in Table 7. As can be seen, for EDBA-PR and EDBA-NONE adjusted and unadjusted p-values are simultaneously less than or equal to 0.05. Therefore, it can be confirmed statistically that the difference in the number of routes for EDBA-PR and EDBA-NONE with respect to EDBA-OR is significant. The same does not happen between the EDBA-FULL and EDBA-OR variants.

Table 7   Adjusted and unadjusted p-values of Holm's test for the number of vehicles

Algorithm	Adjusted p	Unadjusted p
EDBA-PR	0.005235	0.001745
EDBA-NONE	0.005235	0.001745
EDBA-FULL	0.371093	0.371093

Table 8 Adjusted and unadjusted p-values of Holm's test for the total traveled distance

Algorithm	Adjusted p	Unadjusted p
EDBA-PR	0.002389	0.000796
EDBA-NONE	0.002389	0.000796
EDBA-FULL	0.179712	0.179712

To conclude our statistical analysis, new Holm's tests has been performed. In this case the test is related to the traveled distance. The results of this test are depicted in Table 8. In this case, related to the traveled distance, there are significant differences between EDBA-PR and EDBA-NONE with respect to EDBA-OR.

Finally, as a conclusion of the experimentation and the subsequent statistical analysis of the results, it can be ensured that the EDBA-OR variant is the one that obtains the best results. These results are statistically better than those obtained by the EDBA-PR and EDBA-NONE variants. On the contrary, the results obtained by the EDBA-FULL variant are worse than those obtained by EDBA-OR. But the difference in results is not statistically significant.

7 Conclusions

We have presented in this work an Evolutionary Discrete Bat Algorithm for solving the famous Vehicle Routing Problem with Time Windows. The developed method presents some originality, such as the use of the Hamming distance to measure the distance between two bats (solutions) of the swarm, and the application of some recently proposed optimization operators, which have been firstly used in a BA. Specifically, these operators perform selective extractions of nodes in an attempt to minimize the number of routes in the current solution.

With the intention of validating that the proposed EDBA and the used route optimization operators are effective for solving the VRPTW, the results obtained by the EDBA has been compared with the ones obtained by different variants of the technique. For this experimentation, the 56 instances of the well-known Solomon's VRPTW benchmark have been used. Furthermore, two different statistical tests have been performed in order to enrich the conclusions: the non-parametric Friedmans test for multiple comparisons, and the post-hoc Holm's test.

The opportunities for future work related to the research presented in this paper are broad. For example, more complex benchmarks and further comparison of the performance of the proposed EBFA with other metaheuristics can be carried. In addition, it may be useful to apply the route optimization heuristic operators described in this work to other techniques (including classic techniques) such as the Genetic Algorithm or the Tabu Search, in order to test their efficiency. Furthermore, it can be expected that the proposed approach and operators can also used to solve travelling salesman problems and other combinatorial optimization problems.

References

1. Laporte, G.: The vehicle routing problem: an overview of exact and approximate algorithms. Eur. J. Operat. Res. **59**(3), 345–358 (1992)
2. Kirkpatrick, S., Gellat, C., Vecchi, M.: Optimization by simmulated annealing. Science **220**(4598), 671–680 (1983)
3. Glover, F.: Tabu search, part I. ORSA J. Comput. **1**(3), 190–206 (1989)
4. Dorigo, M., Blum, C.: Ant colony optimization theory: a survey. Theor. Comput. Sci. **344**(2), 243–278 (2005)
5. Goldberg, D.: Genetic Algorithms in Search, Optimization, and Machine Learning. Addison-Wesley Professional (1989)
6. De Jong, K.: Analysis of the behavior of a class of genetic adaptive systems. PhD thesis, University of Michigan, Michigan, USA (1975)
7. Kennedy, J., Eberhart, R., et al.: Particle swarm optimization. In: Proceedings of IEEE International Conference on Neural Networks. Vol. 4., Perth, Australia, pp. 1942–1948 (1995)
8. Atashpaz-Gargari, E., Lucas, C.: Imperialist competitive algorithm: an algorithm for optimization inspired by imperialistic competition. In: Evolutionary Computation, 2007. CEC 2007. IEEE Congress on, IEEE pp. 4661–4667 (2007)
9. Karaboga, D., Basturk, B.: A powerful and efficient algorithm for numerical function optimization: artificial bee colony (abc) algorithm. J. Glob. Optim. **39**(3), 459–471 (2007)
10. Yang, X.S.: A new metaheuristic bat-inspired algorithm. In: Nature Inspired Cooperative Strategies for Optimization. Springer pp. 65–74 (2010)
11. Yang, X.S., He, X.: Bat algorithm: literature review and applications. Int. J. Bio-Ins. Comput. **5**(3), 141–149 (2013)
12. Parpinelli, R.S., Lopes, H.S.: New inspirations in swarm intelligence: a survey. Int. J. Bio-Ins. Comput. **3**(1), 1–16 (2011)
13. Dhar, S., Alam, S., Santra, M., Saha, P., Thakur, S.: A novel method for edge detection in a gray image based on human psychovisual phenomenon and bat algorithm. In: Computer, Communication and Electrical Technology. CRC Press, pp. 3–7 (2007)
14. Tharakeshwar, T., Seetharamu, K., Prasad, B.D.: Multi-objective optimization using bat algorithm for shell and tube heat exchangers. Appl. Therm. Eng. **110**, 1029–1038 (2017)
15. Osaba, E., Carballedo, R., Yang, X.S., Diaz, F.: An evolutionary discrete firefly algorithm with novel operators for solving the vehicle routing problem with time windows. In: Nature-Inspired Computation in Engineering. Springer, pp. 21–41 (2016)
16. Lawler, E.L.: The traveling salesman problem: a guided tour of combinatorial optimization. Wiley-Interscience Series in Discrete Mathematics (1985)
17. Christofides, N.: The vehicle routing problem. RAIRO Operat. Res. Recherche Opérationnel. **10**(V1), 55–70 (1976)
18. Wassan, N., Wassan, N., Nagy, G., Salhi, S.: The multiple trip vehicle routing problem with backhauls: formulation and a two-level variable neighbourhood search. Comput. Operat. Res. **78**, 454–467 (2017)
19. Veenstra, M., Roodbergen, K.J., Vis, I.F., Coelho, L.C.: The pickup and delivery traveling salesman problem with handling costs. Eur. J. Operat. Res. **257**(1), 118–132 (2017)
20. Bräysy, O., Gendreau, M.: Vehicle routing problem with time windows, part I: route construction and local search algorithms. Transport. Sci. **39**(1), 104–118 (2005)
21. Potvin, J.Y., Bengio, S.: The vehicle routing problem with time windows part II: genetic search. INFORMS J. Comput. **8**(2), 165–172 (1996)
22. Laporte, G.: The traveling salesman problem: an overview of exact and approximate algorithms. Eur. J. of Oper. Res. **59**(2), 231–247 (1992)
23. Bräysy, O., Gendreau, M.: Vehicle routing problem with time windows, part II: metaheuristics. Transport. Sci. **39**(1), 119–139 (2005)
24. Yang, X.S.: Nature-inspired metaheuristic algorithms. Luniver press (2010)
25. Taha, A., Hachimi, M., Moudden, A.: Adapted bat algorithm for capacitated vehicle routing problem. Int. Rev. Comput. Soft. (IRECOS) **10**(6), 610–619 (2015)

26. Zhou, Y., Luo, Q., Xie, J., Zheng, H.: A hybrid bat algorithm with path relinking for the capacitated vehicle routing problem. In: Metaheuristics and Optimization in Civil Engineering. Springer, pp. 255–276 (2016)
27. Yang, X.S.: Nature-Inspired Metaheuristic Algorithms. Luniver Press, UK (2008)
28. Fister, I., Yang, X.S., Fister, D., Fister Jr, I.: Firefly algorithm: a brief review of the expanding literature. In: Cuckoo Search and Firefly Algorithm. Springer, pp. 347–360 (2014)
29. Fister, I., Fister Jr., I., Yang, X.S., Brest, J.: A comprehensive review of firefly algorithms. Swarm Evolut. Comput. **13**, 34–46 (2013)
30. Jati, G.K., Suyanto. In: Evolutionary Discrete Firefly Algorithm for Travelling Salesman Problem. Springer, Berlin Heidelberg, pp. 393–403 (2011)
31. Alinaghian, M., Naderipour, M.: A novel comprehensive macroscopic model for time-dependent vehicle routing problem with multi-alternative graph to reduce fuel consumption: a case study. Comput. Indust. Eng. **99**, 210–222 (2016)
32. Del Ser, J., Torre-Bastida, A.I., Lana, I., Bilbao, M.N., Perfecto, C.: Nature-inspired heuristics for the multiple-vehicle selective pickup and delivery problem under maximum profit and incentive fairness criteria. In: IEEE Congress on Evolutionary Computation (2017)
33. Yang, X.S., Deb, S.: Cuckoo search via lévy flights. In: World Congress on Nature & Biologically Inspired Computing. IEEE, pp. 210–214 (2009)
34. Ouaarab, A., Ahiod, B., Yang, X.S.: Discrete cuckoo search algorithm for the travelling salesman problem. Neural Comput. Appl. **24**(7–8), 1659–1669 (2014)
35. Alssager, M., Othman, Z.A.: Taguchi-based parameter setting of cuckoo search algorithm for capacitated vehicle routing problem. In: Advances in Machine Learning and Signal Processing. Springer, pp. 71–79 (2016)
36. Teymourian, E., Kayvanfar, V., Komaki, G.M., Zandieh, M.: Enhanced intelligent water drops and cuckoo search algorithms for solving the capacitated vehicle routing problem. Informat. Sci. **334**, 354–378 (2016)
37. Chen, X., Wang, J.: A novel hybrid cuckoo search algorithm for optimizing vehicle routing problem in logistics distribution system. J. Comput. Theor. Nanosci. **13**(1), 114–119 (2016)
38. Geem, Z.W., Kim, J.H., Loganathan, G.: A new heuristic optimization algorithm: harmony search. Simulation **76**(2), 60–68 (2001)
39. Manjarres, D., Landa-Torres, I., Gil-Lopez, S., Del Ser, J., Bilbao, M.N., Salcedo-Sanz, S., Geem, Z.W.: A survey on applications of the harmony search algorithm. Eng. Appl. Artific. Intell. **26**(8), 1818–1831 (2013)
40. Assad, A., Deep, K.: Applications of harmony search algorithm in data mining: a survey. In: Proceedings of Fifth International Conference on Soft Computing for Problem Solving. Springer , pp. 863–874 (2016)
41. Mohd Alia, O., Mandava, R.: The variants of the harmony search algorithm: an overview. Artific. Intell. Rev. **36**(1), 49–68 (2011)
42. Geem, Z.W., Lee, K.S., Park, Y.: Application of harmony search to vehicle routing. Am. J. Appl. Sci. **2**(12), 1552–1557 (2005)
43. Del Ser, J., Bilbao, M.N., Perfecto, C., Salcedo-Sanz, S.: A harmony search approach for the selective pick-up and delivery problem with delayed drop-off. In: Harmony Search Algorithm. Springer, pp. 121–131 (2016)
44. Rashedi, E., Nezamabadi-Pour, H., Saryazdi, S.: Gsa: a gravitational search algorithm. Informat. Sci. **179**(13), 2232–2248 (2009)
45. Precup, R.E., David, R.C., Petriu, E.M., Radac, M.B., Preitl, S.: Adaptive gsa-based optimal tuning of pi controlled servo systems with reduced process parametric sensitivity, robust stability and controller robustness. IEEE Trans. Cybernet. **44**(11), 1997–2009 (2014)
46. Precup, R.E., David, R.C., Petriu, E.M., Preitl, S., Rădac, M.B.: Fuzzy logic-based adaptive gravitational search algorithm for optimal tuning of fuzzy-controlled servo systems. IET Control Theor. Appl. **7**(1), 99–107 (2013)
47. Duman, S., Güvenç, U., Sönmez, Y., Yörükeren, N.: Optimal power flow using gravitational search algorithm. Energy Convers. Manag. **59**, 86–95 (2012)

48. Nodehi, A.N., Fadaei, M., Ebrahimi, P.: Solving the traveling salesman problem using randomized gravitational emulation search algorithm. J. Curr. Res. Sci. **2**, 818 (2016)
49. Hosseinabadi, A.A.R., Kardgar, M., Shojafar, M., Shamshirband, S., Abraham, A.: Gravitational search algorithm to solve open vehicle routing problem. In: Innovations in Bio-Inspired Computing and Applications. Springer, pp. 93–103 (2016)
50. Hosseinabadi, A.A.R., Rostami, N.S.H., Kardgar, M., Mirkamali, S., Abraham, A.: A new efficient approach for solving the capacitated vehicle routing problem using the gravitational emulation local search algorithm. Appl. Mathemat, Model (2017)
51. Desaulniers, G., Errico, F., Irnich, S., Schneider, M.: Exact algorithms for electric vehicle-routing problems with time windows. Les Cahiers du GERAD G-2014-110, GERAD, Montréal, Canada (2014)
52. Belhaiza, S., Hansen, P., Laporte, G.: A hybrid variable neighborhood tabu search heuristic for the vehicle routing problem with multiple time windows. Comput. Operat. Res. **52**, 269–281 (2014)
53. Toklu, N.E., Gambardella, L.M., Montemanni, R.: A multiple ant colony system for a vehicle routing problem with time windows and uncertain travel times. J. Traffic Logist. Eng. **2**(1) (2014)
54. Nguyen, P.K., Crainic, T.G., Toulouse, M.: A hybrid generational genetic algorithm for the periodic vehicle routing problem with time windows. J. Heurist. **20**(4), 383–416 (2014)
55. Yassen, E.T., Ayob, M., Nazri, M.Z.A., Sabar, N.R.: Meta-harmony search algorithm for the vehicle routing problem with time windows. Informat. Sci. **325**, 140–158 (2015)
56. Yang, X.S., Deb, S.: Cuckoo search: recent advances and applications. Neural Comput. Appl. **24**(1), 169–174 (2014)
57. Kallehauge, B., Larsen, J., Madsen, O.B., Solomon, M.M.: Vehicle routing problem with time windows. Springer (2005)
58. Gendreau, M., Tarantilis, C.D.: Solving large-scale vehicle routing problems with time windows: the state-of-the-art. CIRRELT (2010)
59. Afifi, S., Guibadj, R.N., Moukrim, A.: New lower bounds on the number of vehicles for the vehicle routing problem with time windows. In: Integration of AI and OR Techniques in Constraint Programming. Springer, pp. 422–437 (2014)
60. Agra, A., Christiansen, M., Figueiredo, R., Hvattum, L.M., Poss, M., Requejo, C.: The robust vehicle routing problem with time windows. Comput. Operat. Res. **40**(3), 856–866 (2013)
61. Azi, N., Gendreau, M., Potvin, J.Y.: An exact algorithm for a single-vehicle routing problem with time windows and multiple routes. Eur. J. Operat. Res. **178**(3), 755–766 (2007)
62. Bräysy, O., Gendreau, M.: Tabu search heuristics for the vehicle routing problem with time windows. Top **10**(2), 211–237 (2002)
63. Cordeau, J.F., Desaulniers, G., Desrosiers, J., Solomon, M.M., Soumis, F.: Vrp with time windows. Vehicle Rout. Prob. **9**, 157–193 (2001)
64. Glover, F.: Ejection chains, reference structures and alternating path methods for traveling salesman problems. Discr. Appl. Mathemat. **65**(1–3), 223–253 (1996)
65. Osaba, E., Yang, X.S., Diaz, F., Onieva, E., Masegosa, A.D., Perallos, A.: A discrete firefly algorithm to solve a rich vehicle routing problem modelling a newspaper distribution system with recycling policy. Soft Comput. 1–14 (2016)
66. Irnich, S.: A unified modeling and solution framework for vehicle routing and local search-based metaheuristics. INFORMS J. Comput. **20**(2), 270–287 (2008)
67. Campbell, A.M., Savelsbergh, M.: Efficient Insertion heuristics for vehicle routing and scheduling problems. Transport. Sci. **38**(3), 369–378 (2004)
68. Solomon, M.M.: Algorithms for the vehicle routing and scheduling problems with time window constraints. Oper. Res. **35**(2), 254–265 (1987)
69. Derrac, J., García, S., Molina, D., Herrera, F.: A practical tutorial on the use of nonparametric statistical tests as a methodology for comparing evolutionary and swarm intelligence algorithms. Swarm Evol. Computat. **1**(1), 3–18 (2011)

Variants of the Flower Pollination Algorithm: A Review

Zaid Abdi Alkareem Alyasseri, Ahamad Tajudin Khader,
Mohammed Azmi Al-Betar, Mohammed A. Awadallah
and Xin-She Yang

Abstract The flower pollination algorithm (FPA) is a nature-inspired algorithm that imitates the pollination behavior of flowering plants. Optimal plant reproduction strategy involves the survival of the fittest as well as the optimal reproduction of plants in terms of numbers. These factors represent the fundamentals of the FPA and are optimization-oriented. Yang developed the FPA in 2012, which has since shown superiority to other metaheuristic algorithms in solving various real-world problems, such as power and energy, signal and image processing, communications, structural design, clustering and feature selection, global function optimization, computer gaming, and wireless sensor networking. Recently, many variants of FPA have been developed by modification, hybridization, and parameter-tuning to cope with the complex nature of optimization problems. Therefore, this chapter provides a comprehensive review for FPA variants from 2012 to present.

Z.A.A Alyasseri · A.T. Khader
School of Computer Sciences, Universiti Sains Malaysia (USM),
Pulau Pinang, Malaysia
e-mail: zaid.alyasseri@uokufa.edu.iq

A.T. Khader
e-mail: tajudin@cs.usm.my

Z.A.A Alyasseri
ECE Department - Faculty of Engineering, University of Kufa,
P.O. Box 21, Najaf, Iraq

M.A. Al-Betar (✉)
Department of Information Technology, Al-Huson University College,
Al-Balqa Applied University, P.O. Box 50, Irbid, Al-Huson, Jordan
e-mail: mohbetar@bau.edu.jo

M.A. Awadallah
Department of Computer Science, Al-Aqsa University,
P.O. Box 4051, Gaza, Palestine
e-mail: ma.awadallah@alaqsa.edu.ps

X.-S. Yang
School of Science and Technology, Middlesex University,
London NW4 4BT, UK
e-mail: x.yang@mdx.ac.uk

© Springer International Publishing AG 2018
X.-S. Yang (ed.), *Nature-Inspired Algorithms and Applied Optimization*,
Studies in Computational Intelligence 744, https://doi.org/10.1007/978-3-319-67669-2_5

Keywords Algorithm · Flower pollination algorithm · Optimization · Nature-inspired algorithm · Swarm intelligence · Metaheuristics

1 Introduction

Many phenomena in nature have unique characteristics that can be utilized and converted into a mathematical model or even an algorithm to solve real-world problems. Over the last few decades, researchers have developed many nature-inspired algorithms so as to attempt to find the best solutions for various optimization problems. Examples are genetic algorithm (GA) [1], artificial bee colony (ABC) [2], particle swarm optimization (PSO) [3], gray wolf algorithm (GWA) [4], firefly algorithm (FA) [5], bat algorithm (BA) [6], and cuckoo search (CS) [7]. These algorithms have been successfully applied to a wide range of optimization problems and are widely used in the literature of metaheuristics for the last two decades [8, 9]. On the other hand, nature still has many other phenomena that can be utilized to solve different types of problems. One phenomenon is flowering plant reproduction strategy through pollination, which inspired Yang in 2012 to propose a new algorithm called the flower pollination algorithm (FPA) [10].

FPA is a swarm-based optimization technique that has attracted the attention of many researchers in several optimization fields due to its impressive characteristics. FPA has very fewer parameters and has shown a robust performance when applied in various optimization problems. In addition, FPA is a flexible, adaptable, scalable, and simple optimization method. Therefore, FPA, compared with other metaheuristic algorithms, shows good results for solving various real-life optimization problems from different domains such as electrical and power system [11–15], signal and image processing [16–18], wireless sensor networking [19–21], clustering and classification [22, 23], global function optimization [24], computer gaming [25], structural and mechanical engineering optimization [26–28], and many others [29, 30].

Procedurally, FPA is a population-based optimization technique, initiated with a set of provisional or random solutions. At each iteration, either one of the two operators is carried out for each individual population member: local pollination operator and global pollination operator. In a local pollination operator, the decision variables of the current solution attract the other two randomly selected solutions from two population members. In a global pollination operator, the decision variables of the current solution attract to the globally best solution found. The switch operator is responsible for exchanging the improvement loop either locally or globally. This process repeats until a predefined stopping criterion is met.

In recent years, the procedural optimization framework of FPA at its initial version has undergone modification or hybridization to enhance its performance with relation to different types of problem landscapes. Therefore, the original form of FPA is first presented in this review paper in terms of its theoretical aspects. Then, several FPA versions are reviewed and analyzed critically by presenting modified

Fig. 1 Distribution of
published research articles
on FPA

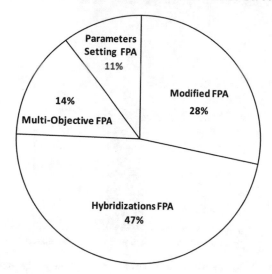

Distribution Of Published Research Articles On FPA

and hybridized versions in detail, including the multi-objective and parameter-less variants. Figure 1 summarizes these variants and their foundations as a pie chart. In addition, the limitations of FPA will also be discussed. Critical analysis concerning the FPA optimization framework is presented to provide new opportunities for interested readers to carry out more research about FPA. Furthermore, some important applications of FPA are comprehensively summarized. Finally, this review chapter concludes with recommending possible future work on FPA.

The Materials and Methods reviewed here have been selected based on their modifications and relevance. Figure 2 shows the main sources of these materials and methods. In the figure, the selected literature and studies are classified, based loosely on the publishers such as IEEE Explorer, ScienceDirect, SpringerLink, Taylor & Francis, and others. There are other ways of presenting the data. For example, Fig. 3 shows the distribution of publications, Materials and Methods, based on the year of publication. As the time progressed, the interest in FPA increased and attracted the attention of the research community in the last 5 years.

Therefore, this chapter is organized as follows. Section 2 describes the flower pollination foundation with Sect. 2.2 describing the flower pollination algorithm. Section 3 provides in detail all the major variants of FPA. Some applications are briefly outlined in Sect. 4, and and critical analysis of FPA is carried out in Sect. 5. Finally, the conclusion will be drawn with some recommendation in Sect. 6.

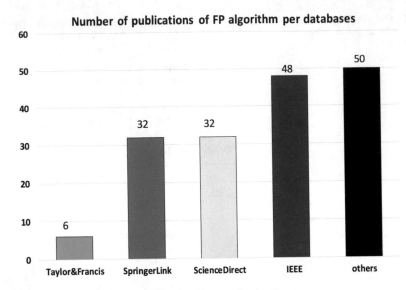

Fig. 2 Number of publications of FPA algorithm per databases

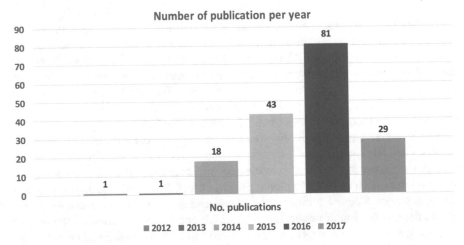

Fig. 3 Number of publications of FPA algorithm per year

2 Flower Pollination and Flower Pollination Algorithm

The majority of plants are flowering plants and there are more than 250,000 species of flowering plants around the world, where pollination represents the main repro-duction strategy of the plants [31, 32]. Pollination is a process of transferring pollen from one flower to another by wind or pollinators such as insects, butterflies, bees, birds and bats. Flowering plants have evolved to produce nectar to attract pollinators and to ensure pollination [33]. In addition, some pollinators and plant species such as hummingbirds and ornithophilous flowering plants form some co-evolutionary flower constancy [31, 32]. Based on the main characteristics of pollination, the flower pollination algorithm has been developed [10].

2.1 FPA in Optimization Context: Nature's Inspiration

Before we describe the flower pollination algorithm in detail, let us briefly review the basic form of pollination in flowering plants. Pollination takes two basic forms: biotic or abiotic.

1. *Biotic pollination*: The main form of pollination is biotic pollination, also called cross-pollination, by pollinators such as insects and birds and others. Almost 90% of flowering plants use this form of pollination. As pollinators move and even fly with various paces and speeds, the motion of pollen can be quite long distant. Such pollination can also be considered as global pollination with potential Lévy flights properties [10, 34, 90]. If pollen is encoded as a solution vector, this action can be equivalent to global search.
2. *Abiotic pollination*: Another form of pollination is abiotic pollination, also called self-pollination, which does not require pollinators. It is estimated that about 10% of floral plants take this form of pollination. As the pollination tends to be local and self-pollination, it can be achieved by wind and diffusion [10, 33]. The distance travelled by such local motion is typically short, and such action can thus be considered as local search.
3. *Flower constancy*: Sometimes, it is advantageous for both plants and pollinators such as hummingbirds to form a partnership to save energy with guaranteed success. Consequently, flower constancy has been evolved. In this case, pollinators only visit a fixed set of flower types without wasting energy for exploring new flower types, while the flower plants evolve to provide sufficient nectar reward to pollinators so as to encourage frequent visits by pollinators and thus maximize their reproduction success [31, 33].

The above characteristics have been used to design an optimization algorithm, called flower pollination algorithm (FPA) [10]. The main characteristics and the algorithm components of FPA can be summarized in Table 1, which shows the relationship or equivalence between optimization terms and flower context.

With these components and characteristics, we can now describe the standard flower pollination algorithm in detail.

Table 1 Pollination and its optimization components

Flower pollination	Optimization components (in FPA)
Pollinators (insects, butterflies, birds)	Moves/modification of variables
Biotic	Global search
Abiotic	Local search
Lévy flight	Step sizes (obeying a power law)
Pollen/flowers	Solution vectors
Flower constancy	Similarity in solution vectors
Evolution of flowers	Iterative evolution of solutions
Optimal flower reproduction	Optimal solution set

2.2 Flower Pollination Algorithm

FPA is a nature-inspired algorithm that mimics the main pollination behavior of flowering plants. The four idealization rules were used by Yang in 2012 [10] and they can be summarized as follows:

Rule 1   Global pollination involves biotic and cross-pollination where pollinators carry the pollen based on Lévy flights.

Rule 2   Local pollination involves abiotic and self-pollination.

Rule 3   Flower constancy can be considered as a reproduction probability that is proportional to the similarity between any two flowers.

Rule 4   Switch probability $p \in [0, 1]$ can be controlled between local pollination and global pollination due to some external factors, such as wind. Local pollination has a significant fraction p in overall pollination activities.

To illustrate the mechanism of the FPA based on these four rules, three key steps can be described in the following three subsections.

2.2.1   Global Search of FPA (Biotic)

As mentioned above, pollinators such as birds and bats can transfer pollen over long distances during biotic pollination, ensuring the diversity and the fittest pollination for reproduction. Therefore, the first (Rule 1) and third (Rule 3) FPA rules can be mathematically formulated as follows:

$$x_i^{t+1} = x_i^t + L(g^* - x_i^t) \tag{1}$$

where x_i^t is the pollen or solution vector at iteration t and g^* is the best solution found among all solutions at the current iteration. The parameter L is the strength of pollination, which is essentially a step size. Because pollinators move over long distances with various distance intervals, the Lévy flight can an efficient simulator for this characteristic [10]; that is, L can be drawn from a Lévy distribution as follows:

$$L \sim \frac{\lambda \Gamma(\lambda) \sin(\pi \lambda/2)}{\pi} \frac{1}{s^{1+\lambda}}, \quad (s \ll 0) \tag{2}$$

where $\Gamma(\lambda)$ denotes the standard gamma function and this distribution is valid for large steps $s > 0$. Normally, it is recommended that $\lambda = 1.5$ can be used [10].

2.2.2 Local Search of FPA (Abiotic)

As abiotic pollination occurs by wind or diffusion without any pollinators, the local pollination (Rule 2) and flower constancy (Rule 3) can be represented as follows:

$$x_i^{t+1} = x_i^t + \varepsilon\, (x_j^t - x_k^t) \tag{3}$$

where x_j^t and x_k^t are pollen from different flowers of the same plant type. This equation essentially mimics the flower constancy in a limited neighborhood. Mathematically speaking, if x_j^t and x_j^k are from the same species that can be selected from the same population, the equation becomes a local random walk if we draw ε from a uniform distribution in [0, 1], and the new solution vector generated will not be too far away from existing solutions.

2.2.3 Switch Probability in FPA

Though we have simulated both biotic and abiotic pollination, we have not considered the percentage and frequency of each pollination type. To mimic this feature, we use a switch probability (Rule 4), where the value of p determines whether the solution modification follows either local or global pollination. Though a naive value of $p = 0.5$ can be used, a more realistic and effective value of $p = 0.8$ gives better performance (than $p = 0.5$) for most applications [10].

Figure 4 shows the flowchart of FPA. Three key steps can be summarized in the FPA pseudocode shown in Algorithm 1.

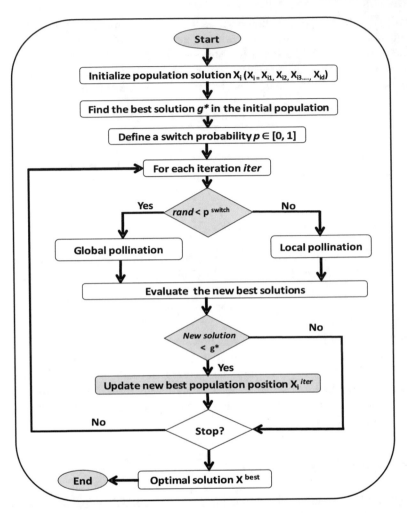

Fig. 4 Flower pollination algorithm flowchart

3 Recent FPA Variants

Though the standard FPA works well for many applications [10], it can still be improved. Given the complex nature of real-world optimization problems, the basic structure of FPA has been modified to enhance its performance. The modification has been done in many parts of the FPA structure, which will be discussed in Sect. 3.1. Furthermore, several FPA hybridization schemes, which will be discussed in Sect. 3.2, have also been introduced to accelerate the convergence and to improve the balance between exploration and exploitation. Multi-objective optimization problems are also a category of problems. Therefore, Sect. 3.3 describes the

---

Algorithm 1 Flower Pollination Algorithm pseudo-code

1: Objective min $f(x)$, $x \in \mathfrak{R}^d$
2: Initialize a population of n flowers/pollen with random solutions
3: Fins the best solution g^* in the initial population
4: Define a switch probability $p \in [0, 1]$
5: Calculate all $f(x)$ for n solutions
6: $t=0$
7: **while** $t \leq MaxGeneration$ **do**
8:   **for** $i = 1, ..., n$ **do**
9:     **if** $rnd \leq p$ **then**
10:       Draw a (d-dimensional) step vector L which obeys a Lévy distribution
11:       Global pollination via $x_i^{t+1} = x_i^t + L * (g^* - x_i^t)$
12:     **else**
13:       Draw from a uniform distribution $\varepsilon \in [0,1]$
14:       Randomly choose j and k among all solutions
15:       Do local pollination via $x_i^{t+1} = x_i^t + \varepsilon (x_j^t - x_k^t)$
16:     **end if**
17:     Calculate all new $f(x^{t+1})$
18:     **if** $f(x^{t+1}) \leq f(x^t)$ **then**
19:       $x^t = x^{t+1}$
20:     **end if**
21:   **end for**
22:   Find the current best solution g^* among all x_i^t
23:   $t = t + 1$
24: **end while**

---

multi-objective version of FPA. Finally, it will be useful if an easy-to-use FPA can be built to deal with parameterless structure, which is discussed in Sect. 3.4.

3.1 Modified Versions of FPA

One way of modifications in FPA structure is to deal with problem complexity or high dimensionality. Initially, the researchers realized that FPA uses two search branches and their fitness values are closely linked to the problem landscape. A possible modification is to reduce some attributes in the search space. Others try to discretize FPA to produce binary FPA to solve combinatorial problems. Various modifications will be explained below.

3.1.1 Modified FPA Based on Operators

Yamany et al. [35] proposed a modified FPA based on an attribute reduction approach. The main objective of their proposed algorithm is to handle a possible large search space. The proposed approach suggests a minimum number of attributes and obtains a comparable or even the best classification accuracy by using all attributes and conventional attribute reduction techniques. The strategy of the proposed algorithm improves three new initialization phases, which are driven by forward selection and backward selection. Their proposed technique utilized eight datasets from the UCI machine learning benchmarks and obtained a better performance than other meta-heuristic algorithms such as GA and PSO.

Zhou et al. [27] proposed an elite opposition-based flower pollination algorithm (EOFPA), which is a new variant for solving function optimization and structure designs. EOFPA showed an improvement in the balance of exploration and exploitation. The authors tested their proposed algorithm using 18 standard benchmark functions, which yielded impressive results.

For economic load dispatch problems in power generation system, Putra et al. [14] proposed a modified FPA (MFPA), which used a dynamic switching probability, the application of real-coded GA (RCGA) as mutation for global and local search, and the differentiation between temporary local search and optimal solution. MFPA was then evaluated for 10 benchmarks of power systems, and their experimental results showed a lower fuel cost than that found by the standard FPA as well as other similarly applicable solutions for similar economic dispatch problems. In another study, Regalado et al. [36] proposed an MFPA for fuel cost value and time required for obtaining a global optimal solution, and the MFPA tested under the IEEE 30-bus test system showed superior results over standard FPA and other metaheuristic optimization algorithms.

To reduce the real power loss and to improve bus voltages, Namachivayam et al. [37] introduced an MFPA for network reconfiguration and optimal placement of shunt capacitors. The proposed algorithm involved the adaptation of the local search of standard FPA and enhanced the global search by using dynamic switching probability approach. Their proposed algorithm (MFPA) was evaluated using 118-bus, 69-bus, and 33-bus radial distribution test feeders, yielding better results than other metaheuristic algorithms, such as HSA, simulated annealing (SA), and IBPSO.

The modified flower pollination algorithm (MFPA) variant proposed by Dubey et al. [38] solved the economic dispatch problems in large power systems where the technique involved two improving phases. The first phase was the addition of a scaling factor to enhance the local pollination of FPA, while the second improvement was the addition of an intensive exploitation step in tuning the best FPA solution. MFPA was then tested using several mathematical benchmarks as well as four large power systems, and MFPA performance was compared with recent methods for economic dispatch problems, revealing the successful outcomes of their proposed algorithm.

3.1.2   Binary FPA

The original FPA was designed to solve continuous optimization problems. To solve discrete and combinatorial optimization problems, proper modifications are needed. Rodrigues et al. [39] developed a binary flower pollination algorithm (BFPA) for feature selection and tested the BFPA on six datasets, and BFPA provided better results than Particle Swarm Optimization (PSO), harmony search (HS), and firefly algorithm (FA). Later, Rodrigues et al. [16] applied BFPA to address the problem of reducing the number of required sensors for person identification based on EEG signals. BFPA was used to select the optimal subset of channels that provides the highest accuracy. BFPA experiments results showed recognition rates of up to 87% based on the Optimum-Path Forest classifier.

Shilaja et al. [40] suggested a technique called CEED to solve 20 photovoltaic and 5 thermal power generation problems. The CEED technique was a combination of Economic Dispatch Euclidean Affine Flower Pollination Algorithm and BFPA. In addition, testing the proposed CEED technique through IEEE 30 bus and IEEE 57 bus systems provided better results than existing methods.

Dahi et al. [41] conducted a systematic study to evaluate BFPA performance in solving the Antenna Positioning Problem (APP). BFPA was tested using realistic, synthetic, and random data with different dimensions and compared with Population-Based Incremental Learning (PBIL) and the Differential Evolution algorithm (DE), which are two of the efficient algorithms in the APP domain. BFPA achieved a more competitive technical finding than PBIL and DE in the APP domain.

3.1.3   Chotic-Based FPA

The standard FPA uses random numbers, and some randomization can also be achieved by using chaotic maps. In mechanical engineering, Meng et al. [28] has developed a modified flower pollination algorithm (MFPA) to solve a design problem. This MFPA involved the adaptive inertia weight and the chaos theory in the enhancement of local search. Evaluating the performance results of MFPA through five mechanical engineering benchmarks, namely, speed reducer, gear train, tubular column design, pressure vessel, and tension/compression spring design, showed better results than other algorithms for mechanical engineering problems.

Metwalli et al. [42] presented a new method for solving fractional programming problems (FPPs) based on development of chaos-based Flower Pollination Algorithm (CFPA). The performance of CFPA has been proven using several FPP benchmarks. The proposed algorithm was compared with metaheuristic solution methods for solving FPPs where the former showed superiority over other fractional programming problem-solving methods.

Many power system techniques have been proposed for wind speed forecasting, but most of these systems do not have any efficient model on data preprocessing. Therefore, Zhang et al. [43] proposed a new model that involved a combination of three short-term techniques for wind speed forecasting. Their novel system included complete ensemble empirical mode decomposition adaptive noise (CEEMDAN), FPA with chaotic local search (CLSFPA), five neural networks, and no negative constraint theory. CLSFPA aimed to choose the optimal weight coefficients of the combined model. It was shown that their combined algorithms could effectively forecast the wind speed at high accuracy after evaluating the 15 min wind speed data from four different farms in eastern China.

3.2 Hybridized Versions of FPA

One of the main problems for any metaheuristic methods is to strike the right balance between global-wide exploration and local neighborhood exploitation during the search process. Some methods are very powerful in exploring several regions of the problem landscape, despite lacking in the exploitation of each region, and these algorithms tend to be population-based or swarm-based algorithms. Other methods are very powerful in exploiting the good elements in a specific region in the search space at the expense of sacrificing the simultaneous exploration of several regions, which is true for gradient-based methods or trust-region methods. Therefore, research communities have been trying to hybridize FPA with other algorithms to improve its performance. Such types of hybridization are summarized below.

3.2.1 Hybridization with Local Search Based Algorithms

The hybridization of FPA with simulated annealing (SA) for engineering optimization problems, named (FPSA), was developed in [44]. In their method, solutions generated by FPA improved locally using the SA algorithm to enhance the search performance and to speed up the convergence. FPSA had a better performance than other methods in the literature.

Jensi and Jiji [45] proposed FPA hybridization with the K-Means algorithm for data clustering and used the K-Means algorithm to enhance the local exploitation of FPA. The performance of their hybrid algorithm was more effective than using classical K-Means or FPA alone.

Sayed et al. [46] introduced a hybrid algorithm called BCFA, which combined Clonal Selection Algorithm (CSA) with FPA to solve feature selection problem. The authors used the Optimum-Path Forest classifier as an objective function, and their proposed hybrid algorithm (BCFA) produced better results than other metaheuristic algorithms.

Emad Nabil [24] proposed an improved variant of FPA hybrid, which was a combination of MFPA and CSA. To evaluate the performance of MFPA, a total of 23 optimization benchmark problems were tested, and MFPA efficiency was compared with SA, GA, FPA, BA, and FA. Results showed that the proposed hybrid MFPA can obtain the better results than standard FPA and the former outperformed the other four metaheuristic algorithms.

A novel FPA hybrid from the integration of FPA and Path Relinking metaheuristic was presented in [30] and was used in the context of generating healthy and nutritional meals for older adults. This hybridization aimed to improve the search for optimal or near-optimal personalized menu recommendations in terms of execution time and quality. The performance test of this hybrid version on real-world dataset showed the superiority of the algorithm to classical FPA in terms of both solution quality and execution time.

Zhang et al. [43] proposed CLSFPA that combined FPA and chaotic local search to enhance the exploration capabilities and to improve the local search ability of FPA. The authors used their version of CLSFPA to solve wind speed forecasting by combining CEEMDAN, which broke down the raw wind speed series into several independent intrinsic mode function components, with corresponding frequencies for easy analysis and forecasting. Later on, CLSFPA was employed for each component to improve the accuracy of the forecast. Experimental results illustrated that the proposed CLSFPA was effective in predicting wind speed with high precision.

3.2.2   Hybridization with Population Based Algorithms

Abdel-Raouf et al. [47] presented a hybrid method, called hybrid FPA for optimization problems, by combining the FPA with the PSO algorithm to improve the search accuracy. Results showed that their method was more accurate, reliable, and efficient in finding the optimal solutions than the other methods in the literature. In another study, the same authors integrated the FPA with chaotic HS algorithm in [25] and used the hybrid algorithm to solve Sudoku puzzles with satisfactory results compared with chaotic HS algorithm.

Nigdeli et al. [48] integrated FPA with HS algorithm for tuning of mass dampers. In their method, four different types of generations were used, the global and local search process of HS and the global and local pollination of FPA. A probability-based method determined the kind of generation utilized in the construction of new solutions. This proposed probability was calculated based on the objective of optimization, and they found that their probability-based FPA performed better than classical FPA in terms of convergence rates.

Recently, Lenin et al. [49] used a hybrid FPA with chaotic HS algorithm to solve the optimal reactive power dispatch problem. The basic idea was to improve the accuracy of the FPA search process. The authors tested the performance of their hybrid FPA using the standard IEEE 57-bus system dataset, and found its effectiveness and robustness in minimizing real power loss.

A novel hybridization of ABC with FPA, called Bee Pollinated Flower Pollination Algorithm (BPFPA), was introduced in Ref. [50] for solar PV parameter estimation. In BPFPA, the bee behavior of discarding pollen combines within FPA, and an elite-based mutation operation replaced the local pollination of FPA. Introducing these modifications in FPA not only enhanced its randomness but the hybrid method also had a faster execution speed and higher robustness than other methods.

Abdel-Baset and Hezam [51] proposed the hybridization of FPA with GA, called FPA-GA, to solve constrained optimization problems. In their method, the GA was triggered after the FPA loop. Introducing the hybridization enhanced the search accuracy of FPA. The authors tested the performance of their method on seven well-known benchmark design problems, and the performance of the proposed FPA-GA was better than the basic GA, the basic FPA, and other algorithms.

Hybridization of DE algorithm with FPA, called DE-FPA, was proposed in [52] to solve benchmark optimization problems. The main idea of their algorithm was to synthesize the strength and power of both algorithms, where this combination provided a right balance between exploration and exploitation capabilities. For their studies, it seemed that DE algorithm was notably a main source of exploration, whereas the exploitation was the main FPA characteristic in this case. Experimental results showed that DE-FPA outperforms classical DE and FPA in terms of performance and convergence rate. Ramadas et al. developed similar hybridization to solve 15 global optimization benchmark problems [53]. Experimental results illustrated that the proposed method was better than or equal to the standard DE almost all test problems.

To solve the wind-thermal dynamic multi-objective optimal dispatch problem, Dubey et al. [54] hybridized FPA with DE and named their algorithm as HFPA. Furthermore, 5-class, 3-step time varying fuzzy selection mechanism (TVFSM) was integrated with HFPA to find a fuzzy selection index (FSI) by aggregating different conflicting objectives. FSI notably helped the decision maker in finding the trade-off solutions by problem specific parameter selection. Based on the experimental results, the HFPA-TVFSM algorithm was more efficient than DE, FPA, and HFPA.

Kalra and Arora [34] improved the FPA performance with the firefly algorithm (FA) to tackle multimodal optimization functions and to overcome the shortcomings of each algorithm. In FPA, the convergence speed improves, and the chances of trapping within local optima decrease by reducing the effect of randomness in FA. Experimentally, the proposed hybrid algorithm obtained better accuracy and faster convergence than FA and FPAs.

A multi-objective optimal power flow problem was tackled through a new hybrid FPA in [15]. The authors combined the FPA with PSO algorithm order to enhance the global search and validated their algorithm using IEEE 30 test bus system and IEEE 118 test bus system. Based from their results, the hybrid algorithm significantly performed better than FPA and PSO alone. A similar study was evaluated on the optimal reactive power dispatch problem in [55]. The authors evaluated the proposed hybrid algorithm evaluated using IEEE 30 and, IEEE 57 bus test systems and found that the algorithm performance of is better in reducing real power loss.

A novel FAP hybridization with PSO algorithm, called HFPA, was proposed in [56], where the PSO was integrated intentionally to enhance the exploration capability of FPA. The authors used their method to design wide-band infinite impulse response digital differentiators and digital integrators. Simulation results illustrated that the proposed HFPA achieved a superior performance in the least number of function evaluations when compared with the other methods in the literature.

Chakraborty et al. [57] integrated the global search capabilities of FPA with the local search behavior of gravitational search algorithm (GSA) for training the feedforward neural network so as to strike a right balance between exploration and exploitation during the search process. Furthermore, dynamic switch probability and adaptive weights of the GSA velocity operator were introduced to avoid trapping in the local minima and to guide the search toward the global minima, respectively. The authors investigated their method using a set of real-world benchmark datasets retrieved from the UCI Machine Learning Repository. These real-world benchmark data sets included cancer, glass, iris, vertebral column, and wine The numerical experiment demonstrated that their method performed significantly better than FPA and GSA for all datasets.

Salgotra and Singh combined FPA with the bat algorithm, called bat flower pollination algorithm (BFA), to avoid getting stuck in the local optima problem and to enhance the convergence speed [58]. Evaluation of the proposed BFA algorithm entailed using thirteen benchmark functions with superior performance and comparison with state-of-the-art algorithms. For more validation, the authors evaluated their method by synthesizing unequally spaced linear antenna arrays for single and multi-objective design. Interestingly, the proposed BFA obtained better synthesis results than other techniques in the literature.

3.2.3 Hybridization with Other Components

Zawbaa et al. [59] presented a new model for multi-objective feature selection based on the combination of FPA and rough set theory to find the optimal feature set for classification. Their model exploited the characteristics of filter-based feature selection and wrapper-based feature selection. Filter-based method served as a data-oriented technique, while wrapper-based method was a classification technique. The authors verified the performance of their method on eight UCI datasets and found that the proposed method was highly competitive when compared with the classical FPA, PSO, and GA.

Abdel-Baset and Hezam combined FPA and conjugate direction (CD) method to solving the ill-conditioned system of linear and nonlinear equations [44]. FPA was used for fast convergence and to find more than one root, whereas CD method was utilized to increase the accuracy of final results and to avoid getting stuck in the local minima. Numerical simulation results showed that the proposed method was very competitive when compared with the other methods in literature.

Valenzuela et al. [60] introduced FFPA a novel hybrid from the integration of FPA with fuzzy inference system. The fuzzy inference system was used in their method to adapt the probability of switching between local and global pollination. The performance of the proposed method was competitive with other approaches when evaluated on eight benchmark mathematical functions.

In 2016, Wang et al. [61] presented a new hybrid variant by integrating FPA with a bee pollinator to solve clustering problems. Elite-based mutation and crossover operators were used in the local search process of FPA to improve convergence speed and population diversity. Furthermore, the discarded pollen operator was used in the global search process of FPA to improve the exploration capability, thus potentially avoiding from getting stuck in the local minima. The authors evaluated their method using ten datasets and the experimental results demonstrated that the proposed method had a higher accuracy, level of stability, and convergence speed, when compared with K-Means, PSO, DE, CS, ABC, and FPA.

Majidpour et al. [62] integrated FPA with Ada-Boost algorithm to enhance the accuracy in text document classification, where the former was used for feature selection, while the latter was used to classify text documents. The authors evaluated the performance of their proposed algorithm by using three standard datasets such as Reuters-21578, WEBKB, and CADE 12. Experimental results showed that the proposed algorithm performed better than the Ada-Boost algorithm and other algorithms.

Jain et al. [63] introduced a multi-objective FPA hybrid to solve the channel allocation problem in optical division multiplexing. The authors divided the population into a set of subpopulations, and combined the mutation strategy of DE with their method to improve search efficiency and to increase population diversity. Simulation demonstrated that the results of the proposed method were satisfactory when compared with other methods.

Xu and Wang proposed a new version of hybrid FPA for estimating the parameters of solar cells and PV modules [64] by integrating the FPA with the Nelder-Mead simplex method to enhance the local search ability of classical FPA. Furthermore, generalized opposition-based learning mechanism was combined with their method to avoid getting stuck in the local optima. The authors evaluated their method using three different solar models including the single diode model, the double diode model, and a PV module. Numerical results clearly demonstrated that the proposed hybrid FPA was better than the other methods in terms of the accuracy of final solutions, convergence speed, and stability.

In another study, FPA was hybridized with the Nelder-Mead simplex method to enhance the local exploitation ability [65]. The authors tested their method using three typical chaotic system parameter estimation problems with three unknown system parameters, including the Lorenz system, the Rossler system, and the Lorenz system under a noise condition. A comparative evaluation conducted with the state-of-the-art methods revealed the proposed algorithm as an effective technique for solving the parameter estimation problem of chaotic systems.

Bensouyad and Saidouni [66] developed a hybrid version of FPA for Graph Coloring Problem by using the efficient constructive method called Recursive Largest First to maintain the feasibility of the solutions during the search process. Furthermore, combining the swap and inversion strategies within their method could keep population diversity and avoid the stagnation problem that occurs during the search process. The experimental results showed that the proposed method was competitive with the state-of-the-art methods.

Combining FPA with a randomized-location modification operator, called a modified randomized-location flower pollination algorithm (MRLFPA), for medical image segmentation was presented in [67]. In MRLFPA, the randomized-location strategy easily overcame the weakness of the classical FPA. The performance of the proposed MRLFPA was tested using eight medical images with different characteristics. A comparative evaluation with other algorithms revealed the effectiveness of MRLFPA in terms of solution quality, stability, and computation efficiency.

3.3   Multi-objective Versions of FPA

Yang et al. [68] proposed the first attempt of extending FPA to solve multi-objective engineering optimization problems (MOFPA) through a random weighted sum method. MOFPA was evaluated using several engineering optimization problems to produce optimal results. Later, the same authors proposed a novel technique for MOFPA [69] by introducing several multi-objective test functions and two bi-objective design benchmarks, and the results of the proposed algorithm were very efficient, compared with other algorithms.

In the radial distribution system, Tamilselvan et al. [70] introduced an MOFPA to calculate the power flow and losses in the system. The proposed algorithm was then evaluated using two standard test cases of IEEE 33 and IEEE 69 radial distribution systems. Dimitrios Gonidakis also implemented MOFPA [71] in solving the environmental/economic dispatch problem. Testing the proposed algorithm from using two power generation systems showed the MOFPA having several advantages over other modern optimization techniques.

Shilaja et al. [72] proposed an enhanced FPA called EFPA to find the best solution for the OPF problem and implemented EFPA for multi-objective of transmission loss and power plant emission, minimization of generating cost, and improvement of voltage stability. Evaluation of EFPA using the standard IEEE 30 test showed the proposed algorithm yielding better results than other optimization algorithms.

For the power loss reduction, Rajaram et al. [73] proposed a multi-objective FPA to improve the load distribution and the voltage profile for distribution network reconfiguration. MOFPA provided better results than other published techniques. Emary et al. [17] implemented a multi-objective FPA for retinal vessel localization with pattern search (PS). The proposed technique utilized the FPA to find the optimal clustering of the given retinal image. In addition, the proposed algorithm applied PS as a local search approach to improve segmentation results. The proposed method was tested using a standard benchmark named as DRIVE dataset. The results of the proposed technique were comparable with other optimization algorithms in terms of accuracy, sensitivity, and specificity with many extendable features.

3.4   Parameters Setting Versions of FPA

To improve the ease of tuning parameters in the standard FPA, Salgotra et al. [74] proposed a new FPA variant called the adaptive-Lévy flower pollination algorithm (ALFPA), which involved new mutation operators, dynamic switching, and adapting local search. ALFPA had been tested using 17 benchmarks and compared with other optimization algorithms such as ABC, FA, BA, DE, and GWO. As a result, they showed ALFPA had superiority in numerical results for standard benchmark functions as well as in statistical tests.

In the manufacturing industry, one challenging task is the multi-pass turning of parameters such as feed rate, cutting speed, and depth of cut. Xu et al. [75] proposed an improved FPA to solve this problem where the proposed algorithm gusseted in keeping the local and global search operator of the standard FPA, while using Deb's heuristic rules to initiate the new population. The proposed algorithm has shown comparative results and outstanding performance.

4 Applications of Flower Pollination Algorithm

When we presented various variants of FPA, we also outlined their applications in a diverse range of real-world scenarios. Here, we briefly introduce other applications that were not mentioned earlier. The FPA is successfully tailored for several domains of optimization problems, including electrical and power system [11–15], signal and image processing [16–18], wireless sensor networking [19–21], clustering and classification [22, 23], global function optimization [24], computer gaming [25], structural and mechanical engineering [26–28], and and many others [29, 30].

Table 2 summarizes these applications. Some applications are high dimensional, complex, and non-convex problems that the classical version of FPA may not easily tackle alone without any amendment. The main focus of this review is not to provide a detailed review of all FPA applications, but rather show the alternative FPA variants and the FPA pros and cons, which can motivate researchers to explore new possibilities in enhancing FPA to solve other application problems. Thus, this review presents a comprehensive but not exhaustive summary of FPA applications.

Figure 5 shows the FPA application domains. Most research areas that use FPA are electrical and power system where most problems can efficiently be solved by FPA, including economic/emission dispatch, load frequency control, optimal power flow, and so on. The FPA in the domain of structural and mechanical engineering has shown powerful results in solving challenging problems such as structural design, tuning of mass dampers, and multi-pass turning parameters. In the clustering and classification domain, the FPA is successfully applicable in several problems such as data clustering and feature selection. Furthermore, FPA has obtained optimal findings in wireless and network system domain where it is applicable in addressing different problems such as wireless sensor networks, antenna positioning, and vehicle path planning problem. Another domain, where using FPA shows its superiority, is signal and image processing, in which the algorithm solves medical image segmentation, retinal vessel localization, EEG channel selection, and multilevel image thresholding. Finally, the FPA has shown good results with the optimization problems of standard benchmark functions.

Table 2 FPA applications

Problem	References
Electrical and power system domain	
Economic/Emission dispatch	Abdelaziz et al. [11], Prathiba et al. [76], Abdelaziz et al. [13], Shilaja et al. [40], Dimitrios Gonidakis [71], Dubey et al. [38], Putra et al. [14], Regalado et al. [36], Dubey et al. [54], Abdelaziz et al. [97], Nigdei et al. [98], Kerta et al. [99], Velamuri et al. [105], Abdelaziz et al. [106]
Reactive power dispatch	Lenin et al. [49], Sakthivel et al. [96]
Optimal power flow	Rajalashmi and Prabha [15], Kanagasabai and RavindhranathReddy [55], Shilaja et al. [72], Shilaja et al. [95]
Transmission laser welding	Acherjee et al. [77]
Optimal design of wideband integrators	Mahata et al. [56]
Solar PV parameter estimation	Ram et al. [50]
Estimation of photovoltaic parameters	Xu and Wang [64], Xu et al. [65]
Forecasting of petroleum consumption	Chiroma et al. [22]
Load frequency control	Lakshmi et al. [78], Jagatheesan et al. [79]
Network reconfiguration	Namachivayam et al. [37], Rajaram et al. [73]
Optimal power flow	Shilaja et al. [72], Mahdad et al. [102]
Optimal parameters of photovoltaic (PV)	Alam et al. [80]
Pi-pd cascade controller	Dash et al. [81]
Wind speed forecasting	Zhang et al. [43]
Wireless and network system domain	
Wireless sensor networks	Sharawi et al. [19], Hajjej et al. [21], Rana et al. [82], Pan et al. [92], Sesli et al. [101], Binh et al. [111], Sharma et al. [117]
Antenna positioning	Dahi et al. [41], Vedula et al. [108], Salgotra et al. [110]
Optical division multiplexing	Jain et al. [63]
Vehicle path planning problem	Zhou et al. [83]
Linear antenna array optimization	Saxena et al. [84]
Synthesis of linear antenna array	Salgotra and Singh [58]
Clustering and classification domain	
Data clustering	Agarwal et al. [23], Łukasik et al. [85], Jensi and Jiji [45], Wang et al. [61], Ramadas et al. [112]
Train the feed forward neural network	Chakraborty et al. [57]
Feature selection	Rodrigues et al. [39], Sayed et al. [46], Zawbaa et al. [59], Majidpour et al. [62]
Structural and mechanical engineering domain	
Frames and and truss systems	Nigdeli et al. [26]
Structure engineering design	Zhou et al. [27], Bekdaş et al. [103], Bekdaş et al. [104]

(continued)

Table 2 (continued)

Problem	References
Tuning of mass dampers	Nigdeli et al. [48]
Multi-pass turning parameters	Xu et al. [75]
Mechanical engineering design	Meng et al. [28]
Assembly sequence optimization	Mishra et al. [86]
Games domain	
Sudoku puzzles	Abdel-Raouf et al. [25]
General healthy domain	
Generating menu recommendations	Pop et al. [30], Pasaribu et al. [93]
Image and signal processing domain	
Shape matching	Zhou et al. [113]
Medical image segmentation	Wang et al. [67], Emary et al. [100]
EEG channel selection	Rodrigues et al. [16]
Retinal vessel localization	Emary et al. [17]
Graph coloring problem	Bensouyad and Saidouni [66]
Multilevel image thresholding	Ouadfel et al. [18]
Standard benchmark functions domain	
Global optimization	Emad Nabil [24], Sakib et al. [91], Draa [94], Łukasik et al. [107], Nasser et al. [114], Hegazy et al. [115], Rathasamuth et al. [116]
Ill-conditioned set of equations	Abdel-Baset and Hezam [44]
Roots identified	Platt et al. [87]
Fractional programming problems	Metwalli et al [42]

5   Critical Analysis of FPA Variants

Almost all metaheuristic algorithms can be modified by hybridizing with other algorithms or adding new components. FPA has undergone various modifications and hybridization with other techniques to address the complexity nature of optimization problems. As we have summarized above, the research community has investigated many different ways to improve the convergence of FPA and to overcome its potential weakness or drawbacks. Here we choose a few variants and analyze them critically so as to gain a better understanding and insight into different variants and try to understand why certain hybrid and additional components can improve the performance.

The first issue is related to the exploration ability of FPA by Lévy flights which can be too aggressive by generating large steps, which may lead to the case when newly generated solutions can be potentially outside of the search domain and thus reduce the true exploration ability. Some variants try to enhance the exploration capability. For example, Wang et al. [61] related the exploration capability of FPA to the pollination operator. Their modification of classical FPA used three steps: use of dynamic switching probability, application of the real-coded GA (RCGA) as

Fig. 5 FPA applications

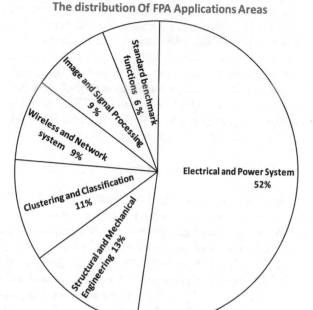

The distribution Of FPA Applications Areas

Standard benchmark functions 6 %

Image and Signal Processing 9 %

Wireless and Network system 9%

Clustering and Classification 11%

Structural and Mechanical Engineering 13%

Electrical and Power System 52%

mutation for global and local search, and the differentiation between the temporary local search and optimal solution [14]. In addition, exploitation in the standard FPA is mainly by selection and use of g^* (the best solution found so far). Some variants try to enhance this ability by using gradients or simplex method. Hybridization has been carried out by improving its performance in exploration as FPA has a stronger ability in exploitation [47, 50, 61]. Further research can try to understand why these hybrid algorithms work both numerically and theoretically.

Another issue is about discretization of continuous FPA because theclassical FPA cannot deal with the binary optimization problem directly. To handle this type of problems, Rodrigues et al. [16] produced a new binary FPA version where the local pollination operator has been modified. The search space is modeled as an n-dimensional Boolean lattice, in which the solutions are updated across the corners of a hypercube.

In addition, many optimization problems in real-world applications are multi-objective. Thus, some variants have focused on the extension of FPA to solve challenging multi-objective problems [68, 69]. This area requires further research because multi-objective optimization can be very computationally extensive in higher dimensions. Effective methods should be sorted to generate high quality Pareto fronts for multiobjective optimization.

Like all metaheuristic algorithms, the parameter values of an algorithm may affect the performance of that algorithm. Therefore, some variants of FPA attempted to tune FPA parameters. For example, FPA parameters are tunable through a dynamic switching probability, application of the mutation for global and local search, and identification of the difference between the temporary local search and optimal solution [14, 74, 75]. In addition, the standard FPA has not been applied to solve large-scale problem. Therefore, several researchers have attempted to use FPA with modifications to solve a large spectrum of optimization problems [38, 74]. FPA has a good robust ability in solving continuous optimization problems, and this ability has been tested for discrete optimization problems [88].

Even with the above various modifications and variants, there are still more research opportunities in this area. We will highlight some of research directions in the next section.

6 Conclusions and Future Directions

This review has summarized the most recent FPA variants, based on the main optimization framework of FPA initially proposed by Yang [10]. The variations of FPA variants have been discussed based on four classifications, namely, modified variants, hybridized variants, multi-objective variants, and adaptive parameter variants. Modified FPA variants have been further classified into three classes: modified FPAs based on operators, binary FPAs, and chaotic-based FPAs. Furthermore, the hybridized FPA versions have also been categorized into local search-based hybridization, population-based hybridization, and hybridization with other components. Finally, a critical analysis of some FPA variants has been carried out show to gain some insight.

Even the FPA performance has been enhanced in many places and in many ways, future research can focus on the following areas:

- **Theoretical Analysis**: As FPA has successfully been applied into a wide variety of optimization problem, there is one mathematical study on its global convergence by He et al. [89] that has proved that FPA can have guaranteed global convergence. However, this study does not provide any information about the rate of convergence. Therefore, further studies can focus on the theoretical analysis of the convergence rate, stability and robustness of FPA. Such methodology will also be useful to analyze other metaheuristic algorithms.
- **Adaptivity of Parameters**: Almost all algorithms have algorithm-dependent parameters and such parameters can influence their performance. However, the tuning, control and variations of such parameters can be difficult. Ideally, an algorithm should tune its parameters automatically and adapt their values according to the type of problem under consideration.
- **Large-Scale Combinatorial Optimization**: FPA has been applied to many problems including some combinatorial optimization problems with good results. However, like almost all other metaheuristic algorithms, it has not been applied to truly large-scale combinatorial problems with thousands of design variables. Therefore, future research should focus on the application of FPA and other algorithms to solve large-scale problems that are important in real-world applications.
- **Population Structure**: Currently, almost all population-based algorithms including FPA and its variants use a simple structure of updating the population. All

solutions in the population are either updated in series or in parallel. It would be useful to investigate the possibility of updating them in a non-synchronized, unstructured way. It will also be useful to see if they work for a mixed parallel-series population with random mixing, even with island models such as cellular automata type of structures.

In summary, this review has justified that FPA is a potentially powerful and useful tool for solving different optimization problems in a diverse range of applications. New modifications and improvements can enhance its performance even further. The authors hope that this chapter can inspire interested researchers and practitioners to carry out more research in this area and to solve more complex and challenging problems in practice.

Acknowledgements The first author would like to thank the University Science Malaysia (USM) and The World Academic Science (TWAS) for supporting his Ph.D. study which is under (USM-TWAS Postgraduate Fellowship, FR number: 3240287134).

References

1. Holland, J.H.: Adaptation in Natural and Artificial Systems. An Introductory Analysis With Application to Biology, Control, and Artificial Intelligence. University of Michigan Press, Ann Arbor, MI (1975)
2. Karaboga, D.: An idea based on honey bee swarm for numerical optimization. Technical report-tr06, Erciyes university, engineering faculty, computer engineering department (2005)
3. Kennedy, J.: Particle swarm optimization. Encyclopedia of Machine Learning, pp. 760–766. Springer (2011)
4. Mirjalili, S., Mirjalili, S.M., Lewis, A.: Grey wolf optimizer. Adv. Eng. Softw. **69**, 46–61 (2014)
5. Yang, X.-S.: Firefly algorithm, stochastic test functions and design optimisation. Int. J. Bio-Inspired Comput. **2**(2), 78–84 (2010). Inderscience Publishers
6. Yang, X.-S.: A new metaheuristic bat-inspired algorithm. In: Nature Inspired Cooperative Strategies for Optimization (NICSO 2010), pp. 65–74. Springer (2010)
7. Gandomi, A.H., Yang, X.-S., Alavi, A.H.: Cuckoo search algorithm: a metaheuristic approach to solve structural optimization problems. Eng. Comput. **29**(1), 17–35 (2013)
8. Yang, X.S.: Nature-Inspired Metaheuristic Algorithms. Luniver Press (2008)
9. Yang, X.S.: Engineering Optimization: An Introduction with Metaheuristic Applications. Wiley (2010)
10. Yang, X.-S.: Flower pollination algorithm for global optimization. In: International Conference on Unconventional Computing and Natural Computation, pp. 240–249. Springer (2012)
11. Abdelaziz, A., Ali, E., Elazim, S.A.: Combined economic and emission dispatch solution using flower pollination algorithm. Int. J. Electr. Power Energy Syst. **80**, 264–274 (2016)
12. Singh, U., Salgotra, R.: Synthesis of linear antenna array using flower pollination algorithm. Neural Comput. Appl., 1–11 (2016)
13. Abdelaziz, A., Ali, E., Elazim, S.A.: Implementation of flower pollination algorithm for solving economic load dispatch and combined economic emission dispatch problems in power systems. Energy **101**, 506–518 (2016)
14. Putra, P.H., Saputra, T.A., et al.: Modified flower pollination algorithm for nonsmooth and multiple fuel options economic dispatch. In: 2016 8th International Conference on Information Technology and Electrical Engineering (ICITEE), pp. 1–5. IEEE (2016)

15. Rajalashmi, K., Prabha, S.: A hybrid algorithm for multiobjective optimal power flow problem using particle swarm algorithm and enhanced flower pollination algorithm. Asian J. Res. Social Sci. Humanit. **7**(1), 923–940 (2017)
16. Rodrigues, D., Silva, G.F., Papa, J.P., Marana, A.N., Yang, X.-S.: Eeg-based person identification through binary flower pollination algorithm. Expert Syst. Appl. **62**, 81–90 (2016)
17. Emary, E., Zawbaa, H.M., Hassanien, A.E., Parv, B.: Multi-objective retinal vessel localization using flower pollination search algorithm with pattern search. Adv. Data Anal. Class., 1–17 (2016)
18. Ouadfel, S., Taleb-Ahmed, A.: Social spiders optimization and flower pollination algorithm for multilevel image thresholding: a performance study. Expert Syst. Appl. **55**, 566–584 (2016)
19. Sharawi, M., Emary, E., Saroit, I.A., El-Mahdy, H.: Flower pollination optimization algorithm for wireless sensor network lifetime global optimization. Int. J. Soft Comput. Eng. **4**(3), 54–59 (2014)
20. Shankar, T., James, T., Mageshvaran, R., Rajesh, A.: Lifetime improvement in wsn using flower pollination meta heuristic algorithm based localization approach. Indian J. Sci. Technol. **9**(37)
21. Hajjej, F., Ejbali, R., Zaied, M.: An efficient deployment approach for improved coverage in wireless sensor networks based on flower pollination algorithm, pp. 117–129 (2016). doi:10.5121/csit.2016.61511
22. Chiroma, H., Khan, A., Abubakar, A.I., Saadi, Y., Hamza, M.F., Shuib, L., Gital, A.Y., Herawan, T.: A new approach for forecasting opec petroleum consumption based on neural network train by using flower pollination algorithm. Appl. Soft Comput. **48**, 50–58 (2016)
23. Agarwal, P., Mehta, S.: Enhanced flower pollination algorithm on data clustering. Int. J. Comput. Appl. **38**(2–3), 144–155 (2016)
24. Nabil, E.: A modified flower pollination algorithm for global optimization. Expert Syst. Appl. **57**, 192–203 (2016)
25. Abdel-Raouf, O., El-Henawy, I., Abdel-Baset, M.: A novel hybrid flower pollination algorithm with chaotic harmony search for solving sudoku puzzles. Int. J. Mod. Educ. Comput. Sci. **6**(3), 38 (2014)
26. Nigdeli, S.M., Bekdaş, G., Yang, X.-S.: Application of the flower pollination algorithm in structural engineering. In: Metaheuristics and Optimization in Civil Engineering, pp. 25–42. Springer (2016)
27. Zhou, Y., Wang, R., Luo, Q.: Elite opposition-based flower pollination algorithm. Neurocomputing **188**, 294–310 (2016)
28. Meng, O.K., Pauline, O., Kiong, S.C., Wahab, H.A., Jafferi, N.: Application of modified flower pollination algorithm on mechanical engineering design problem. In: IOP Conference Series: Materials Science and Engineering, vol. 165, p. 012032. IOP Publishing (2017)
29. Pant, S., Kumar, A., Ram, M.: Flower pollination algorithm development: a state of art review. Int. J. Syst. Assur. Eng. Manag., 1–9 (2017)
30. Pop, C.B., Chifu, V.R., Salomie, I., Racz, D.S., Bonta, R.M.: Hybridization of the flower pollination algorithma ase study in the problem of generating healthy nutritional meals for older adults. In: Nature-Inspired Computing and Optimization, pp. 151–183. Springer (2017)
31. Bell, A.: Plant Form: An Illustrated Guide to Flowering Plant Morphology. Oxford University Press, Oxford (1991)
32. Cronquist, A.: An Integrated System of Calssificaiton of Flowering Plants. Columbia University Press, New York (1981)
33. Glover, B.J.: Understanding Flowers and Flowering: An Integrated Approach. Oxford University Press (2007)
34. Kalra, S., Arora, S.: Firefly algorithm hybridized with flower pollination algorithm for multimodal functions. In: Proceedings of the International Congress on Information and Communication Technology, pp. 207–219. Springer (2016)
35. Yamany, W., Zawbaa, H.M., Emary, E., Hassanien, A.E.: Attribute reduction approach based on modified flower pollination algorithm. In: 2015 IEEE International Conference on Fuzzy Systems (FUZZ-IEEE), pp. 1–7. IEEE (2015)

36. Regalado, J.A., Emilio, B.E., Cuevas, E.: Optimal power flow solution using modified flower pollination algorithm. In: 2015 IEEE International Autumn Meeting on Power, Electronics and Computing (ROPEC), pp. 1–6. IEEE (2015)
37. Namachivayam, G., Sankaralingam, C., Perumal, S.K., Devanathan, S.T.: Reconfiguration and capacitor placement of radial distribution systems by modified flower pollination algorithm. Electr. Power Compon. Syst. 44(13), 1492–1502 (2016)
38. Dubey, H.M., Pandit, M., Panigrahi, B.K.: A biologically inspired modified flower pollination algorithm for solving economic dispatch problems in modern power systems. Cogn. Comput. 7(5), 594–608 (2015)
39. Rodrigues, D., Yang, X.-S., De Souza, A.N., Papa, J.P.: Binary flower pollination algorithm and its application to feature selection. In: Recent Advances in Swarm Intelligence and Evolutionary Computation, pp. 85–100. Springer (2015)
40. Shilaja, C., Ravi, K.: Optimization of emission/economic dispatch using euclidean affine flower pollination algorithm (efpa) and binary fpa (bfpa) in solar photo voltaic generation. Renew. Energy 107, 550–566 (2017)
41. Dahi, Z.A.E.M., Mezioud, C., Draa, A.: On the efficiency of the binary flower pollination algorithm: application on the antenna positioning problem. Appl. Soft Comput. 47, 395–414 (2016)
42. Metwalli, M.A.-B., Hezam, I., Yardım, D., Ozkan, I.A., Saritas, I., Aslam, D.M.: A modified flower pollination algorithm for fractional programming problems. Int. J. Intell. Syst. Appl. Eng. 3(3) (2015)
43. Zhang, W., Qu, Z., Zhang, K., Mao, W., Ma, Y., Fan, X.: A combined model based on ceemdan and modified flower pollination algorithm for wind speed forecasting. Energy Convers. Manag. 136, 439–451 (2017)
44. Abdel-Baset, M., Hezam, I.: A hybrid flower pollination algorithm for engineering optimization problems. Int. J. Comput. Appl. 140(12) (2016)
45. Jensi, R., Jiji, G.W.: Hybrid data clustering approach using k-means and flower pollination algorithm (2015). arXiv:1505.03236
46. Sayed, S.A.-F., Nabil, E., Badr, A.: A binary clonal flower pollination algorithm for feature selection. Pattern Recogn. Lett. 77, 21–27 (2016)
47. Abdel-Raouf, O., Abdel-Baset, M., et al.: A new hybrid flower pollination algorithm for solving constrained global optimization problems. Int. J. Appl. Oper. Res.-An Open Access J. 4(2), 1–13 (2014)
48. Nigdeli, S.M., Bekdaş, G., Yang, X.-S.: Optimum tuning of mass dampers by using a hybrid method using harmony search and flower pollination algorithm. In: International Conference on Harmony Search Algorithm, pp. 222–231. Springer (2017)
49. Lenin, K., Ravindhranath, R., Surya, K.: Shrinkage of active power loss by hybridization of flower pollination algorithm with chaotic harmony search algorithm. Control Theory Inf. 4, 31–38 (2014)
50. Ram, J.P., Babu, T.S., Dragicevic, T., Rajasekar, N.: A new hybrid bee pollinator flower pollination algorithm for solar pv parameter estimation. Energy Convers. Manag. 135, 463–476 (2017)
51. Abdel-Baset, M., Hezam, I.M.: An effective hybrid flower pollination and genetic algorithm for constrained optimization problems. Adv. Eng. Technol. Appl. Int. J. 4, 27–27 (2015)
52. Chakraborty, D., Saha, S., Dutta, O.: De-fpa: A hybrid differential evolution-flower pollination algorithm for function minimization. In: 2014 International Conference on High Performance Computing and Applications (ICHPCA), pp. 1–6. IEEE (2014)
53. Ramadas, M., Pant, M., Abraham, A., Kumar, S.: ssfpa/de: an efficient hybrid differential evolution–flower pollination algorithm based approach. Int. J. Syst. Assur. Eng. Manag., 1–14 (2016)
54. Dubey, H.M., Pandit, M., Panigrahi, B.: Hybrid flower pollination algorithm with time-varying fuzzy selection mechanism for wind integrated multi-objective dynamic economic dispatch. Renew. Energy 83, 188–202 (2015)

55. Kanagasabai, L., RavindhranathReddy, B.: Reduction of real power loss by using fusion of flower pollination algorithm with particle swarm optimization. J. Inst. Ind. Appl. Eng. **2**(3), 97–103 (2014)

56. Mahata, S., Saha, S.K., Kar, R., Mandal, D.: Optimal design of wideband digital integrators and differentiators using hybrid flower pollination algorithm. Soft Comput., 1–27 (2017)

57. Chakraborty, D., Saha, S., Maity, S.: Training feedforward neural networks using hybrid flower pollination-gravitational search algorithm. In: 2015 International Conference on Futuristic Trends on Computational Analysis and Knowledge Management (ABLAZE), pp. 261–266. IEEE (2015)

58. Kusuma, I., Ma'sum, M.A., Sanabila, H., Wisesa, H., Jatmiko, W., Arymurthy, A., Wiweko, B.: Fetal head segmentation based on gaussian elliptical path optimize by flower pollination algorithm and cuckoo search. In: 2016 International Conference on Advanced Computer Science and Information Systems (ICACSIS), pp. 564–571. IEEE (2016)

59. Zawbaa, H.M., Hassanien, A.E., Emary, E., Yamany, W., Parv, B.: Hybrid flower pollination algorithm with rough sets for feature selection. In: 2015 11th International Computer Engineering Conference (ICENCO), pp. 278–283. IEEE (2015)

60. Valenzuela, L., Valdez, F., Melin, P.: Flower pollination algorithm with fuzzy approach for solving optimization problems. In: Nature-Inspired Design of Hybrid Intelligent Systems, pp. 357–369. Springer (2017)

61. Wang, R., Zhou, Y., Qiao, S., Huang, K.: Flower pollination algorithm with bee pollinator for cluster analysis. Inf. Process. Lett. **116**(1), 1–14 (2016)

62. Majidpour, H., Soleimanian Gharehchopogh, F.: An improved flower pollination algorithm with adaboost algorithm for feature selection in text documents classification. J. Adv. Comput. Res

63. Jain, P., Bansal, S., Singh, A.K., Gupta, N.: Golomb ruler sequences optimization for fwm crosstalk reduction: multi-population hybrid flower pollination algorithm. In: Progress in Electromagnetics Research Symposium (PIERS), Prague, Czech Republic, pp. 2463–2467 (2015)

64. Xu, S., Wang, Y.: Parameter estimation of photovoltaic modules using a hybrid flower pollination algorithm. Energy Convers. Manag. **144**, 53–68 (2017)

65. Xu, S., Wang, Y., Liu, X.: Parameter estimation for chaotic systems via a hybrid flower pollination algorithm. Neural Comput. Appl. 1–17 (2017)

66. Bensouyad, M., Saidouni, D.E.: A hybrid discrete flower pollination algorithm for graph coloring problem. In: Proceedings of the The International Conference on Engineering & MIS 2015, p. 22. ACM (2015)

67. Wang, R., Zhou, Y., Zhao, C., Wu, H.: A hybrid flower pollination algorithm based modified randomized location for multi-threshold medical image segmentation. Bio-Med. Mater. Eng. **26**(s1), S1345–S1351 (2015)

68. Yang, X.-S., Karamanoglu, M., He, X.: Multi-objective flower algorithm for optimization. Proc. Comput. Sci. **18**, 861–868 (2013)

69. Yang, X.-S., Karamanoglu, M., He, X.: Flower pollination algorithm: a novel approach for multiobjective optimization. Eng. Optim. **46**(9), 1222–1237 (2014)

70. Tamilselvan, V., Jayabarathi, T.: Multi objective flower pollination algorithm for solving capacitor placement in radial distribution system using data structure load flow analysis. Arch. Electr. Eng. **65**(2), 203–220 (2016)

71. Gonidakis, D.: Application of flower pollination algorithm to multi-objective environmental/economic dispatch. Int. J. Manag. Sci. Eng. Manag. **11**(4), 213–221 (2016)

72. Shilaja, C., Ravi, K.: Multi-objective optimal power flow problem using enhanced flower pollination algorithm. Gazi Univ. J. Sci. **30**(1), 79–91 (2017)

73. Rajaram, R., Kumar, K.S.: Multiobjective power loss reduction using flower pollination algorithm. **8**(5), 2239–2245 (2015)

74. Salgotra, R., Singh, U.: Application of mutation operators to flower pollination algorithm. Expert Syst. Appl. **79**, 112–129 (2017)

75. Xu, S., Wang, Y., Huang, F.: Optimization of multi-pass turning parameters through an improved flower pollination algorithm. Int. J. Adv. Manuf. Technol., 1–12 (2016)
76. Prathiba, R., Moses, M.B., Sakthivel, S.: Flower pollination algorithm applied for different economic load dispatch problems. Int. J. Eng. Technol. (IJET) 6(2), 1009–16 (2014)
77. Acherjee, B., Maity, D., Kuar, A.S.: Parameters optimisation of transmission laser welding of dissimilar plastics using rsm and flower pollination algorithm integrated approach. Int. J. Math. Modell. Num. Optim. 8(1), 1–22 (2017)
78. Lakshmi, D., Fathima, A.P., Muthu, R., et al.: A novel flower pollination algorithm to solve load frequency control for a hydro-thermal deregulated power system. Circuits Syst. 7(04), 166 (2016)
79. Jagatheesan, K., Anand, B., Samanta, S., Dey, N., Santhi, V., Ashour, A.S., Balas, V.E.: Application of flower pollination algorithm in load frequency control of multi-area interconnected power system with nonlinearity. Neural Comput. Appl., 1–14 (2016)
80. Alam, D., Yousri, D., Eteiba, M.: Flower pollination algorithm based solar pv parameter estimation. Energy Convers. Manag. 101, 410–422 (2015)
81. Dash, P., Saikia, L.C., Sinha, N.: Flower pollination algorithm optimized pi-pd cascade controller in automatic generation control of a multi-area power system. Int. J. Electr. Power Energy Syst. 82, 19–28 (2016)
82. Rana, D., Arora, M.: Energy efficient cluster-based routing protocol in wireless sensor network using flower pollination algorithm. Int. J. Control Theory Appl. 10(10), 119–133 (2017)
83. Zhou, Y., Wang, R.: An improved flower pollination algorithm for optimal unmanned undersea vehicle path planning problem. Int. J. Pattern Recognit. Artif. Intell. 30(04), 1659010 (2016)
84. Saxena, P., Kothari, A.: Linear antenna array optimization using flower pollination algorithm. SpringerPlus 5(1), 306 (2016)
85. Łukasik, S., Kowalski, P.A., Charytanowicz, M., Kulczycki, P.: Clustering using flower pollination algorithm and calinski-harabasz index. In: 2016 IEEE Congress on Evolutionary Computation (CEC), pp. 2724–2728. IEEE (2016)
86. Mishra, A., Deb, S.: Assembly sequence optimization using a flower pollination algorithm-based approach. J. Intell. Manuf., 1–22 (2016)
87. Platt, G.: Application of the flower pollination algorithm in nonlinear algebraic systems with multiple solutions. Eng. Optim. 2014, 117 (2014)
88. Wang, R., Zhou, Y., Zhou, Y., Bao, Z.: Local greedy flower pollination algorithm for solving planar graph coloring problem. J. Comput. Theor. Nanosci. 12(11), 4087–4096 (2015)
89. He, X.S., Yang, X.S., Karamanoglu, M., Zhao, Y.X.: Global convegence analysis of the flower pollination algorithm: a discrete-time Markov chain approach. Proc. Comput. Sci. 108, 1354–1363 (2017)
90. Pavlyukevich, I.: Lévy flights, non-local search and simulated annealing. J. Comput. Phys. 226, 1830–1844 (2007)
91. Sakib, N., Kabir, M.W.U., Subbir, M., Alam, S.: A comparative study of flower pollination algorithm and bat algorithm on continuous optimization problems. Int. J. Appl. Inf. Syst. 7(9), 13–19 (2014)
92. Pan, J.-S., Dao, T.-K., Pan, T.-S., Chu, S.-C., Roddick, J.F.: An improvement of flower pollination algorithm for node localization optimization in wsn. J. Inf. Hiding Multimed. Signal Process. 8(2), 486–499 (2017)
93. Pasaribu, U.S., al Mashumah, F., Permana, D.: Estimation of the transition matrix in Markov chain model of customer lifetime value using flower pollination algorithm. Appl. Math. Sci. 9(69), 3409–3418 (2015)
94. Draa, A.: On the performances of the flower pollination algorithm-qualitative and quantitative analyses. Appl. Soft Comput. 34, 349–371 (2015)
95. Shilaja, C., Ravi, K.: Optimal line flow in conventional power system using euclidean affine flower pollination algorithm. Int. J. Renew. Energy Res. C. 6(1)
96. Sakthivel, S., Manopriya, P., Venus, S., Ranjitha, S., Subhashini, R.: Optimal reactive power dispatch problem solved by using flower pollination algorithm. Int. J. Appl. Eng. Res. 11(6), 4387–4391 (2016)

97. Abdelaziz, A., Ali, E., Elazim, S.A.: Optimal sizing and locations of capacitors in radial distribution systems via flower pollination optimization algorithm and power loss index. Eng. Sci. Technol. Int. J. **19**(1), 610–618 (2016)

98. Nigdei, S.M., Bekdaş, G., Yang, X.: Optimum tuning of mass dampers for seismic structures using flower pollination algorithm. Int. J. Theor. Appl. Mech

99. Kerta, S., Hamid, Z., Musirin, I.: Real power generation tracing for deregulated power system using the flower pollination algorithm technique. J. Theor. Appl. Inf. Technol. **81**(3), 564 (2015)

100. Emary, E., Zawbaa, H.M., Hassanien, A.E., Tolba, M.F., Snášel, V.: Retinal vessel segmentation based on flower pollination search algorithm. In: Proceedings of the Fifth International Conference on Innovations in Bio-Inspired Computing and Applications IBICA 2014, pp. 93–100. Springer (2014)

101. Sesli, E., Hacıoğlu, G.: RSSI and flower pollination algorithm based location estimation for wireless sensor networks. Int. J. Intell. Syst. Appl. Eng., 13–17 (2016)

102. Mahdad, B., Srairi, K.: Security constrained optimal power flow solution using new adaptive partitioning flower pollination algorithm. Appl. Soft Comput. **46**, 501–522 (2016)

103. Bekdaş, G., Nigdei, S.M., Yang, X.: Size optimization of truss structures employing flower pollination algorithm without grouping structural members. Int. J. Theor. Appl. Mech. **1**, 269–273 (2016)

104. Bekdaş, G., Nigdeli, S.M., Yang, X.-S.: Sizing optimization of truss structures using flower pollination algorithm. Appl. Soft Comput. **37**, 322–331 (2015)

105. Velamuri, S., Sreejith, S., Ponnambalam, P.: Static economic dispatch incorporating wind farm using flower pollination algorithm. Perspect. Sci. **8**, 260–262 (2016)

106. Abdelaziz, A.Y., Ali, E.S.: Static var compensator damping controller design based on flower pollination algorithm for a multi-machine power system. Electr. Power Compon. Syst. **43**(11), 1268–1277 (2015)

107. Łukasik, S., Kowalski, P.A.: Study of flower pollination algorithm for continuous optimization. In: Intelligent Systems' 2014, pp. 451–459. Springer (2015)

108. Vedula, V., Paladuga, S., Prithvi, M.R.: Synthesis of circular array antenna for sidelobe level and aperture size control using flower pollination algorithm. Int. J. Antennas Propag. (2015)

109. Preethi, C., Vanathi, P.: Attribute selection using binary flower pollination algorithm with greedy crossover and one to allinitialisation. Electron. Lett. **52**(21), 1757–1759 (2016)

110. Salgotra, R., Singh, U.: A novel bat flower pollination algorithm for synthesis of linear antenna arrays. Neural Comput. Appl., 1–14 (2016)

111. Binh, H.T.T., Hanh, N.T., Dey, N., et al.: Improved cuckoo search and chaotic flower pollination optimization algorithm for maximizing area coverage in wireless sensor networks. Neural Comput. Appl., 1–13 (2016)

112. Ramadas, M., Abraham, A., Kumar, S.: Using data clustering on ssfpa/de-a search strategy flower pollination algorithm with differential evolution. In: International Conference on Hybrid Intelligent Systems, pp. 539–550. Springer (2016)

113. Zhou, Y., Zhang, S., Luo, Q., Wen, C.: Using flower pollination algorithm and atomic potential function for shape matching. Neural Comput. Appl., 1–20 (2016)

114. Nasser, A.B., Alsewari, A.A., Muazu, A.A., Kamal, Z., et al.: Comparative performance analysis of flower pollination algorithm and harmony search based strategies: a case study of applying interaction testing in the real world. Int. J. Eng. Lang. Educ., 1–5 (2016)

115. Hegazy, O., Soliman, O.S., Salam, M.A.: Comparative study between fpa, ba, mcs, abc, and pso algorithms in training and optimizing of ls-svm for stock market prediction. Int. J. Adv. Comput. Res. **5**(18), 35 (2015)

116. Rathasamuth, W., Nootyaskool, S.: Comparison solving discrete space on flower pollination algorithm, pso and ga. In: 2016 8th International Conference on Knowledge and Smart Technology (KST), pp. 18–21. IEEE (2016)

117. Sharma, S., Rana, A.: Power system loss minimization using flower pollination algorithm (fpa)-a comparative study. Int. J. Adv. Res. Ideas Innov. Technol., 374–378 (2017)

On the Hypercomplex-Based Search Spaces for Optimization Purposes

João Paulo Papa, Gustavo Henrique de Rosa and Xin-She Yang

Abstract Most applications can be modeled using real-valued algebra. Neverthe-less, certain problems may be better addressed using different mathematical tools. In this context, complex numbers can be viewed as an alternative to standard algebra, where imaginary numbers allow a broader collection of tools to deal with different types of problems. In addition, hypercomplex numbers extend naïve complex alge-bra by means of additional imaginary numbers, such as quaternions and octonions. In this work, we will review the literature concerning hypercomplex spaces with an emphasis on the main concepts and fundamentals that build the quaternion and octonion algebra, and why they are interesting approaches that can overcome some potential drawbacks of certain optimization techniques. We show that quaternion-and octonion-based algebra can be used to different optimization problems, allow-ing smoother fitness landscapes and providing better results than those represented in standard search spaces.

Keywords Meta-heuristic · Hypercomplex numbers · Optimization · Quaternions · Octonions

1 Introduction

Optimization problems are relevant in many situations and applications, that range from engineering [13, 24] to medicine [20], among others. Very often, one must deal with different challenges when working on optimization problems, including

J.P. Papa (✉)
School of Sciences, São Paulo State University, Bauru, Brazil
e-mail: papa@fc.unesp.br

G.H. de Rosa
Department of Computing, São Paulo State University, Bauru, Brazil
e-mail: gth.rosa@uol.com.br

X.-S. Yang
School of Science and Technology, Middlesex University London, London, UK
e-mail: x.yang@mdx.ac.uk

© Springer International Publishing AG 2018
X.-S. Yang (ed.), *Nature-Inspired Algorithms and Applied Optimization*,
Studies in Computational Intelligence 744, https://doi.org/10.1007/978-3-319-67669-2_6

the selection of designing variables, establishing their proper bounds, setting up hyper-parameters [15, 16, 18, 21], and obviously obtaining the appropriate solutions without be potentially trapped at local optima. The potential trap of local optima can be of great concern if the optimal solutions are needed, which somehow fosters the research on different approaches to handle the problem of optimizing complex fitness functions. For example, hybrid variants [8], aging mechanisms [1], and fitness landscape analysis [19] are among the different approaches that are often used to deal with the issues related to local optima.

Another direction is an approach that attempts to embed the search space into different representations of those that are normally used. Fister et al. [4], for instance, presented a Firefly Algorithm embedded in a quaternion-based space [2]. The idea is to map each possible solution encoded as an n-dimensional firefly as an $n \times 4$ tensor, where each decision variable is now encoded as a hypercomplex number, which contains four parts, being one real-valued and the remaining three parts concerning imaginary numbers. Later on, Fister et al. [3] proposed a quaternion-based Bat Algorithm, and Papa et al. [14] introduced the well-known Harmony Search in the context of Deep Belief Network fine-tuning using quaternion representations.

Normalized quaternions, also known as versors, are widely used to represent the orientation of objects in three-dimensional spaces, and thus can be efficient to perform rotations in such spaces. The idea behind using quaternionic search spaces concerns the possibility of having smoother fitness landscapes, although it has not been mathematically demonstrated. However, the results obtained previously support that assumption [3, 4, 14]. Another interesting extension of quaternions is referred to as octonions, which are composed of eight dimensions, twice the number of a quaternion dimension [6]. Although they are not well known in the literature, they have interesting properties that make them suitable to be used in special relativity and quantum logic, among other research fields.

In this chapter, we present some insights and recent results concerning optimization on hypercomplex spaces. In addition, we also consider a toy example to help enthusiasts implementing their own technique on top of LibOPT [17], an open-source library for the implementation of meta-heuristic-based optimization techniques.[1] The experiments also comprise quaternion- and octonion-based implementations of some well-known techniques in the literature, such as Firefly Algorithm (FA) [25], Artificial Bee Colony (ABC) [10], Bat Algorithm (BA) [27] and Particle Swarm Optimization (PSO) [11].

In light of the experiments, the reader can observe that hypercomplex representations are useful to achieve reasonable results in higher-dimensional problems, as well as such spaces can allow a faster convergence in a number of situations, which are addressed here by means of benchmarking functions. We hope the reader can benefit from such initial study concerning hypercomplex search spaces and their applications to the context of meta-heuristic techniques.

---

[1] https://github.com/jppbsi/LibOPT.

2 Hypercomplex Representations

2.1 Complex Numbers

Mathematicians often have to face some challenging problems that seem to be unsolvable. For example, a seemingly simple problem as follows:

$$x^2 + 1 = 0, \tag{1}$$

does not have any real solution. The solution $x^2 = -1$ does not appear as reasonable, since the square of any number must be positive, $x \in \Re$.

In order to deal with the aforementioned problem, the term *imaginary number* was coined. Such a number is usually represented as follows:

$$i^2 = -1, \tag{2}$$

which does not have any physical meaning in practice. The imaginary numbers compose the so-called *complex numbers*, which have a real and an imaginary term, as follows:

$$c = a + bi, \tag{3}$$

where $a, b \in \Re$ and $i^2 = -1$. Roughly speaking, one can observe we can obtain a real number by just setting $b = 0$, or even to obtain an imaginary number by setting $a = 0$. Therefore, the complex numbers generalize both real and imaginary numbers.

Complex numbers have interesting properties when performing rotations in two-dimensional spaces. In order to clarify that, we can map the complex numbers onto a two-dimensional plane called *complex plane*, where the real part (**Re**) is mapped on the horizontal axis, and the imaginary part (**Im**) is mapped on the vertical axis, as displayed in Fig. 1.

Fig. 1 Complex plane used to map complex numbers onto a two-dimensional representation

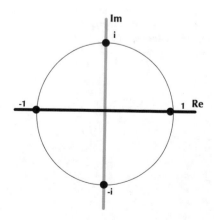

The interesting point in using complex numbers concerns the fact we can rotate them through the complex plane by 90° by using simple multiplication by i. In order to show that, let us consider an arbitrary point $p = 1 + i$ in the complex plane. Let r be the result of the multiplication of p by i, as follows:

$$r = pi = (1 + i)i = i + i^2 = -1 + i. \tag{4}$$

Now, let us multiply r by the very same quantity i, thus obtaining s:

$$s = ri = (-1 + i)i = -i + i^2 = -1 - i. \tag{5}$$

Repeating the very same computation once more, we can obtain t as follows:

$$t = si = (-1 - i)i = -i - i^2 = 1 - i. \tag{6}$$

Finally, by multiplying t by i, we can obtain:

$$u = ti = (+1 - i)i = i - i^2 = 1 + i, \tag{7}$$

which is the very same position we started, i.e., $p = u$. Figure 2 depicts the above calculations.

2.2 Hypercomplex Numbers

Roughly speaking, hypercomplex numbers extend complex numbers by adding more imaginary terms, which allow them to perform rotations in higher-dimensional complex spaces. In this chapter, we consider two well-known hypercomplex representations: quaternions and octonions.

Fig. 2 Rotating complex numbers through the complex plane

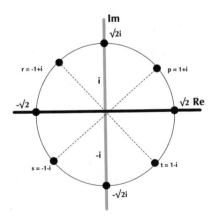

2.2.1 Quaternions

A quaternion q is composed of real and complex numbers, i.e., $q = x_0 + x_1 i + x_2 j + x_3 k$, where $x_0, x_1, x_2, x_3 \in \Re$ and i, j, k are imaginary numbers (also known as "fundamental quaternions units") obeying the next set of equations:

$$ij = k, \tag{8}$$

$$jk = i, \tag{9}$$

$$ki = j, \tag{10}$$

$$ji = -k, \tag{11}$$

$$kj = -i, \tag{12}$$

$$ik = -j, \tag{13}$$

and

$$i^2 = j^2 = k^2 = -1. \tag{14}$$

Roughly speaking, a quaternion q is represented in a 4-dimensional space over the real numbers, i.e., $\Re^4$.

Given two quaternions $q_1 = x_0 + x_1 i + x_2 j + x_3 k$ and $q_2 = y_0 + y_1 i + y_2 j + y_3 k$, the quaternion algebra defines a set of main operations [2]. The addition, for instance, can be defined by:

$$
\begin{aligned}
q_1 + q_2 &= (x_0 + x_1 i + x_2 j + x_3 k) + (y_0 + y_1 i + y_2 j + y_3 k) \\
&= (x_0 + y_0) + (x_1 + y_1)i + (x_2 + y_2)j + (x_3 + y_3)k,
\end{aligned} \tag{15}
$$

while the subtraction is defined as follows:

$$
\begin{aligned}
q_1 - q_2 &= (x_0 + x_1 i + x_2 j + x_3 k) - (y_0 + y_1 i + y_2 j + y_3 k) \\
&= (x_0 - y_0) + (x_1 - y_1)i + (x_2 - y_2)j + (x_3 - y_3)k.
\end{aligned} \tag{16}
$$

Another important operation is the norm, which maps a given quaternion to a real-valued number, as follows:

$$
\begin{aligned}
N(q_1) &= N(x_0 + x_1 i + x_2 j + x_3 k) \\
&= \sqrt{x_0^2 + x_1^2 + x_2^2 + x_3^2}.
\end{aligned} \tag{17}
$$

Finally, Fister et al. [3, 4] introduced two other operations, *qrand* and *qzero*. The former initializes a given quaternion with values drawn from a Gaussian distribution, and it can be defined as follows:

$$qrand() = \{x_i = \mathcal{N}(0, 1) | i \in \{0, 1, 2, 3\}\}. \tag{18}$$

The latter function initialized a quaternion with zero values, as follows:

$$qzero() = \{x_i = 0 | i \in \{0, 1, 2, 3\}\}. \tag{19}$$

2.2.2 Octonions

Roughly speaking, octonions can be modeled as a natural extension of quaternions, and they were discovered independently by John T. Graves and Arthur Cayley around by 1843. An octonion has seven complex parts, and one real-valued term, being defined as follows:

$$o = x_0 e_0 + x_1 e_1 + x_2 e_2 + \cdots + x_7 e_7, \tag{20}$$

where $x_i \in \mathfrak{R}$ and e_i concerns the imaginary number, $i = 0, \ldots, 7$. Usually, $e_0 = 1$ in order to obtain the real-valued term of the octonion.

The addition, subtraction and norm operations are computed similarly to the quaternions' formulae, which can lead to an easy implementation framework for handling different hypercomplex representations.

3 LibOPT—A Library for Optimization Purposes

In this section, we first present LibOPT [17] for further working on a toy example using hypercomplex-based optimization.

3.1 *Hypercomplex Tools*

LibOPT is an open-source optimization library available at GitHub, where a home-page presents all techniques[2] and benchmarking functions currently available. To date, LibOPT implements the following approaches concerning quaternion- and octonion-based representations:

---

[2]https://github.com/jppbsi/LibOPT/wiki.

- Particle Swarm Optimization [11];
- Particle Swarm Optimization with Adaptive Inertia Weight [12];
- Bat Algorithm [27];
- Flower Pollination Algorithm [23];
- Firefly Algorithm [25];
- Cuckoo Search [26];
- Black Hole Algorithm [7];
- Artificial Bee Colony [10];
- Harmony Search [14];
- Improved Harmony Search [14]; and
- Parameter-setting-free Harmony Search [5].

Additionally, LibOPT implements 112 benchmarking functions[3] [9], which are not displayed here for the sake of space.

3.2   Installation

The library was implemented and tested to work under Unix- and MacOS-based operational systems, and can be quickly installed by executing the make command right after decompressing the compressed file. On MacOS, if there is any problem, a possible solution is to use the GNU/gcc compiler.[4]

3.3   Data Structures

Despite other directories, LibOPT is composed of two main folders, such as LibOPT \include and LibOPT\src, where the first one is in charge of the header files, and the latter is responsible for the source files and main implementations.

The library was created based on a fast prototyping ideal, where a main structure, called Agent, controls all the common information shared among the implemented techniques, as implemented below:

```
typedef struct Agent_{
    /* common definitions */
    int n; /* number of decision variables */
    double *x; /* position */
    double fit; /* fitness value */
    double **t; /* tensor */
}Agent;
```

---

[3]https://github.com/jppbsi/LibOPT/wiki/Benchmarking-functions.
[4]https://github.com/jppbsi/LibOPT/wiki/Installation.

As stated in the above code-snippet, n is the number of decision variables to be optimized, and **x** stands for an array that encodes the current position of the agent when working under standard search spaces. Further, variable `fit` stores the fitness value, and **t** stands for a matrix-like structure that is used to implement the hypercomplex-based versions of the naïve techniques, and it works similarly to **x**, but in another search space representation.

Nonetheless, there is another main structure which depicts the whole search space, including additional information about the optimization problem not described in the `Agent` structure:

```
typedef struct SearchSpace_{
    /* common definitions */
    int m; /* number of agents (solutions) */
    int n; /* number of decision variables */
    int iterations; /* number of iterations */
    Agent **a; /* array of pointers to agents */
    double *LB; /* lower boundaries */
    double *UB; /* upper boundaries */
    double *g; /* global best agent */
    double **t_g; /* global best tensor */
    int best; /* index of the best agent */
    double gfit; /* global best fitness */
    int is_integer_opt; /* integer-valued problem? */
}SearchSpace;
```

Notice the library contains a quite detailed explanation about every attribute information in order to avoid possible misunderstandings, thus leading the user to the maximum advantages of LibOPT. Figure 3 depicts an outline of how a hypercomplex search space structure works under LibOPT definitions. In this example, there is only one decision variable to be optimized, which is represented by a tensor $\mathbf{t} \in \mathfrak{R}^k$, where $k \in \{4, 8\}$. One can observe a `SearchSpace` structure and three agents allocated, which encode each decision variable as a quaternion ($k = 4$) or an octonion ($k = 8$).

3.4 Model Files

As aforementioned, albeit most techniques have something in common (e.g., number of decision variables, current position, and maybe velocity), they may also differ in the number of parameters. Such occasion led us to design a model file-based implementation, which means all parameter setting up required for a given optimization technique must be provided in a single text file, hereinafter called "model file".

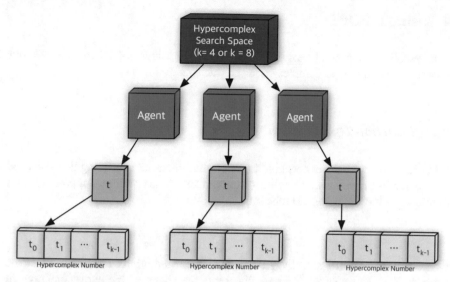

Fig. 3 Hypercomplex search space structure on top of LibOPT

For the sake of explanation, let us consider the model file of the Firefly Algorithm.[5] Roughly speaking, the user must input all information required by that technique, as follows:

```
10 2 100 #<n_particles> <dimension> <max_iterations>
0.2 1 1 #<alpha> <beta_0> <gamma>
-10 10 #<LB> <UB> x[0]
-10 10 #<LB> <UB> x[1]
```

The first line contains three integers: number of agents (particles), number of decision variables (dimension) and number of iterations. Notice everything right after the caracter # is considering a comment, thus not taking into account by the parser. The next line configures FA parameters α, β_0 and γ. The last two lines aim at setting up the range of each decision variable. Since we have two dimensions in this example, each line stands for one variable, say $x[0]$ and later $x[1]$. In the above example, we have a problem with 10 particles, 2 decision variables and 100 iterations for convergence. Also, we used $\alpha = 0.2$, $\beta_0 = 1$, $\gamma = 1$ and $x[i] \in [-10, 10]$, $i \in \{0, 1\}$.

---

[5] Detailed information concerning the model files of the techniques implemented in LibOPT can be found at https://github.com/jppbsi/LibOPT/wiki/Model-files.

4 Using LibOPT

In this section, we present a toy example concerning using LibOPT to optimize your own function on a hypercomplex-based search space.

4.1 Function Optimization

LibOPT works with the concept of "function minimization", meaning that you need to take that into account when trying to "maximize" some function. Suppose we want to minimize the following 2D function:

$$f(x, y) = x^4 + y^2 + 10, \tag{21}$$

where $x, y \in [-10, 10]$ and $x, y \in \Re$. Note that for simplicity reasons, we will be using x as x_0 and y as x_1. Figure 4 illustrates the shape of the above function, in which one can observe a global minimum as of $f(x, y) = 10$.

Since all functions are implemented in both LibOPT/include/function.h (header) and LibOPT/src/function.c files, one must add the function's signature in the first file, and the function's implementation in the second one. In LibOPT/include/function.h, the following line of code must be added: double MyFunction(Agent *a, va_list arg);. With respect to the file LibOPT/src/function.c, one should implement the function as follows:

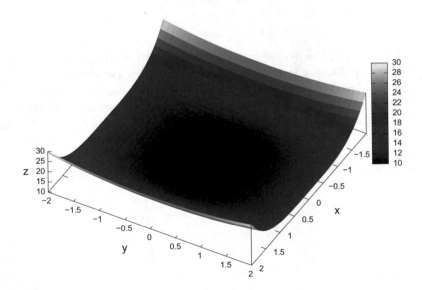

Fig. 4 Plot representing $f(x, y) = x^4 + y^2 + 10$

```
double MyFunction(Agent *a, va_list arg){
    double output;

    if(!a){
        fprintf(stderr,"\nAgent not allocated @MyFunction.");
        return DBL_MAX;
    }
    if(a->n < 1){
        fprintf(stderr,"\nInvalid number of decision variables
            @MyFunction. It must be equal or greater than one.\n")
            ;
        return DBL_MAX;
    }
    output = pow(a->x[0], 4) + pow(a->x[1], 2) + 10; /* Equation
        (1) */

    return output;
}
```

In the above source-code, the first two conditional structures verify whether the Agent has been allocated or not, and if the number of decision variables is greater than 1. The next line implements the function itself: since $\mathbf{x} \in \mathfrak{R}^2$, each agent has two dimensions only, i.e., a->x[0] and a->x[1]. Notice that LibOPT uses double precision for the data types.

Although the user can implement any function to be optimized, we need to consider the guidelines implemented in LibOPT/include/common.h by the following function: typedef double (*prtFun)(Agent *, va_list arg). This signature tells us the function to be minimized should return a double value, as well as its first parameter should be an Agent, followed by a list of arguments, which depends on the function.

4.2 Toy Example

In our example, suppose we want to use a hypercomplex quaternion-based Firefly Algorithm to minimize MyFunction. For the sake of explanation, we will use the same parameters defined by the model described in Sect. 3.4:

```
10 2 100 #<n_particles> <dimension> <max_iterations>
0.2 1 1 #<alpha> <beta_0> <gamma>
-10 10 #<LB> <UB> x[0]
-10 10 #<LB> <UB> x[1]
```

Let fa_model.txt be the file name concerning the above model. Basically, one needs to create a main file to call hypercomplex quaternion-based FA procedure as follows:

```
#include "common.h"
#include "function.h"
```

```
#include "fa.h"

int main(){
    SearchSpace *s = NULL;
    int i;

    s = ReadSearchSpaceFromFile("fa_model.txt", _FA_);

    s->t_g = AllocateTensor(s->n, _QUATERNION_);
    for (i = 0; i < s->m; i++)
        s->a[i]->t = AllocateTensor(s->n, _QUATERNION_);

    InitializeTensorSearchSpace(s, _QUATERNION_);

    if(CheckSearchSpace(s, _FA_))
        runTensorFA(s, _QUATERNION_, MyFunction);

    DeallocateTensor(&s->t_g, s->n);

    for (i = 0; i < s->m; i++)
        DeallocateTensor(&s->a[i]->t, s->n);

    DestroySearchSpace(&s, _FA_);

    return 0;
}
```

As one can observe, it is quite simple to execute hypercomplex quaternion-based FA, since we only need to to call seven main functions:

- `ReadSearchSpaceFromFile`: it reads the model file and creates a search space;
- `AllocateTensor`: it allocates a tensor to the desired variable. Note that we need to allocate the global best tensor `s->t_g` and each agent's tensor `s->a[i]->t`;
- `InitializeTensorSearchSpace`: it initializes the hypercomplex search space;
- `CheckSearchSpace`: it checks whether the search space is valid or not;
- `runTensorFA`: it minimizes function `MyFunction`;
- `DeallocateTensor`: it deallocates the used tensors; and
- `DestroySearchSpace`: it deallocates the search space.

Notice one can find a number of similar examples in `LibOPT/examples`. Figure 5 displays a convergence plot along the iterations concerning the function defined by Eq. (21) and considered in this toy example.

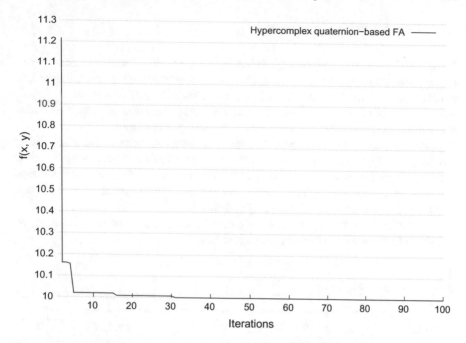

Fig. 5 Convergence plot regarding hypercomplex quaternion-based FA considering Eq. 21

5 Methodology and Experiments

As aforementioned in Sect. 2, quaternions and octonions employ a wider search space, where each parameter is encoded by $\mathbb{R}^4$ and $\mathbb{R}^8$ spaces, respectively, and then mapped into an $\mathbb{R}$ space within the chosen limits.

The main issue when working with hypercomplex representations concerns with the boundaries of the real-valued term x_i associated to each term. Therefore, we have devised a function to tackle this boundary issue, mapping x_i to plausible boundaries, as follows:

$$span(x_i) = (U_i - L_i)\left(\frac{N(q)}{\sqrt{M}}\right) + L_i, \tag{22}$$

where U_i and L_i stand for the upper and lower bounds of decision variable x_i, respectively, and $N(q)$ denotes the norm over quaternion q (similarly to octonions). In addition, M corresponds to the number of dimensions of the hypercomplex representation (i.e., $M = 4$ for quaternions and $M = 8$ for octonions).

5.1 Benchmarking Functions

Nevertheless, we need a practical solution in order to validate and compare the performance of optimization techniques. One interesting approach is to test these algorithms against some benchmarking functions. However, it is important to include a wide variety of test functions, such as unimodal, multimodal, separable, non-separable, regular, irregular or even multi-dimensional functions.

Thus, we have selected 10 different benchmarking functions in order to validate our proposed approach. Table 1 depicts the employed functions, where the identifier column stands for their names, the function column stands for their mathematical formulations, the bounds column stands for their variables' lower and upper bounds, and the $f(x^*)$ column stands for their optimum values.

In addition, in order to provide a more comprehensive view of each function, we outline below some of their main characteristics:

Table 1 Benchmarking functions

Identifier	Function	Bounds	$f(x^*)$
Csendes	$f_1(x) = \sum_{i=1}^{D} x_i^6(2 + \sin(\frac{1}{x_i}))$	$-1 \leq x_i \leq 1$	0
Lévy	$f_2(x) = \sin^2(\pi w_1) + \sum_{i=1}^{D-1} (w_i - 1)^2[1 + 10\sin^2(\pi w_i + 1)] + \\ +(w_d - 1)^2[1 + \sin^2(2\pi w_d)]$, where $w_i = 1 + \frac{x_i-1}{4}$	$-10 \leq x_i \leq 10$	0
Quartic	$f_3(x) = \sum_{i=1}^{D} ix_i^4 + \text{random}[0, 1)$	$-1.28 \leq x_i \leq 1.28$	0
Rastrigin	$f_4(x) = 10n + \sum_{i=1}^{D}[x_i^2 - 10\cos(2\pi x_i)]$	$-5.12 \leq x_i \leq 5.12$	0
Rosenbrock	$f_5(x) = \sum_{i=1}^{D-1} [100(x_{i+1} - x_i^2)^2 + (x_i - 1)^2]$	$-30 \leq x_i \leq 30$	0
Salomon	$f_6(x) = 1 - \cos\left(2\pi\sqrt{\sum_{i=1}^{D} x_i^2}\right) + 0.1\sqrt{\sum_{i=1}^{D} x_i^2}$	$-100 \leq x_i \leq 100$	0
Schewefel	$f_7(x) = (\sum_{i=1}^{D} x_i^2)^{\sqrt{\pi}}$	$-100 \leq x_i \leq 100$	0
Sphere	$f_8(x) = \sum_{i=1}^{D} x_i^2$	$-10 \leq x_i \leq 10$	0
Xin-She Yang #1	$f_9(x) = \sum_{i=1}^{D} \epsilon_i |x_i|^i$	$-5 \leq x_i \leq 5$	0
Zakharov	$f_{10}(x) = \sum_{i=1}^{D} x_i^2 + \left(\frac{1}{2}\sum_{i=1}^{D} ix_i\right)^2 + \left(\frac{1}{2}\sum_{i=1}^{D} ix_i\right)^4$	$-5 \leq x_i \leq 10$	0

- Csendes (f_1)—continuous, differentiable, non-separable, non-scalable and uni-modal;
- Lévy (f_2)—continuous, differentiable and multimodal;
- Quartic (f_3)—continuous, differentiable, separable, scalable;
- Rastrigin (f_4)—continuous, differentiable, separable, scalable and multimodal;
- Rosenbrock (f_5)—continuous, differentiable, non-separable, scalable and unimodal;
- Salomon (f_6)—continuous, differentiable, non-separable, scalable and multimodal;
- Schewefel (f_7)—continuous, differentiable, partially-separable, scalable and uni-modal;
- Sphere (f_8)—continuous, differentiable, separable, scalable and multimodal;
- Xin-She Yang #1 (f_9)—separable;
- Zakharov (f_{10})—continuous, differentiable, non-separable, scalable and unimodal.

5.2 Experimental Setup

In this work, we compared four meta-heuristic approaches against their quaternion and octonion versions, including

- Artificial Bee Colony: ABC, QABC (Quaternion ABC) and OABC (Octonion ABC);
- Bat Algorithm: BA, QBA (Quaternion BA) and OBA (Octonion BA);
- Flower Pollination Algorithm: FPA, QFPA (Quaternion FPA) and OFPA (Octonion FPA);
- Particle Swarm Optimization: PSO, QPSO (Quaternion PSO) and OPSO (Octonion PSO).

To evaluate the robustness of parameter fine-tuning, we have used four distinct dimensions $D = 10, 25, 50$ and 100.

In order to provide an in-depth analysis, we have conducted experiments with 25 runs for each algorithm and used the following metrics: best fitness values, means of best fitness values and standard deviations of best fitness values. We also used a population size of 100 with $2000D$ iterations for all techniques. Therefore, this means we have 20,000 iterations for a $D = 10$ space, 50,000 iterations for a $D = 25$ space, 100,000 iterations for a $D = 50$ space, and 200,000 iterations for a $D = 100$ space. Table 2 presents the parameter configuration for each optimization technique.[6]

Regarding ABC, the number of trials limit stands for the amount of trials that a solution can be improved by an employee bee. Considering BA, we have the minimum and maximum frequency ranges, f_{min} and f_{max}, respectively, as well as the loudness parameter A and pulse rate r. In FPA, β is used to compute the Lévy distribution, while p is the probability of local pollination. Finally, PSO defines w as the inertia weight, and c_1 and c_2 as the control parameters.

---

[6]Notice that these values have been empirically setup.

Table 2 Parameter configuration

Technique	Parameters
ABC	Trials limit $= 1000$
BA	$f_{min} = 0, f_{max} = 2, A = 0.5, r = 0.5$
FPA	$\beta = 1.5, p = 0.8$
PSO	$c_1 = 1.7, c_2 = 1.7, w = 0.7$

5.3 Experiments

The results are presented for each employed dimension ($D = 10, 25, 50$ and 100) within the following format: (BF, MBF, SDBF), where BF stands for "best fitness value", MBF for "mean of best fitness value", and SDBF for "standard deviation of best fitness value". Tables 3, 4, 5, 6, 7, 8, 9, 10, 11 and 12 present the results concerning functions f_1 to f_{10}, respectively, being the best values in bold according to the Wilcoxon signed-rank statistical test with significance as of 0.05 [22]. Note that the statistical test evaluated the MBF measure only.

Considering the 10-dimensional search space, hypercomplex-based approaches achieved the best results in 4 out of 10 situations, meanwhile in 25-dimensional problems, quaternionic- and octonic-based representations obtained the best results in 5 out of 10 functions. If one considers higher dimensional search spaces, i.e., 50 and 100 dimensions, hypercomplex approaches obtained the best results in 6 out of 10 problems. Therefore, we can clearly observe the benefits in using such alternative representations in optimization problems.

Looking at all the results for all the ten functions, one can observe that only three functions did not obtain any best result by means of hypercomplex-based search spaces, and these three functions are f_3, f_6 and f_{10}. However, the quaternionic and octonic variants of the optimization techniques considered in this work obtained the best results or equally best results when compared to the standard versions in most cases for the aforementioned functions in general. For both functions f_3 and f_{10}, the modality is either unimodal or dominated by a big mode, the optimal solutions are relatively easily achievable compared to other more complex functions, while f_6 the variable of modal height is not much (due to the cosine term). Therefore, the more complicated hypercomplex representations may not benefit much because the landscape are already sufficiently smooth for these functions.

It can be expected that the benefit of using hypercomplex representations may be higher for the objective functions with highly complex modality and such benefit seems to be slightly significant for more higher-dimension problems.

ABC technique seems to be the one that did not benefit from hypercomplex search spaces as much as the others considered in this work. As a matter of fact, ABC was the technique that obtained the best results in the situations the hypercomplex spaces did not improve standard search spaces. In order to have a deeper look into this, we conducted extra experiments to analyze such behavior.

Table 3 Results concerning Csendes (f_1) function

Technique	10D	25D	50D	100D
ABC	(0.00e + 00, 0.00e + 00, 0.00e + 00)	(0.00e + 00, 0.00e + 00, 0.00e + 00)	(0.00e + 00, 0.00e + 00, 0.00e + 00)	(0.00e + 00, 0.00e + 00, 0.00e + 00)
QABC	(0.00e + 00, 0.00e + 00, 0.00e + 00)	(0.00e + 00, 0.00e + 00, 0.00e + 00)	(0.00e + 00, 0.00e + 00, 0.00e + 00)	(1.00e − 06, 1.35e − 02, 0.00e + 00)
OABC	(0.00e + 00, 0.00e + 00, 0.00e + 00)	(0.00e + 00, 4.20e − 05, 0.00e + 00)	(1.33e − 04, 5.41e − 04, 0.00e + 00)	(9.36e − 04, 1.07e − 03, 0.00e + 00)
BA	(0.00e + 00, 0.00e + 00, 0.00e + 00)	(0.00e + 00, 0.00e + 00, 0.00e + 00)	(0.00e + 00, 0.00e + 00, 0.00e + 00)	(0.00e + 00, 1.00e − 06, 1.00e − 06)
QBA	(0.00e + 00, 1.00e − 06, 1.00e − 06)	(3.00e − 06, 2.00e − 06, 3.00e − 06)	(6.00e − 06, 3.00e − 06, 4.00e − 06)	(8.00e − 06, 8.00e − 06, 1.10e − 05)
OBA	(0.00e + 00, 0.00e + 00, 1.00e − 06)	(3.00e − 06, 2.00e − 06, 3.00e − 06)	(7.00e − 06, 6.00e − 06, 8.00e − 06)	(1.10e − 05, 1.00e − 05, 1.50e − 05)
FPA	(0.00e + 00, 0.00e + 00, 0.00e + 00)	(0.00e + 00, 0.00e + 00, 0.00e + 00)	(0.00e + 00, 0.00e + 00, 0.00e + 00)	(1.00e − 06, 1.00e − 06, 1.00e − 06)
QFPA	(0.00e + 00, 0.00e + 00, 0.00e + 00)	(0.00e + 00, 0.00e + 00, 0.00e + 00)	(0.00e + 00, 0.00e + 00, 0.00e + 00)	(0.00e + 00, 0.00e + 00, 0.00e + 00)
OFPA	(0.00e + 00, 0.00e + 00, 0.00e + 00)	(0.00e + 00, 0.00e + 00, 0.00e + 00)	(0.00e + 00, 0.00e + 00, 0.00e + 00)	(0.00e + 00, 0.00e + 00, 0.00e + 00)
PSO	(0.00e + 00, 0.00e + 00, 0.00e + 00)	(0.00e + 00, 0.00e + 00, 0.00e + 00)	(0.00e + 00, 1.00e − 06, 2.00e − 06)	(1.00e − 06, 0.00e + 00, 0.00e + 00)
QPSO	(0.00e + 00, 0.00e + 00, 0.00e + 00)	(0.00e + 00, 0.00e + 00, 0.00e + 00)	(1.00e − 06, 0.00e + 00, 0.00e + 00)	(1.00e − 06, 1.00e − 06, 2.00e − 06)
OPSO	(0.00e + 00, 0.00e + 00, 0.00e + 00)	(0.00e + 00, 0.00e + 00, 0.00e + 00)	(0.00e + 00, 0.00e + 00, 0.00e + 00)	(1.00e − 06, 1.00e − 06, 3.00e − 06)

Table 4 Results concerning Levy's (f_2) function

Technique	$D = 10$	$D = 25$	$D = 50$	$D = 100$
ABC	(7.77e−03, 1.12e+00, 0.00e+00)	(3.00e+00, 3.00e+00, 0.00e+00)	(6.12e+00, 6.12e+00, 0.00e+00)	(1.24e+01, 1.24e+01, 0.00e+00)
QABC	(0.00e+00, 2.65e−03, 0.00e+00)	(1.13e−01, 1.09e+00, 0.00e+00)	(2.16e+00, 3.59e+00, 0.00e+00)	(1.03e+01, 1.25e+01, 0.00e+00)
OABC	(3.80e−05, 8.79e−03, 0.00e+00)	(9.54e−01, 2.27e+00, 0.00e+00)	(4.09e+00, 4.85e+00, 0.00e+00)	(1.03e+01, 1.38e+01, 0.00e+00)
BA	(0.00e+00, 2.72e−01, 3.85e−01)	(1.82e+00, 5.72e+00, 8.08e+00)	(7.45e+00, 1.26e+01, 1.79e+01)	(1.33e+01, 1.14e+01, 1.61e+01)
QBA	(2.69e−01, 5.44e−01, 7.69e−01)	(8.95e−01, 9.92e−01, 1.40e+00)	(3.24e+00, 3.72e+00, 5.26e+00)	(8.59e+00, 7.39e+00, 1.05e+01)
OBA	(3.00e−06, 1.79e−01, 2.53e−01)	(3.58e−01, 1.04e+00, 1.47e+00)	(1.52e+00, 2.17e+00, 3.07e+00)	(2.51e+00, 2.43e+00, 3.43e+00)
FPA	(0.00e+00, 0.00e+00, 0.00e+00)	(0.00e+00, 0.00e+00, 0.00e+00)	(8.95e−02, 1.49e−01, 2.58e−01)	(1.07e+00, 1.04e+00, 1.81e+00)
QFPA	**(0.00e+00, 0.00e+00, 0.00e+00)**	**(0.00e+00, 0.00e+00, 0.00e+00)**	**(0.00e+00, 2.98e−02, 5.17e−02)**	**(8.95e−02, 2.98e−02, 5.17e−02)**
OFPA	**(0.00e+00, 0.00e+00, 0.00e+00)**	**(0.00e+00, 0.00e+00, 0.00e+00)**	**(0.00e+00, 0.00e+00, 0.00e+00)**	(2.69e−01, 1.19e−01, 2.07e−01)
PSO	(0.00e+00, 6.71e−02, 1.34e−01)	(1.57e+00, 9.74e−01, 1.95e+00)	(4.40e+00, 7.13e+00, 1.43e+01)	(8.56e+00, 2.49e+00, 4.98e+00)
QPSO	(0.00e+00, 1.58e−01, 3.17e−01)	(2.04e−01, 5.11e−02, 1.02e−01)	(1.62e+00, 7.67e−01, 1.53e+00)	(3.84e+00, 1.10e+00, 2.20e+00)
OPSO	(0.00e+00, 2.24e−02, 4.48e−02)	(1.81e−01, 2.59e−01, 5.17e−01)	(6.98e−01, 2.45e−01, 4.90e−01)	(2.82e+00, 8.02e−01, 1.60e+00)

Table 5 Results concerning Quartic (f_3) function

Technique	10D	25D	50D	100D
ABC	(0.00e + 00, 2.00e − 06, 0.00e + 00)	(0.00e + 00, 0.00e + 00, 0.00e + 00)	(0.00e + 00, 0.00e + 00, 0.00e + 00)	(0.00e + 00, 0.00e + 00, 0.00e + 00)
QABC	(2.44e − 04, 2.33e − 03, 0.00e + 00)	(1.12e − 02, 7.73e − 02, 0.00e + 00)	(8.36e − 02, 1.43e − 01, 0.00e + 00)	(1.36e + 00, 8.80e + 00, 0.00e + 00)
OABC	(2.19e − 03, 1.47e − 02, 0.00e + 00)	(6.66e − 02, 2.87e − 01, 0.00e + 00)	(5.22e − 01, 9.47e − 01, 0.00e + 00)	(3.06e + 00, 5.45e + 00, 0.00e + 00)
BA	(5.22e − 04, 1.82e − 03, 2.57e − 03)	(4.70e − 03, 2.35e − 03, 3.32e − 03)	(1.68e − 02, 1.71e − 02, 2.42e − 02)	(6.87e − 02, 5.42e − 02, 7.66e − 02)
QBA	(9.02e − 03, 1.57e − 02, 2.22e − 02)	(1.08e − 02, 1.10e − 02, 1.55e − 02)	(7.48e − 03, 4.68e − 03, 6.62e − 03)	(3.60e − 03, 1.91e − 03, 2.70e − 03)
OBA	(7.03e − 03, 1.20e − 02, 1.70e − 02)	(2.15e − 02, 1.97e − 02, 2.78e − 02)	(1.66e − 02, 1.16e − 02, 1.64e − 02)	(8.79e − 03, 6.33e − 03, 8.95e − 03)
FPA	(1.42e − 04, 9.90e − 05, 1.71e − 04)	(3.01e − 04, 1.25e − 04, 2.16e − 04)	(1.54e − 03, 1.37e − 03, 2.37e − 03)	(2.82e − 02, 2.08e − 02, 3.61e − 02)
QFPA	(2.37e − 04, 1.93e − 04, 3.34e − 04)	(6.08e − 04, 4.32e − 04, 7.49e − 04)	(9.22e − 04, 8.39e − 04, 1.45e − 03)	(5.30e − 03, 3.03e − 03, 5.25e − 03)
OFPA	(5.90e − 05, 1.54e − 04, 2.67e − 04)	(5.86e − 04, 3.96e − 04, 6.85e − 04)	(9.40e − 04, 3.13e − 04, 5.43e − 04)	(2.08e − 03, 1.85e − 03, 3.21e − 03)
PSO	(1.48e − 04, 7.02e − 04, 1.40e − 03)	(2.08e − 02, 7.59e − 03, 1.52e − 02)	(1.98e − 01, 1.89e − 01, 3.77e − 01)	(1.40e + 00, 4.89e − 01, 9.79e − 01)
QPSO	(1.51e − 04, 4.22e − 04, 8.44e − 04)	(3.34e − 03, 2.24e − 03, 4.48e − 03)	(3.67e − 02, 3.21e − 02, 6.43e − 02)	(2.48e − 01, 7.97e − 02, 1.59e − 01)
OPSO	(2.30e − 05, 6.00e − 05, 1.20e − 04)	(2.51e − 03, 9.19e − 04, 1.84e − 03)	(1.99e − 02, 2.07e − 02, 4.14e − 02)	(7.17e − 02, 4.59e − 02, 9.19e − 02)

Table 6 Results concerning Rastrigin (f_4) function

Technique	10D	25D	50D	100D
ABC	**(0.00e + 00, 0.00e + 00, 0.00e + 00)**	**(0.00e + 00, 0.00e + 00, 0.00e + 00)**	**(0.00e + 00, 0.00e + 00, 0.00e + 00)**	**(0.00e + 00, 0.00e + 00, 0.00e + 00)**
QABC	(2.76e − 01, 3.01e + 00, 0.00e + 00)	(1.03e + 01, 2.34e + 01, 0.00e + 00)	(4.12e + 01, 6.53e + 01, 0.00e + 00)	(1.22e + 02, 1.62e + 02, 0.00e + 00)
OABC	(2.01e + 00, 6.15e + 00, 0.00e + 00)	(1.41e + 01, 3.19e + 01, 0.00e + 00)	(4.29e + 01, 8.55e + 01, 0.00e + 00)	(1.43e + 02, 1.70e + 02, 0.00e + 00)
BA	(7.96e + 00, 1.54e + 01, 2.18e + 01)	(2.59e + 01, 3.78e + 01, 5.35e + 01)	(5.57e + 01, 4.48e + 01, 6.33e + 01)	(1.01e + 02, 9.40e + 01, 1.33e + 02)
QBA	(5.97e + 00, 1.04e + 01, 1.48e + 01)	(2.69e + 01, 2.99e + 01, 4.22e + 01)	(3.19e + 01, 6.92e + 01, 9.78e + 01)	(1.13e + 02, 1.21e + 02, 1.72e + 02)
OBA	(3.98e + 00, 6.47e + 00, 9.15e + 00)	(7.96e + 00, 1.79e + 01, 2.53e + 01)	(3.68e + 01, 2.69e + 01, 3.80e + 01)	(6.97e + 01, 6.22e + 01, 8.80e + 01)
FPA	**(0.00e + 00, 0.00e + 00, 0.00e + 00)**	(5.97e + 00, 8.95e + 00, 1.55e + 01)	(3.58e + 01, 1.43e + 01, 2.47e + 01)	(4.97e + 01, 2.12e + 01, 3.68e + 01)
QFPA	(2.98e + 00, 9.95e − 01, 1.72e + 00)	(1.49e + 01, 7.30e + 00, 1.26e + 01)	(3.58e + 01, 1.63e + 01, 2.81e + 01)	(5.67e + 01, 2.95e + 01, 5.11e + 01)
OFPA	(1.99e + 00, 1.99e + 00, 3.45e + 00)	(9.95e + 00, 5.31e + 00, 9.19e + 00)	(1.99e + 01, 1.09e + 01, 1.90e + 01)	(4.68e + 01, 1.56e + 01, 2.70e + 01)
PSO	(4.97e + 00, 3.98e + 00, 7.96e + 00)	(2.87e + 01, 1.55e + 01, 3.09e + 01)	(8.95e + 01, 4.23e + 01, 8.46e + 01)	(2.35e + 02, 7.82e + 01, 1.56e + 02)
QPSO	(4.97e + 00, 2.98e + 00, 5.97e + 00)	(1.31e + 01, 1.12e + 01, 2.24e + 01)	(4.34e + 01, 1.66e + 01, 3.33e + 01)	(1.09e + 02, 4.78e + 01, 9.55e + 01)
OPSO	(2.98e + 00, 9.95e − 01, 1.99e + 00)	(6.98e + 00, 1.11e + 01, 2.22e + 01)	(3.26e + 01, 1.15e + 01, 2.30e + 01)	(6.04e + 01, 4.09e + 01, 8.18e + 01)

Table 7 Results concerning Rosenbrock's (f_5) function

Technique	$D = 10$	$D = 25$	$D = 50$	$D = 100$
ABC	(4.91e − 04, 4.91e − 04, 0.00e + 00)	(2.42e − 03, 2.02e + 00, 0.00e + 00)	(3.40e − 03, 4.95e + 00, 0.00e + 00)	(8.27e + 01, 9.28e + 01, 0.00e + 00)
QABC	(2.62e − 03, 9.43e − 03, 0.00e + 00)	(6.94e − 01, 2.49e + 01, 0.00e + 00)	(8.30e + 01, 3.28e + 02, 0.00e + 00)	(4.80e + 02, 4.92e + 02, 0.00e + 00)
OABC	(3.80e − 04, 8.86e − 02, 0.00e + 00)	(4.53e − 02, 6.88e + 01, 0.00e + 00)	(1.84e + 05, 5.51e + 05, 0.00e + 00)	(7.31e + 05, 2.26e + 06, 0.00e + 00)
BA	(1.10e − 05, 3.63e − 04, 5.13e − 04)	(2.43e − 02, 3.95e + 01, 5.59e + 01)	(1.32e − 02, 5.03e + 00, 7.11e + 00)	(5.62e + 01, 4.79e + 01, 6.77e + 01)
QBA	(1.21e − 01, 2.13e + 00, 3.01e + 00)	(7.24e + 00, 7.82e + 00, 1.11e + 01)	(4.05e + 01, 2.43e + 01, 3.43e + 01)	(8.10e + 01, 4.72e + 01, 6.68e + 01)
OBA	(1.01e − 01, 7.53e − 02, 1.06e − 01)	(9.04e − 01, 6.85e + 00, 9.68e + 00)	(4.13e + 01, 4.91e + 01, 6.95e + 01)	(9.14e + 01, 7.62e + 01, 1.08e + 02)
FPA	**(0.00e + 00, 0.00e + 00, 0.00e + 00)**	**(0.00e + 00, 0.00e + 00, 0.00e + 00)**	**(0.00e + 00, 0.00e + 00, 0.00e + 00)**	**(0.00e + 00, 1.35e + 00, 2.35e + 00)**
QFPA	(0.00e + 00, 7.00e − 06, 1.20e − 05)	(0.00e + 00, 0.00e + 00, 0.00e + 00)	(0.00e + 00, 0.00e + 00, 0.00e + 00)	**(0.00e + 00, 1.61e − 02, 2.79e − 02)**
OFPA	(0.00e + 00, 1.76e − 02, 3.05e − 02)	(0.00e + 00, 0.00e + 00, 0.00e + 00)	(0.00e + 00, 0.00e + 00, 0.00e + 00)	**(0.00e + 00, 0.00e + 00, 0.00e + 00)**
PSO	(8.97e − 01, 2.07e + 00, 4.14e + 00)	(2.48e + 01, 6.07e + 01, 1.21e + 02)	(1.83e + 02, 3.28e + 02, 6.57e + 02)	(1.35e + 02, 1.29e + 03, 2.58e + 03)
QPSO	(0.00e + 00, 1.42e + 00, 2.84e + 00)	(3.04e + 01, 1.85e + 01, 3.70e + 01)	(4.81e + 02, 2.49e + 02, 4.99e + 02)	(2.50e + 03, 1.89e + 03, 3.78e + 03)
OPSO	(7.24e − 02, 6.45e − 01, 1.29e + 00)	(4.02e + 01, 2.18e + 01, 4.36e + 01)	(3.06e + 02, 4.52e + 02, 9.05e + 02)	(8.52e + 02, 3.02e + 02, 6.04e + 02)

Table 8 Results concerning Salomon's (f_6) function

Technique	10D	25D	50D	100D
ABC	**(0.00e + 00, 0.00e + 00, 0.00e + 00)**	**(0.00e + 00, 0.00e + 00, 0.00e + 00)**	**(0.00e + 00, 0.00e + 00, 0.00e + 00)**	**(0.00e + 00, 0.00e + 00, 0.00e + 00)**
QABC	(9.99e − 02, 1.02e − 01, 0.00e + 00)	(2.00e − 01, 7.00e − 01, 0.00e + 00)	(1.10e + 00, 2.40e + 00, 0.00e + 00)	(6.70e + 00, 7.30e + 00, 0.00e + 00)
OABC	(9.99e − 02, 4.00e − 01, 0.00e + 00)	(8.00e − 01, 2.40e + 00, 0.00e + 00)	(2.60e + 00, 5.40e + 00, 0.00e + 00)	(1.06e + 01, 1.12e + 01, 0.00e + 00)
BA	(2.10e + 00, 2.00e + 00, 2.83e + 00)	(6.80e + 00, 3.75e + 00, 5.30e + 00)	(1.06e + 01, 8.60e + 00, 1.22e + 01)	(1.67e + 01, 8.40e + 00, 1.19e + 01)
QBA	(2.20e + 00, 1.45e + 00, 2.05e + 00)	(5.70e + 00, 3.40e + 00, 4.81e + 00)	(9.60e + 00, 5.50e + 00, 7.78e + 00)	(1.48e + 01, 8.50e + 00, 1.20e + 01)
OBA	(1.50e + 00, 1.25e + 00, 1.77e + 00)	(4.30e + 00, 2.55e + 00, 3.61e + 00)	(7.10e + 00, 3.55e + 00, 5.02e + 00)	(1.07e + 01, 6.10e + 00, 8.63e + 00)
FPA	(9.99e − 02, 3.33e − 02, 5.77e − 02)	(9.99e − 02, 6.66e − 02, 1.15e − 01)	(9.00e − 01, 6.00e − 01, 1.04e + 00)	(3.80e + 00, 1.33e + 00, 2.31e + 00)
QFPA	(9.99e − 02, 3.33e − 02, 5.77e − 02)	(9.99e − 02, 3.33e − 02, 5.77e − 02)	(6.00e − 01, 2.67e − 01, 4.62e − 01)	(2.00e + 00, 7.33e − 01, 1.27e + 00)
OFPA	(9.99e − 02, 3.33e − 02, 5.77e − 02)	(2.00e − 01, 6.66e − 02, 1.15e − 01)	(6.00e − 01, 3.00e − 01, 5.20e − 01)	(1.70e + 00, 6.00e − 01, 1.04e + 00)
PSO	(3.00e − 01, 1.50e − 01, 3.00e − 01)	(1.70e + 00, 1.10e + 00, 2.20e + 00)	(4.60e + 00, 1.15e + 00, 2.30e + 00)	(9.20e + 00, 2.65e + 00, 5.30e + 00)
QPSO	(2.00e − 01, 7.50e − 02, 1.50e − 01)	(1.30e + 00, 6.75e − 01, 1.35e + 00)	(2.80e + 00, 8.75e − 01, 1.75e + 00)	(5.10e + 00, 1.70e + 00, 3.40e + 00)
OPSO	(2.00e − 01, 1.00e − 01, 2.00e − 01)	(1.20e + 00, 3.00e − 01, 6.00e − 01)	(2.40e + 00, 7.50e − 01, 1.50e + 00)	(3.60e + 00, 1.12e + 00, 2.25e + 00)

Table 9 Results concerning Schewefel's (f_7) function

Technique	$D = 10$	$D = 25$	$D = 50$	$D = 100$
ABC	(0.00e + 00, 0.00e + 00, 0.00e + 00)	(0.00e + 00, 0.00e + 00, 0.00e + 00)	(0.00e + 00, 0.00e + 00, 0.00e + 00)	(0.00e + 00, 0.00e + 00, 0.00e + 00)
QABC	(0.00e + 00, 1.80e − 02, 0.00e + 00)	(1.82e − 03, 5.04e + 00, 0.00e + 00)	(1.81e − 02, 6.55e + 04, 0.00e + 00)	(1.91e + 05, 7.73e + 05, 0.00e + 00)
OABC	(2.00e − 06, 8.11e − 01, 0.00e + 00)	(1.83e − 03, 6.75e + 04, 0.00e + 00)	(6.01e + 02, 2.51e + 05, 0.00e + 00)	(2.32e + 06, 3.35e + 06, 0.00e + 00)
BA	(0.00e + 00, 0.00e + 00, 0.00e + 00)	(0.00e + 00, 0.00e + 00, 0.00e + 00)	(0.00e + 00, 0.00e + 00, 0.00e + 00)	(0.00e + 00, 0.00e + 00, 0.00e + 00)
QBA	(2.00e − 06, 1.00e − 06, 1.00e − 06)	(5.20e − 04, 4.94e − 04, 6.98e − 04)	(8.14e − 03, 4.81e − 03, 6.80e − 03)	(1.00e − 01, 6.75e − 02, 9.55e − 02)
OBA	(0.00e + 00, 0.00e + 00, 1.00e − 06)	(8.10e − 05, 1.02e − 04, 1.44e − 04)	(1.78e − 03, 1.48e − 03, 2.10e − 03)	(3.55e − 02, 2.11e − 02, 2.98e − 02)
FPA	(0.00e + 00, 0.00e + 00, 0.00e + 00)	(0.00e + 00, 0.00e + 00, 0.00e + 00)	(0.00e + 00, 0.00e + 00, 0.00e + 00)	(0.00e + 00, 0.00e + 00, 0.00e + 00)
QFPA	(0.00e + 00, 0.00e + 00, 0.00e + 00)	(0.00e + 00, 0.00e + 00, 0.00e + 00)	(0.00e + 00, 0.00e + 00, 0.00e + 00)	(0.00e + 00, 0.00e + 00, 0.00e + 00)
OFPA	(0.00e + 00, 0.00e + 00, 0.00e + 00)	(0.00e + 00, 0.00e + 00, 0.00e + 00)	(0.00e + 00, 0.00e + 00, 0.00e + 00)	(0.00e + 00, 0.00e + 00, 0.00e + 00)
PSO	(0.00e + 00, 0.00e + 00, 0.00e + 00)	(7.13e − 02, 1.16e + 00, 2.33e + 00)	(1.44e − 01, 1.02e + 00, 2.05e + 00)	(5.00e − 06, 2.09e − 03, 4.18e − 03)
QPSO	(0.00e + 00, 0.00e + 00, 0.00e + 00)	(2.95e − 01, 2.33e + 00, 4.65e + 00)	(5.94e + 02, 1.49e + 02, 2.97e + 02)	(4.95e + 03, 7.54e + 03, 1.51e + 04)
OPSO	(0.00e + 00, 0.00e + 00, 0.00e + 00)	(4.55e − 03, 2.17e − 02, 4.33e − 02)	(3.33e + 01, 4.71e + 01, 9.41e + 01)	(1.36e + 03, 7.87e + 02, 1.57e + 03)

Table 10 Results concerning Sphere (f_8) function

Technique	$D = 10$	$D = 25$	$D = 50$	$D = 100$
ABC	**(0.00e + 00, 0.00e + 00, 0.00e + 00)**	**(0.00e + 00, 0.00e + 00, 0.00e + 00)**	**(0.00e + 00, 0.00e + 00, 0.00e + 00)**	**(0.00e + 00, 0.00e + 00, 0.00e + 00)**
QABC	(1.00e − 06, 1.51e − 03, 2.18e − 03)	(5.30e − 05, 7.70e − 03, 1.57e − 02)	(1.89e − 03, 1.28e + 00, 1.69e + 00)	(6.53e − 02, 7.87e + 00, 4.52e + 00)
OABC	(4.81e − 01, 8.53e − 01, 2.01e − 01)	(3.31e + 00, 7.60e + 00, 1.64e + 00)	(1.78e + 01, 2.35e + 01, 2.10e + 00)	(5.80e + 01, 6.66e + 01, 5.48e + 00)
BA	**(0.00e + 00, 0.00e + 00, 0.00e + 00)**	(1.00e − 06, 4.00e − 06, 2.00e − 06)	(1.60e − 05, 3.20e − 05, 1.10e − 05)	(1.28e − 04, 1.80e − 04, 3.10e − 05)
QBA	(2.00e − 06, 3.00e − 06, 1.00e − 06)	(2.80e − 05, 4.20e − 05, 6.00e − 06)	(1.65e − 04, 2.00e − 04, 1.60e − 05)	(7.70e − 04, 8.55e − 04, 4.10e − 05)
OBA	(1.00e − 06, 2.00e − 06, 0.00e + 00)	(1.80e − 05, 2.10e − 05, 2.00e − 06)	(8.10e − 05, 1.02e − 04, 9.00e − 06)	(3.81e − 04, 4.18e − 04, 1.80e − 05)
FPA	**(0.00e + 00, 0.00e + 00, 0.00e + 00)**	**(0.00e + 00, 0.00e + 00, 0.00e + 00)**	**(0.00e + 00, 0.00e + 00, 0.00e + 00)**	**(0.00e + 00, 0.00e + 00, 0.00e + 00)**
QFPA	(0.00e + 00, 3.79e − 04, 1.00e − 03)	**(0.00e + 00, 0.00e + 00, 0.00e + 00)**	**(0.00e + 00, 0.00e + 00, 0.00e + 00)**	(2.78e + 01, 3.44e + 01, 3.83e + 00)
OFPA	**(0.00e + 00, 0.00e + 00, 0.00e + 00)**	**(0.00e + 00, 0.00e + 00, 0.00e + 00)**	**(0.00e + 00, 0.00e + 00, 0.00e + 00)**	**(0.00e + 00, 0.00e + 00, 0.00e + 00)**
PSO	**(0.00e + 00, 0.00e + 00, 0.00e + 00)**	(4.24e − 04, 1.07e − 02, 1.09e − 02)	(8.85e − 04, 3.23e − 02, 6.50e − 02)	(4.00e − 06, 9.20e − 05, 1.20e − 04)
QPSO	(0.00e + 00, 1.00e − 06, 4.00e − 06)	(5.35e − 04, 1.35e − 02, 1.25e − 02)	(3.62e − 02, 2.77e − 01, 1.53e − 01)	(2.09e − 01, 7.46e − 01, 3.17e − 01)
OPSO	**(0.00e + 00, 0.00e + 00, 0.00e + 00)**	(3.64e − 04, 6.34e − 03, 7.64e − 03)	(1.45e − 02, 1.02e − 01, 8.46e − 02)	(1.08e − 01, 2.48e − 01, 1.49e − 01)

Table 11 Results concerning Xin-She Yang's #1 (f_6) function

Technique	$D = 10$	$D = 25$	$D = 50$	$D = 100$
ABC	(0.00e + 00, 0.00e + 00, 0.00e + 00)	(0.00e + 00, 0.00e + 00, 0.00e + 00)	(0.00e + 00, 0.00e + 00, 0.00e + 00)	(0.00e + 00, 0.00e + 00, 0.00e + 00)
QABC	(0.00e + 00, 1.05e − 02, 0.00e + 00)	(4.74e − 01, 1.49e + 00, 0.00e + 00)	(6.01e + 03, 2.80e + 05, 0.00e + 00)	(2.75e+15, 1.04e+19, 0.00e + 00)
OABC	(0.00e + 00, 3.77e − 02, 0.00e + 00)	(3.59e − 02, 2.59e − 01, 0.00e + 00)	(3.66e + 01, 4.68e + 02, 0.00e + 00)	(9.17e + 09, 6.67e+11, 0.00e + 00)
BA	(7.26e − 03, 2.68e − 02, 3.79e − 02)	(5.01e − 01, 2.41e + 02, 3.41e + 02)	(7.80e + 02, 2.34e + 09, 3.31e + 09)	(2.28e−13, 3.67e+23, 5.19e+23)
QBA	(2.36e − 03, 4.81e − 02, 6.81e − 02)	(8.31e − 02, 4.16e − 02, 5.88e − 02)	(4.64e + 00, 3.56e + 04, 5.04e + 04)	(1.78e+11, 1.26e+16, 1.78e+16)
OBA	(2.05e − 03, 2.66e − 02, 3.76e − 02)	(1.93e − 02, 2.74e − 01, 3.88e − 01)	(9.90e − 02, 3.00e + 02, 4.24e + 02)	(2.20e + 02, 1.66e+11, 2.35e+11)
FPA	(0.00e + 00, 0.00e + 00, 0.00e + 00)	(0.00e + 00, 0.00e + 00, 0.00e + 00)	(0.00e + 00, 0.00e + 00, 0.00e + 00)	(0.00e + 00, 1.00e − 06, 2.00e − 06)
QFPA	(0.00e + 00, 4.00e − 06, 6.00e − 06)	(0.00e + 00, 0.00e + 00, 1.00e − 06)	(0.00e + 00, 0.00e + 00, 0.00e + 00)	(0.00e + 00, 0.00e + 00, 0.00e + 00)
OFPA	(1.00e − 06, 3.00e − 06, 5.00e − 06)	(0.00e + 00, 1.00e − 06, 1.00e − 06)	(0.00e + 00, 1.00e − 06, 0.00e + 00)	(0.00e + 00, 0.00e + 00, 0.00e + 00)
PSO	(3.60e − 05, 1.10e − 05, 2.10e − 05)	(2.92e − 03, 6.08e − 03, 1.22e − 02)	(9.96e − 02, 1.53e − 01, 3.07e − 01)	(2.77e + 03, 7.01e + 09, 1.40e+10)
QPSO	(0.00e + 00, 4.10e − 05, 8.20e − 05)	(3.20e − 05, 1.02e − 03, 2.03e − 03)	(6.00e − 05, 2.21e − 03, 4.42e − 03)	(8.62e − 04, 3.49e − 03, 6.98e − 03)
OPSO	(0.00e + 00, 1.30e − 05, 2.50e − 05)	(1.00e − 06, 1.22e − 04, 2.44e − 04)	(4.70e − 05, 1.27e − 04, 2.54e − 04)	(5.20e − 05, 1.91e − 04, 3.82e − 04)

Table 12 Results concerning Zakharov's (f_{10}) function

Technique	$D = 10$	$D = 25$	$D = 50$	$D = 100$
ABC	**(0.00e + 00, 0.00e + 00, 0.00e + 00)**	**(0.00e + 00, 0.00e + 00, 0.00e + 00)**	**(0.00e + 00, 0.00e + 00, 0.00e + 00)**	**(0.00e + 00, 0.00e + 00, 0.00e + 00)**
QABC	(6.04e + 00, 3.21e + 01, 0.00e + 00)	(1.64e + 02, 2.15e + 02, 0.00e + 00)	(4.63e + 02, 6.06e + 02, 0.00e + 00)	(1.19e + 03, 1.34e + 03, 0.00e + 00)
OABC	(1.72e + 01, 3.20e + 01, 0.00e + 00)	(1.21e + 02, 1.96e + 02, 0.00e + 00)	(3.46e + 02, 6.04e + 05, 0.00e + 00)	(8.07e + 02, 1.87e+12, 0.00e + 00)
BA	(3.00e − 06, 7.00e − 06, 1.00e − 05)	(1.76e − 02, 5.02e + 01, 7.09e + 01)	(2.95e + 01, 2.22e + 02, 3.14e + 02)	(3.37e + 03, 2.03e + 03, 2.87e + 03)
QBA	(8.00e − 06, 6.00e − 06, 9.00e − 06)	(8.10e − 05, 9.10e − 05, 1.29e − 04)	(6.10e − 04, 3.40e − 04, 4.81e − 04)	(2.74e − 03, 1.60e − 03, 2.27e − 03)
OBA	(4.00e − 06, 3.00e − 06, 4.00e − 06)	(6.10e − 05, 3.80e − 05, 5.40e − 05)	(3.07e − 04, 1.59e − 04, 2.25e − 04)	(1.32e − 03, 7.24e − 04, 1.03e − 03)
FPA	**(0.00e + 00, 0.00e + 00, 0.00e + 00)**	**(0.00e + 00, 0.00e + 00, 0.00e + 00)**	**(0.00e + 00, 0.00e + 00, 0.00e + 00)**	**(0.00e + 00, 0.00e + 00, 0.00e + 00)**
QFPA	(0.00e + 00, 2.50e − 05, 4.30e − 05)	(6.01e − 04, 2.02e − 03, 3.50e − 03)	(1.97e − 03, 1.27e − 02, 2.21e − 02)	(1.77e − 01, 2.30e + 00, 3.98e + 00)
OFPA	(0.00e + 00, 1.80e − 05, 3.10e − 05)	(1.50e − 04, 1.05e − 03, 1.81e − 03)	(1.39e − 02, 3.29e − 02, 5.69e − 02)	(3.99e − 01, 4.22e − 01, 7.30e − 01)
PSO	**(0.00e + 00, 0.00e + 00, 0.00e + 00)**	(6.14e + 00, 4.79e + 01, 9.57e + 01)	(9.20e + 01, 2.92e + 02, 5.84e + 02)	(1.73e + 03, 9.44e + 02, 1.89e + 03)
QPSO	(0.00e + 00, 1.62e + 01, 3.23e + 01)	(8.72e + 01, 4.84e + 05, 9.68e + 05)	(5.19e + 02, 1.80e + 02, 3.61e + 02)	(8.82e + 02, 5.93e + 02, 1.19e + 03)
OPSO	(1.00e − 06, 2.50e − 01, 4.99e − 01)	(2.50e + 01, 1.13e + 02, 2.26e + 02)	(9.64e + 01, 1.63e + 02, 3.27e + 02)	(1.68e + 02, 8.31e + 01, 1.66e + 02)

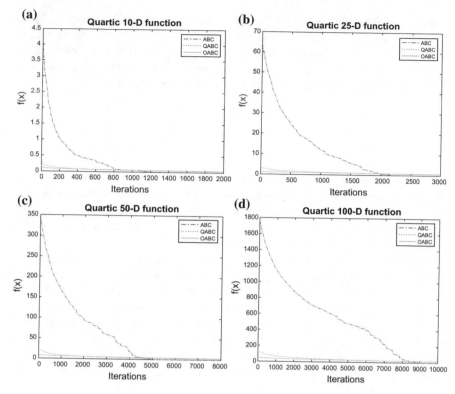

Fig. 6 Convergence plot considering ABC, QABC and OABC over Salomon's function. **a** 10-D, **b** 25-D, **c** 50-D and **d** 100-D

Figure 6 depicts a convergence study of ABC variants over Quartic's function (f_3). Interestingly, although the hypercomplex-based approaches did not obtain the best results in this dataset, one can observe their faster convergence when compared to standard ABC considering the first 800 iterations. As a matter of fact, their final results were considerably close to each other. It can be hypothesized that the standard ABC uses higher trial limits (here 1000), which may lead to higher computational costs. Such computational costs may be even higher when represented in hypercomplex spaces for ABC, and the benefit can be reduced.

The convergence plots for other functions (f_6 and f_{10}) are similar. In this situation, standard ABC converged faster than its hypercomplex-based version for all dimensions. It is worth pointing out that the observations that ABC may not benefit from the hypercomplex representations for the selected benchmarks do not necessarily apply to other functions.

Since both BA and FPA obtained better and best results when formulated in hypercomplex representations, it can be expected that the hypercomplex representations can in general lead to smoother landscapes, especially for highly nonlinear, multimodal landscapes. This point requires further investigation so as to figure out what types of functions should be represented in hypercomplex spaces.

6 Conclusion

Hypercomplex numbers extend standard complex spaces by adding imaginary numbers in their formulation. In this work, we considered mapping search spaces into hypercomplex ones defined by quaternions and octonions. Roughly speaking, each decision variable is encoded by a hypercomplex number, which leads us to work with tensors instead of vectors in the search space.

We have considered 10 benchmarking functions with different dimensions in order to evaluate hypercomplex spaces, and we observed that higher-dimensional problems seem to benefit more than lower-dimensional search spaces. Additionally, a study concerning the convergence highlighted that quaternionic- and octonic-based approaches seem to converge faster than standard ones for some benchmarking function. As a take-home message, we can observe hypercomplex-based search spaces may provide better optimization environments, as well as techniques based on quaternions and octonions can obtain results so accurate as standard ones, but usually with a faster convergence.

For future works, it will be useful to carry out more detailed tests using more benchmarks so as to find the true benefit of hypercomlex representations. In addition, it will be also useful to consider even higher-dimensional hypercomplex spaces, such as representations based on sedenions. Furthermore, it can be expected that quaternions and octonions can be used in the context of hyperparameter fine-tuning in machine learning techniques and image processing.

Acknowledgements The authors are grateful to FAPESP grants #2014/16250-1, #2014/12236-1 and #2015/25739-4, as well as CNPq grant #306166/2014-3.

References

1. Chen, W.N., Zhang, J., Lin, Y., Chen, N., Zhan, Z.H., Chung, H.S.H., Li, Y., Shi, Y.H.: Particle swarm optimization with an aging leader and challengers. IEEE Trans. Evol. Comput. **17**(2), 241–258 (2013)
2. Eberly, D.: Quaternion Algebra and Calculus. Tech. rep, Magic Software (2002)
3. Fister, I., Brest Jr., J., Yang, X.S.: I.F.: Modified bat algorithm with quaternion representation. In: IEEE Congress on Evolutionary Computation, pp. 491–498 (2015)
4. Fister, I., Yang, X.S., Brest Jr., J.: I.F.: Modified firefly algorithm using quaternion representation. Expert Syst. Appl. **40**(18), 7220–7230 (2013)

5. Geem, Z.W., Sim, K.B.: Parameter-setting-free harmony search algorithm. Appl. Math. Comput. **217**(8), 3881–3889 (2010)
6. Graves, J.T.: On a connection between the general theory of normal couples and the theory of complete quadratic functions of two variables. Philos. Mag. **26**(173), 315–320 (1845)
7. Hatamlou, A.: Black hole: a new heuristic optimization approach for data clustering. Inform. Sci. **222**, 175–184 (2013)
8. Hu, X.M., Zhang, J., Yu, Y., Chung, H.S.H., Li, Y.L., Shi, Y.H., Luo, X.N.: Hybrid genetic algorithm using a forward encoding scheme for lifetime maximization of wireless sensor networks. IEEE Trans. Evol. Comput. **14**(5), 766–781 (2010)
9. Jamil, M., Yang, X.S.: A literature survey of benchmark functions for global optimization problems. Int. J. Math. Modell. Numer. Optim. **4**(2), 150–194 (2013)
10. Karaboga, D., Basturk, B.: A powerful and efficient algorithm for numerical function optimization: Artificial bee colony (ABC) algorithm. J. Global Optim. **39**(3), 459–471 (2007)
11. Kennedy, J., Eberhart, R.C.: Swarm Intell. Morgan Kaufmann Publishers Inc., San Francisco, USA (2001)
12. Nickabadi, A., Ebadzadeh, M.M., Safabakhsh, R.: A novel particle swarm optimization algorithm with adaptive inertia weight. Appl. Soft Comput. **11**, 3658–3670 (2011)
13. Oh, B.K., Kim, K.J., Park, Y.K.H.S., Adeli, H.: Evolutionary learning based sustainable strain sensing model for structural health monitoring of high-rise buildings. Appl. Soft Comput. **58**, 576–585 (2017)
14. Papa, J.P., Pereira, D.R., Baldassin, A., Yang, X.S.: On the harmony search using quaternions. In: Schwenker, F., Abbas, H.M., El-Gayar, N., Trentin, E. (eds.) Artificial Neural Networks in Pattern Recognition: 7th IAPR TC3 Workshop, ANNPR, pp. 126–137. Springer International Publishing, Cham (2016)
15. Papa, J.P., Rosa, G.H., Costa, K.A.P., Marana, A.N., Scheirer, W., Cox, D.D.: On the model selection of bernoulli restricted Boltzmann machines through harmony search. In: Proceedings of the Genetic and Evolutionary Computation Conference, GECCO '15, pp. 1449–1450. ACM, New York, USA (2015)
16. Papa, J.P., Rosa, G.H., Marana, A.N., Scheirer, W., Cox, D.D.: Model selection for discriminative restricted boltzmann machines through meta-heuristic techniques. J. Comput. Sci. **9**, 14–18 (2015)
17. Papa, J.P., Rosa, G.H., Rodrigues, D., Yang, X.S.: LibOPT: an open-source platform for fast prototyping soft optimization techniques. ArXiv e-prints (2017). http://adsabs.harvard.edu/abs/2017arXiv170405174P
18. Papa, J.P., Scheirer, W., Cox, D.D.: Fine-tuning deep belief networks using harmony search. Appl. Soft Comput. **46**, 875–885 (2016)
19. Pitzer, E., Affenzeller, M.: A Comprehensive Survey on Fitness Landscape Analysis. Springer, Berlin, Heidelberg (2012)
20. Rodrigues, D., Silva, G.F.A., Papa, J.P., Marana, A.N., Yang, X.S.: EEG-based person identification through binary flower pollination algorithm. Expert Syst. Appl. **62**, 81–90 (2016)
21. Rosa, G.H., Papa, J.P., Marana, A.N., Scheirer, W., Cox, D.D.: Fine-tuning convolutional neural networks using harmony search. In: Progress in Pattern Recognition, Image Analysis, Computer Vision, and Applications. Springer International Publishing (2015)
22. Wilcoxon, F.: Individual comparisons by ranking methods. Biom. Bull. **1**(6), 80–83 (1945)
23. Yang, S.S., Karamanoglu, M., He, X.: Flower pollination algorithm: a novel approach for multiobjective optimization. Eng. Optim. **46**(9), 1222–1237 (2014)
24. Yang, X.S.: Engineering Optimization: An Introduction with Metaheuristic Applications, 1st edn. Wiley Publishing (2010)
25. Yang, X.S.: Firefly algorithm, stochastic test functions and design optimisation. Int. J. Bio-Inspir. Comput. **2**(2), 78–84 (2010)
26. Yang, X.S., Deb, S.: Engineering optimisation by cuckoo search. Int. J. Math. Modell. Numer. Optim. **1**, 330–343 (2010)
27. Yang, X.S., Gandomi, A.H.: Bat algorithm: a novel approach for global engineering optimization. Eng. Comput. **29**(5), 464–483 (2012)

Lévy Flight-Driven Simulated Annealing for B-spline Curve Fitting

Carlos Loucera, Andrés Iglesias and Akemi Gálvez

Abstract Point cloud approximation by spline models, also called *curve and surface reconstruction*, is an active research field in computer-aided design and manufacturing (CAD/CAM). Due to the physical and mechanical processes used to obtain the data, the measurements are often affected by noise and other distortions. Obtaining a suitable spline model to reconstruct the underlying shape of the data while maintaining a low design complexity leads to a multivariate and highly non-linear optimization problem, also known to be non-convex and multi-modal. In this work, we propose a method to fit a given point cloud by means of a B-spline curve model. Our approach to solve the optimization problem is based on a powerful thermodynamics-driven metaheuristic known as the Simulated Annealing. We compute the model parameters by combining traditional SA techniques with Lévy flights (random walks based on the Lévy distribution). The ability to perform such a *flight* allows the algorithm to escape from local minima and energy plateaus, a strong requirement when dealing with highly multi-modal problems. The performance and robustness of our algorithm is tested against three illustrative examples. Our experimental results show that our method is able to reconstruct the underlying shape of the data, even in the presence of noise, with acceptable accuracy and in a completely automated way.

Keywords Data fitting · Curve reconstruction · Reverse engineering · CAD/CAM · B-spline functions · Metaheuristic techniques · Simulated annealing · Lévy flights

C. Loucera
Department of Communications Engineering, Universidad de Cantabria,
Avda. de Los Castros, s/n, E-39005 Santander, Spain
e-mail: loucerac@unican.es

A. Iglesias (✉) · A. Gálvez
Faculty of Sciences, Department of Information Science, Toho University,
2-2-1 Miyama, Narashino Campus, Funabashi 274-8510, Japan
e-mail: iglesias@unican.es

A. Iglesias · A. Gálvez
Department of Applied Mathematics and Computational Sciences,
University of Cantabria, Avda. de Los Castros, s/n, 39005 Santander, Spain
e-mail: galveza@unican.es

© Springer International Publishing AG 2018
X.-S. Yang (ed.), *Nature-Inspired Algorithms and Applied Optimization*,
Studies in Computational Intelligence 744, https://doi.org/10.1007/978-3-319-67669-2_7

1 Introduction

Curve fitting is a major issue in many scientific fields, such as statistics, numerical analysis, data visualization, geometric modeling, image processing, and meteorology, to mention just a few. Data fitting is also a key problem in computer-aided design and manufacturing (CAD/CAM), where the ability to obtain a digital model from a 3D-scanned real-world piece (a process generally known as *reverse engineering*) plays a crucial role in many current manufacturing industries. This process usually involves the fitting of a massive and noisy point cloud obtained with modern data-acquisition technologies such as 3D laser scanning and other devices [1, 2].

B-splines are an industry standard in the CAD/CAM field for data storage and representation. There are several reasons for this choice: B-splines are highly flexible and easy to manipulate; they are also able to describe very complex shapes with a minimal set of parameters. These good properties are due to the particular functional structure of B-splines: they consist of a linear combination of non-linear functions, known as the *basis functions*. The coefficients of this linear combination are usually called the *poles* or *control points*, whereas the local shape parameters are known as the *breakpoints* or *knots*. Finally the parameterization deals with the locations where the B-splines are evaluated to fit the data points. The problem of reconstructing a dataset by means of an optimal B-spline is a multivariate and highly non-linear optimization problem. It is also known to be a non-convex and multi-modal problem. Owing to these challenging features, the general shape reconstruction problem remains largely unsolved so far (in other words, no closed analytical form for the general solution can be automatically obtained). Even powerful and well-tested mathematical optimization methods tend to get trapped in one of the many local minima [3].

Recently, *nature-inspired computation* (a broad set of different computational methods based on mimicking certain biological or social processes from the natural world), has evolved into one of the most fruitful areas of scientific research and knowledge. In fact, bio-inspired optimization has been applied with great success to many difficult engineering problems [4–7]. These methods, often of derivative-free nature, consist of the implementation of search strategies providing a trade-off between local and global optimization without assuming any a priori knowledge about the problem. This ability to combine exploitation and exploration (i.e., conducting exhaustive search in the most promising regions while simultaneously maintaining the possibility to explore the whole fitness landscape seeking for the best global solution) within an unified framework make bio-inspired computation methods excellent tools for solving the data approximation problem. Over the years many methods have solved particular instances of the problem, either with Bézier models [8–16] or local-support curves [17–30]. See also [31] for a recent and detailed review on curve and surface fitting with nature-inspired methods and some recent trends in the field.

In this chapter we apply a variant of the classical simulated annealing algorithm that introduces a Lévy-based re-annealing (restarting the annealing cycle) method coupled with a local exploitation phase driven by the constrained optimization by linear approximations algorithm (COBYLA) [32]. This SA variant searches the B-spline parameterization associated with the problem data, while a suitable breakpoint sequence is found by means of the knot averaging method [33]. Finally the poles are computed by solving traditional linear least-squares problem.

The structure of this chapter is as follows: in Sect. 2 we provide the basic definitions about parametric B-spline curves, along with the mathematical background required to understand in detail the data fitting problem for this type of curves. In Sect. 3 we focus on the Lévy flight-driven simulated annealing, the metaheuristics proposed in this paper. The discussion starts with the description of the classical simulated annealing algorithm. Then, we describe our new variant in detail. The main components of our method are also discussed in detail in this section. Section 4 describes our Lévy flight-driven simulated annealing-based method for B-spline curve fitting. Firstly, a brief outline of the method is presented; then, a more detailed description of each individual component is given. The performance of this new method is illustrated in Sect. 5 through three examples. The experimental results are presented both graphically and numerically. Finally, Sect. 6 summarizes the main conclusions of this chapter and provides some hints about future work in the field.

2 Mathematical Formulation

2.1 Basic Definitions

Mathematically, a *parametric B-spline curve* $\mathbf{C}(t) \subset \mathbb{R}^d$ of order p is a piecewise function expressed as:

$$\mathbf{C}(t) = \sum_{i=0}^{n} \mathbf{P}_i N_{i,p}(t) \tag{1}$$

where $t \in [\alpha, \beta]$ represents the data parameterization, $\{\mathbf{P}_i\}$ are the control net of the curve and $\{N_{i,p}(t)\}_i$ are the so called B-spline basis functions of order p defined on a knot vector $\mathscr{U} = \{u_0 = \alpha, u_1, u_2, \dots, u_{n+p} = \beta\}$, comprised of non-decreasing real numbers u_i called *knots*. The B-spline basis functions $N_{j,p}(t)$ can be computed through the Cox de-Boor recursive formula (see [34] for details):

$$N_{j,p}(t) = \frac{t - u_j}{u_{j+p-1} - u_j} N_{j,p-1}(t) + \frac{u_{j+p} - t}{u_{j+p} - u_{j+1}} N_{j+1,p-1}(t) \tag{2}$$

for $p > 1$, while for $p = 1$ we have:

$$N_{j,1}(t) = \begin{cases} 1 & \text{if } u_j \le t < u_{j+1} \\ 0 & \text{otherwise} \end{cases} \qquad (j = 0, \dots, n+p-1) \tag{3}$$

In this paper we consider only splines clamped at the edges, so $u_0 = \cdots = u_{p-1} = \alpha$, and $u_{n+1} = \cdots = u_{n+p} = \beta$. Each blending function is a local support function, those that vanish outside a certain interval, since each basis is only defined in the corresponding interval. Therefore, perturbations on a given interval only affect a part of the curve, a very useful property in the CAD/CAM industry. For this and other fundamental properties of the B-spline space of functions, see [33].

2.2   Data Fitting

Let $\left\{\mathbf{Q}_k\right\}_{k=1}^M$ be a set of points in $\mathbb{R}^d$. The aim of this paper consists in finding a B-spline curve that approximates the given data, by taking into account both, the fidelity of the reconstruction and its complexity.

In order to reconstruct the underlying shape of the data with a clamped B-spline curve $\mathbf{C}(t)$ of degree p, our method must perform the parameterization, i.e. find the $\{t_k\}$ associated to the original data, compute the poles $\mathbf{P}_j$ with the corresponding breakpoints $\mathscr{U}$ and, finally, the method must deal with the model complexity: how to minimize the number of free parameters of the system. Due to the constrains imposed on the boundary knots, we can assume $\mathbf{C}(t_1) = \mathbf{Q}_1$ and $\mathbf{C}(t_M) = \mathbf{Q}_M$. As a result the equation to minimize in a least-squares sense is given by:

$$E = \sum_{k=2}^{M-1} \left\| \mathbf{Q}_k - \sum_{i=0}^{n} \mathbf{P}_i N_{i,p}(t_k) \right\|_2^2 \tag{4}$$

where $||.||_2$ indicates the Euclidean norm. Note that for known degree, parameterization and knot vector, a solution of the previous linear system can always be computed through standard numerical methods for polynomial system solving, obtaining as a result the B-spline control net. However, in many real-world problems, neither the parameterization nor the knot vector are generally known. In general, even the optimal value for the curve degree is unknown. In such a case, the least-squares minimization problem (4) becomes highly nonlinear, continuous, and multivariate. In addition, the computation of the knot vector has been proved to be a non-convex and multi-modal optimization problem [34, 35]. To overcome such difficulties we propose an optimization simulated annealing schema that deals with each sub-problem sequentially: data parameterization, knot vector and pole computation and finally, model complexity.

3 Lévy-Driven Simulated Annealing

3.1 Basic Principles

Simulated annealing (SA) is a family of stochastic optimization algorithms belonging to the emergent field of nature-inspired computation. Since its inception in the early eighties by Kirkpatrick et al. [36], it has been used on a large number of real-world and synthetic problems [37]. One of the defining features of nature-inspired algorithms is the assumption of a metaphor driving the search for a global optimum. In this particular case, the SA algorithm computationally mimics the thermodynamical processes behind the annealing of a metal: how to improve the material through a set of intervals of rapid heating and slow cooling cycles.

At the initial stages of the physical process, the material is heated to very high temperatures, which leads to free-moving particles. Then, a slow cooling phase follows, when the particles tend to loose part of their mobility. All along the process, the atoms tend to move towards configurations that minimize the system energy, although it may lead to occasional rises of the overall energy. The thermal equilibrium for a certain temperature is reached when these transitions tend to stabilize the energy of the system. This heating-cooling procedure is repeated until the there are almost no particle movement, which coincides with the minimal energy state of the system. The resulting material has a better inner structure at the end of the thermodynamic process.

The SA algorithm constructs a computational metaphor of this thermodynamical process in order to minimize a functional that replicates the energy of the physical process. The procedure starts from a random state/solution, and iteratively generates new solutions sampled from a candidate distribution. This sampling takes into account the temperature of the system, an artificial parameter controlled by the cooling schedule, how and when the temperature is updated, and the previous visited solutions. As the system freezes, the sampled points are closer to the previous solutions (which closely follows loss of mobility of the atoms). In order to guarantee that the system can avoid local minima, each transition is accepted according to a certain probability. If the energy is minimized, the new candidate is always accepted. Otherwise, the chance to discard a worse transition is increased as the system evolves.

Random walks based on the Lévy distribution have already been used as an enhancement to various global search algorithms. Relevant examples are described in [38] for the cuckoo search algorithm, where the flight of cuckoo birds is simulated by Lévy flights, in [39] where that technique is applied to the reconstruction of outline curves of computer fonts with rational Bézier curves, in [40], where Lévy flights are used to maintain the population diversity of particle swarm optimization, and [41] for the flower pollination optimization algorithm, where Lévy random walks are used to mimic the long distances taken by insects during the pollination.

3.2 SA Algorithm

The SA algorithm is designed to minimize a real-valued fitness function (usually called the *system energy*) $f : \mathscr{D} \subseteq \mathbb{R}^d \longrightarrow \mathbb{R}$, within a problem domain $\mathscr{D}$, assumed to be continuous in this paper. Each point $\mathbf{x} \in \mathscr{D}$ is a *state* of the physical system. Given an initial (usually random) state $\mathbf{x}_0$, the algorithm performs an iterative process; at each iteration step, a new state $\mathbf{x}_{new}$ is generated from the current one, $\mathbf{x}_{old}$, through a *neighborhood function*, denoted by $\mathfrak{N} : \mathscr{D} \longrightarrow \mathscr{D}$, i.e., $\mathbf{x}_{new} = \mathfrak{N}(\mathbf{x}_{old})$. Let now $f_{old} \equiv f(\mathbf{x}_{old})$, $f_{new} \equiv f(\mathbf{x}_{new})$ be their associated energies, respectively. The algorithm probabilistically decides between moving the system to the new state $\mathbf{x}_{new}$ or staying in the current state $\mathbf{x}_{old}$. This new state is chosen with a probability function $\mathfrak{P} : \mathscr{D} \times \mathscr{D} \longrightarrow [0,1]$, called the *acceptance function*, which depends on two factors:

1. the difference $\triangle = f_{old} - f_{new}$ of the energy values; and
2. a global parameter called *temperature*, denoted by T, which varies according to a strictly decreasing function $\mathfrak{T} : \mathbb{R}^+ \longrightarrow \mathbb{R}^+$ called the *cooling function*.

In addition, two more conditions are required. The first one is that $\mathfrak{P} > 0$ if $\triangle < 0$, meaning that the system may move to the new state even if it is worse than the current one. This condition is imposed with the goal to prevent *stagnation* (when the system gets trapped in the neighborhood of local optima, leading to premature convergence). The second one is that the lower the temperature, the easier to reject a worse solution. In fact, in the particular case $T = 0$, the procedure will only allow downhill moves, meaning that the algorithm reduces to a greedy search algorithm. The interested reader is referred to [42] for further details about the algorithm and the corresponding pseudocode.

3.3 Lévy SA Algorithm

The description in previous paragaphs refers to the Classical Simulated Annealing (CSA), a combinatorial optimization algorithm that follows as closely as possible the physical annealing process [36]. The algorithm was soon adapted to deal with continuous optimization problems [43, 44], with an impressive capacity to resolve very hard problems. However, the computational costs of these first approaches were prohibitive in many cases and required some heavy fine tuning to guarantee the convergence.

Over the years, many SA variants have been proposed with the aim to overcome the aforementioned difficulties. A complete list is beyond the scope of this work; instead, we will introduce two of the most influential variants. On one hand, the Fast Simulated Annealing (FSA) [45] where a Cauchy visiting distribution is coupled with a time-inversely cooling schedule. On the other hand, the Adaptive Simulated Annealing (ASA) [46], a major milestone in the physics-inspired optimization field, consists in pairing a fitness-sensitive visiting distribution with two sets of temperatures that are raised and lowered thought the live of the algorithm in a cycle called re-annealing.

One of the major drawbacks of most SA variants (and many other nature-inspired optimizers) is the parameter setup: a necessary step that involves knowledge about the problem being solved. The initial temperature is a critical parameter, as it needs to be high enough to let the solutions move freely at the initial stages, but no so high that most computation time is wasted performing random walks aimless on the solution space. If the start temperature is too low, the system may become a greedy search too early, effectively losing its global search capabilities.

Algorithm 1: Lévy Simulated Annealing (by linear approximations)

Input: An initial guess $\mathbf{x}_0$, fitness f, lower $\mathbf{l}$ and upper bounds $\mathbf{u}$
Output: The final solution $\mathbf{x}$
$constraints \leftarrow \texttt{generateLinearConstraints}(\mathbf{x}_0);$
$T_0 \leftarrow \texttt{computeInitialTemperature}(f, \mathbf{l}, \mathbf{u});$
$\mathbf{x} \leftarrow \mathbf{x}_0$ and $f_\mathbf{x} \leftarrow f(\mathbf{x});$
$T \leftarrow T_0;$
while *The System is not Frozen* **do**
    $acceptedFlag \leftarrow False;$
    while *Thermal Equilibrium is not Reached* **do**
        $\mathbf{x}_{new} \leftarrow \mathfrak{N}(\mathbf{x}, T)$ and $f_{new} \leftarrow f(\mathbf{x}_{new});$
        if $\mathfrak{A}(f_{new}, f_\mathbf{x}, T)$ **then**
            $\mathbf{x} \leftarrow \mathbf{x}_{new}$ and $f_\mathbf{x} \leftarrow f_{new};$
            $acceptedFlag \leftarrow True;$
        end
    end
    $T \leftarrow \mathfrak{T}(T);$
    if *not acceptedFlag* **then**
        $[\mathbf{x}, f_\mathbf{x}, T] \leftarrow \texttt{learnFitnessLandscape}(\mathbf{x}, \mathbf{x}_{best}, \mathbf{l}, \mathbf{u}, constraints);$
    end
end
return $\mathbf{x}$

Our SA proposal, as summarized in Algorithm 1, draws its inspiration from the FSA and ASA algorithms, while maintaining the classical structure. The algorithm starts with a random point in the search space, then it iteratively tries to improve it by means of a temperature-driven d-dimensional Cauchy visiting distribution. When no new solutions are accepted for a given thermal cycle, a local search is performed. If successful, the algorithm continues, as it may be a basin to exploit. In the case of an unsuccessful exploitation phase, the algorithm considers that it has been trapped in either a correctly exploited basin or an energy plateau. In both cases, a Lévy flight is performed in order to *escape* from the current situation, and the temperature is re-restarted (mimicking the ASA re-annealing).

The main components of our method are discussed in detail in next paragraphs.

3.3.1 Initial Temperature

To compute the initial temperature we try to approximate the average of the temperatures needed to raise the fitness with a probability of χ_0, as proposed in [47, 48]. Let $X_0^+ = \{\Delta(\mathbf{x}_k)\}_k$ a set of randomly chosen positive transitions, and χ_0 the desired acceptance ratio, typically 0.8. Then, the initial temperature is given by:

$$T_0 = -\frac{\text{mean}\left(X_0^+\right)}{\log\left(\chi_0\right)} \tag{5}$$

This step is done with the `computeInitialTemperature` procedure in Algorithm 1.

3.3.2 Cooling Schedule

The law governing the cooling strategy is given by the following formula:

$$\mathfrak{T} = \frac{T_0}{k_{outer}} \tag{6}$$

where T_0 is computed as in the previous paragraph and k_{outer} is the annealing index, updated after each thermal cycle. This cooling law has been proven to be slow enough to guarantee an stable Cauchy visiting distribution [45]. See [49] for a thoughtful discussion on this and other SA schedules.

3.3.3 Candidate Distribution

The neighborhood function $\mathfrak{N}$ is based on sampling points from the following Cauchy distribution:

$$\mathbf{x}_{new} \leftarrow \mathbf{x}_{old} + \Delta\mathbf{x}_{old} \tag{7}$$

that depends on the previous point and the current temperature of the system, where:

$$\Delta\mathbf{x}_{old} \sim \frac{T}{\left(\|\Delta\mathbf{x}_{old}\|^2 + T^2\right)^{\frac{D+1}{2}}} \tag{8}$$

$\sim$ indicates *sampled from*, and D is the dimension of the solution space. See [45] for a more detailed discussion on the topic.

3.3.4 Acceptance Function

The law governing the probability of accepting a given transition follows the modified Metropolis criterion [50]:

$$\mathfrak{A} \leftarrow \min \left\{ 1, \left(1 + \exp \left(\frac{\Delta f}{T} \right) \right)^{-1} \right\} \tag{9}$$

3.3.5 Learn Fitness Landscape

Our memetic approach tries to learn the shape of the fitness landscape after each thermal equilibrium phase. To do so, if no solution has been accepted during a given inner cycle (i.e., the *acceptedFlag* remains *False*) we assume that we are either trapped in a local basin or traversing an energy plateau. To overcome such a difficult scenario, the algorithm first tries to exploit the neighborhood of the current solution by performing a local search by means of the constrained by linear approximations algorithm (COBYLA). On one hand, if the direct search is successful, we update the SA solution with the new one and continue with the main algorithm. On the other hand, we conclude that the algorithm is traversing an energy plateau so we perform a Lévy flight in order to escape the flat zone. The temperature is restarted after each random walk.

3.3.6 Lévy Flights

Let $\mathbf{x}$ and $\mathbf{x}_{best}$ the current and best-found solution. We simulate the capacity to make a long jump in the solution space by the following Lévy flight:

$$\mathbf{x}_{new} = \mathbf{x} + L \left(\mathbf{x} - \mathbf{x}_{best} \right) \tag{10}$$

where $L > 0$ is the step size, which follows the Lévy distribution given by:

$$L \sim \frac{\lambda \Gamma (\lambda) \sin \left(\lambda \frac{\pi}{2} \right)}{\pi} \frac{1}{s^{1+\lambda}} \quad \text{with} \quad s \gg s_0 > 0 \tag{11}$$

where Γ represents the Gamma function. The given distribution is valid for $s \gg s_0 > 0$, where s_0 is the smallest step. Typically, it is enough to use $s_0 \approx 0.1$ (although the limits are problem-dependent). To compute the step size s, i.e. to draw s from the Lévy distribution given by Eq. (11), we use *Mantegna's algorithm* [51], a well established approach in the nature-inspired optimization field [6]. The procedure can be summarized as follows:

$$s = \frac{u}{|v|^{\frac{1}{\lambda}}}$$

where $u \sim \mathcal{N}\left(0, \sigma_u^2\right)$ and $v \sim \mathcal{N}\left(0, 1\right)$, with:

$$\sigma_u^2 = \left(\frac{\Gamma\left(1+\lambda\right) \sin\left(\frac{\lambda}{2}\pi\right)}{\lambda \Gamma\left(\frac{1+\lambda}{2}\right) 2^{\frac{\lambda-1}{2}}} \right)^{\frac{1}{\lambda}}$$

where $\mathcal{N}\left(\mu, \sigma^2\right)$ represents the normal distribution of mean μ and variance σ^2. Therefore, a Lévy flight can be computed by fixing λ and s_0 using the approximation $L \approx s \cdot s_0$. See [6], pp. 11–20 for a more detailed discussion on Lévy flights.

4   The Method

In this section we will discuss the main parts of our method in detail. Firstly, we introduce the overall outline of the methodology. Then, we continue with a more detailed description of each individual component.

4.1   Outline of the Method

Our method is based on converting the geometric problem of how to fit a B-spline curve to a point cloud in a completely automatic way into a multivariate optimization problem. This is done through the least-squares technique given by Eq. (4), leading to a non-convex, highly non-linear optimization problem with three sets of unknowns, namely; the poles, data parameters, and knots. To overcome such difficulties we propose a method that sequentially computes each set of unknowns, using the previous set as the input for the next step.

Figure 1 shows a diagram that summarizes the workflow of the proposed methodology. Central and right parts of the image indicates the different steps of the method, from top to bottom. Left part of the diagram also shows the different sets of variables computed at each specific step. Before we discuss each component in further detail, we will briefly summarize our workflow diagram (the reader is kindly referred to Fig. 1 for a visual explanation of the different steps):

- *Complexity Parameters*: it refers to the free variables of the model. In this problem, they are the data parameters, the knots and the poles. At the initialization step, they are randomly chosen and then labelled as $\mathcal{T}_0$, $\mathcal{U}_0$, and $\mathcal{P}_0$, respectively. They are the initial input of our data fitting problem, along with the given data points.
- *SA-LSQ*: in this step, we compute the model parameterization by means of the SA algorithm introduced in this paper. The obtained data parameters at step j of our iterative process, $\mathcal{T}_j$, are used as input data for the kAvgKnots method to

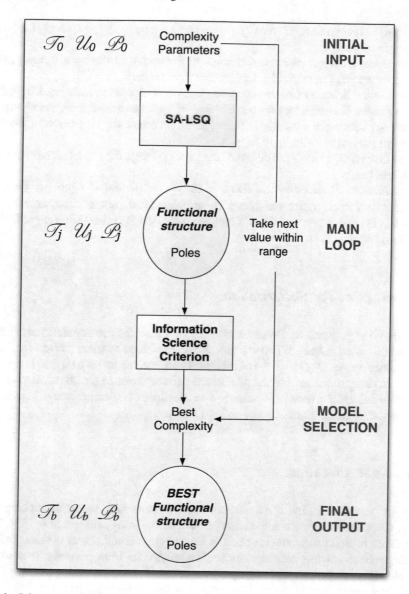

Fig. 1 Schematic workflow of our method (see the main text for further details)

obtain the breakpoints, $\mathcal{U}_j$. Finally, we compute the poles, $\mathcal{P}_j$, by solving the least-squares system (4).

- *Functional structure and poles*: it refers to the output of the current step j to be used as input for next step $j + 1$ of our iterative process.
- *Information Sciences Criterion*: at this step, we compute the associated Bayesian Information Criterion for each model. This will give us a good indicator about the trade-off between the quality of fitting and the complexity of the model, so that we can prevent over-fitting to happen.
- *Best Complexity*: Using the output of the previous step, the model with the lowest *BIC* is chosen.
- *Best Functional Structure and Poles*: at the end of the iterative process, the best values for the sets of free variables of the problem are computed. They are labelled as $\mathcal{T}_b$, $\mathcal{U}_b$, and $\mathcal{P}_b$, respectively. This yields the best B-spline fitting curve for the given data points.

4.2 B-spline Parameterization

The B-spline parameterization is computed by means of the simulated annealing variant proposed in Sect. 3. The solution encoding is done by using a M-dimensional real-valued vector $\mathcal{T} \in (0, 1)^M$ with the first and last values set to 0 and 1, respectively. In order to assure that the parameterization is done in a convenient way, i.e. a strictly increasing vector, we impose a set of linear constrains derived from the parameterization structure: $\mathbf{t}_i < \mathbf{t}_{i+1}, \forall i \neq \{1, M\}$.

4.3 Knot Allocation

Given a parameterization $\mathcal{T}$, we search for a suitable knot vector by making use of an adaptive averaging method based on the methodology presented in [33]. The main idea behind this method is that the knot vector should reflect the distribution of the parameters while keeping control of its length. To do so, we define in advance the number of inner knots needed, then we select an appropriate segmentation of the parameter vector, and finally we apply the knot averaging technique from [33]. The knot allocation procedure is summarized in Algorithm 2.

The averaging method has been successfully used in many data fitting problems [33]. Its mains advantages are the easiness of computation and a resulting acceptable knot sequence that follows the distribution of the data. The i-th term of the k averaging knot sequence from input vector $\mathcal{T}$, is constructed as follows:

$$\mathbf{u}_i = \frac{1}{k-1} \sum_{j=i+1}^{i+k-1} \mathbf{t}_j \tag{12}$$

Algorithm 2: kAverageKnots Knot allocation procedure

Input: Parameterization $\mathcal{T}$, spline order p, number of inner knots k
Output: Knot vector $\mathcal{U}$
Get the number of segments:
$nSeg \leftarrow k + p - 1$;
Get index of partition of unity in nSeg:
$tIndex \leftarrow$ round(linearPartition$(0, 1, nSeg)$);
Filter $\mathcal{T}$:
$\mathcal{T}^* \leftarrow \mathcal{T}[tIndex]$;
Compute the averaging knot sequence over $\mathcal{T}^*$:
$\mathcal{U} \leftarrow$ averageKnots$(\mathcal{T}^*, p)$;
return $\mathcal{U}$

---

where $\mathcal{T} = \{t_j\}_j$. Note that, in order to guarantee the end point interpolation condition, we append p equal knots at both extremes. Without loss of generality, we can assume that $[\alpha, \beta] = [0, 1]$. Therefore, for the problem at hand, this means appending zeros and ones at the left and right end of the knot vector, respectively.

4.4 Control Net Computation

The spline poles are computed by solving the least-squares problem defined in Eq. (4). Note that this system of equations is numerically solvable by means of traditional least-squares methods as it becomes an over-constrained linear system once the data parameters and knot vector are known.

More precisely, the system given by Eq. (4) can be rewritten in the following matrix form:

$$\mathbf{Q} = \mathbf{F} \cdot \mathbf{P} \tag{13}$$

where $\mathbf{Q} = \text{vec}\left(\{\mathbf{Q}_i\}_i\right)$, $\mathbf{P} = \text{vec}\left(\{\mathbf{P}_j\}_j\right)$ and $\mathbf{F}$ is constructed by the column-wise stacking of the B-spline basis functions of order p, with knot vector $\mathcal{U}$, and evaluated at $\mathcal{T}$. Therefore, given that $M \gg n$ and the only set of unknowns in Eq. (13) are the poles, it is indeed an over-constrained linear system, which can be trivially transformed as:

$$\mathbf{F}^T \cdot \mathbf{Q} = \mathbf{F}^T \cdot \mathbf{F} \cdot \mathbf{P}$$

which can be solved by classical least-squares methods. In this work we have opted for the *SVD* decomposition, by means of the Moore-Penrose pseudo-inverse, denoted by $\cdot^\dagger$, of $\mathbf{F}$:

$$\mathbf{P} = \mathbf{F}^\dagger \cdot \mathbf{Q} \tag{14}$$

4.5 Model Selection

In order to maintain a good compromise between data-fidelity and complexity we make use the Bayesian Information Criterion, given by Eq. (15):

$$BIC = \xi \log(E) + d \log(\xi) \tag{15}$$

where ξ denotes the total number of parameters to be fitted, E represents the fitting error (given by the residual sum of squares, RSS) and d accounts for the total number of free parameters of the proposed model. From Eq. (15) we can see that, for a fixed error, the BIC penalizes models with higher complexity, whereas for a given d, the criterion favors those models with higher fidelity.

Note that, in our problem, on one hand the length of the parameter vector is fixed as it is equal to the number of points in the dataset. On the other hand, the number of control points is fixed for a given order p, and k inner knots. Thus, the number of inner knots uniquely determines the model complexity.

5 Experimental Results

In this section we present the evaluation of the proposed Lévy flight-driven simulated annealing-based method for B-spline curve fitting over a set of three illustrative examples. All the examples exhibit challenging features from a geometrical point of view, such as self-intersections and strong changes of slope and curvature. All the fitted datasets correspond to real-world instances: the first one represents a famous logo digitized with noise of medium intensity, while the other two examples are different views of the same dataset: the silhouette of a cat with a high-density uniform sampling and a irregular low-density sampling, respectively.

Regarding the parameter values used in our simulations, the employed SA parameter setup is as follows: the thermal equilibrium cycle runs for $N_{inner} = 50$ iterations. For the logo dataset we choose $\sigma \in \{1, \ldots, 40\}$, while for the other examples we employ $\sigma \in \{1, \ldots, 300\}$. For all examples in our benchmark we consider $\lambda = 1.5, s_0 = 0.1$ as the Lévy flight coefficients and $\chi_0 = 0.8$ for the initial acceptance probability. Note that χ_0 is the only parameter needed to automatically compute the initial temperature. For each example we compute the centripetal parameterization and run the knot averaging method for the given σ range, then we select the model with the lowest BIC (BIC_{best}) and we run our SA implementation for the chosen σ_{best}, starting from a random feasible point, until improving the BIC_{best}. Once this threshold is exceeded, the stop/frozen indicator is raised, marking the end of the algorithm. The optimization phase has been carried out 26 times, discarding the three best and worst results, in order to provide statistical evidence for the results presented and assert the experiment reproducibility. Note that the non-optimization phase is completely stochastic-free, therefore, one run for each σ in the chosen range is enough.

5.1 Example 1: A Famous Logo

The first dataset in our benchmark represents a noisy scanned famous logo, which is also affected by a non-uniform sampling. In addition, the example exhibits some difficult geometrical features, such as several auto-intersections and changes of concavity. However our method is able to reconstruct the overall shape of the curve as shown in Fig. 2. In this figure, the data points are represented by × symbols in black. The picture on the left shows the fitting B-spline curve with a deterministic parameterization, while the picture on the right shows the fitting B-spline curve with the optimized parameterization obtained with our method. They are represented by solid lines in blue and magenta colors, respectively. As the reader can see, the optimized parameterization presents a better visual quality. This is also confirmed by our numerical results, reported in Table 1 (see our discussion about the numerical results in Sect. 5.4).

5.2 Example 2: High-Density Silhouette of a Cat

In this example we reconstruct the silhouette of a cat, sampled by a high density dataset with subtle changes on the curvature and some sharp peaks around the ears. Our experimental results are graphically depicted in Fig. 3, where the meaning of the pictures in this figure is similar to those in Fig. 2. As shown in Fig. 3, due to the high density and uniform sampling, the cases of deterministic parameterization and optimized parameterization are almost visually indistinguishable from each other for this example. This fact is also confirmed by very similar numerical values for both cases in Table 1.

Fig. 2 Best fitting B-spline curve for the famous logo example: (left) with a deterministic parameterization; (right) with the optimized parameterization of our method

Table 1 Summary table of our numerical results for the three examples in the paper. See the main text for the meaning of the symbols in the left column

	Example I	*Example II*	*Example III*
nparam	380	1062	402
dim	188	529	199
σ	31	188	128
	Deterministic parameterization		
BIC	−2677.9178	1316.7934	815.8298
RSS	0.0000	0.0025	0.0011
xNRMSE	0.9919	0.9987	0.9986
yNRMSE	0.9925	0.9983	0.9981
	Optimized parameterization		
BIC	−2682.5495	1221.2876	661.9920
RSS	0.0000	0.0023	0.0008
xNRMSE	0.9917	0.9986	0.9986
yNRMSE	0.9927	0.9984	0.9986

Fig. 3 Best fitting B-spline curve for the high-density silhouette of a cat: (left) with a deterministic parameterization; (right) with the optimized parameterization of our method

5.3 *Example 3: Low-Density Silhouette of a Cat*

The dataset of this example presents the same geometrical features as the previous high-density one, as it corresponds simply to a different sampling of the same original silhouette. The low-density scenario provides a challenge at its own, as there are fewer points to represent the geometrical features and hence, capture the subtle details of this shape. Furthermore, the non-uniform sampling adds additional difficulties when searching for the optimal parameterization and its associated breakpoint sequence.

Figure 4 shows the graphical results for the best (in the BIC sense) deterministic (left) and optimized (right) approximation for the low-density cat dataset. As we

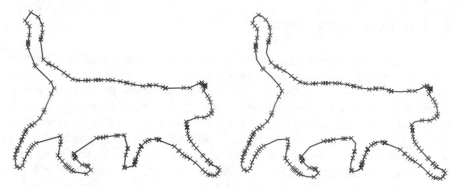

Fig. 4 Best fitting B-spline curve for the low-density silhouette of a cat: (left) with a deterministic parameterization; (right) with the optimized parameterization of our method

can see, both methods can reconstruct with good accuracy the shape of the data. However, it is clear that the optimized version (on the right) is more visually pleasant, as it is smoother and more polished. By contrast, the picture on the left shows some strongly linear parts (compare, for instance, the end parts of the two central legs as well as the end part of the cat tail, for two illustrative examples of this visual appearance in both cases).

5.4 Numerical Results

Although the figures above are helpful to figure out the goodness of our method, its accuracy should still be determined numerically. Table 1 summarizes the numerical results for all the reconstructed examples. The three examples in our benchmark are arranged in columns. For each example, we report (in rows): the total number of parameters, *nparam*, the dimension of the search space, *dim*, the best value of σ parameter (in the BIC sense), and the values of the Bayesian Information Criterion score (BIC), the residual sum of squares of the errors (RSS) and normalized root-mean-square error (NRMSE). The latter value is shown for each coordinate. Two sets of values are reported for each example, corresponding to the deterministic parameterization and the optimized parameterization, respectively.

5.5 Computational Issues

All the computations have been carried out in an `Intel i7-6700` quad core processor with 16 GB of RAM. The source code has been implemented by the authors in the native programming language of MATLAB, version 2014b. We have make use of the COBYLA implementation in the `NLopt` library [52]. Our implementation of the Lévy flight follows Mantegna's algorithm [51] as described in [6].

6   Conclusions and Future Work

In this paper we have presented a methodology for data fitting with local-support curves that merges classical least-squares and knot allocation techniques with a modern metaheuristic approach. Our method computes the spline parameterization by means of a new variant of the thermodynamics-driven simulated annealing method. Instead of using a single candidate distribution function, our approach maintains two sets of distributions: on one hand, new candidates are typically generated with a temperature-sensitive Cauchy distribution, which was explored in a previous article from the authors for rational Bézier surface reconstruction [53]. On the other hand, in order to escape from local minima and energy plateaus, a particle is allowed to take a Lévy flight (a random walk controlled by the Lévy distribution). Promising regions are explored by means of the derivative-free COBYLA algorithm, as previously outlined in the memetic simulated annealing (MeSA) used by the authors in [54].

This technique has been tested against a set of data fitting problems that present interesting features from the optimization and the geometrical points of view. The results are quite promising, as our technique is able to capture with great precision the overall shape of the data, while maintaining the complexity of the model under control. Furthermore, we can produce better results, from an optimization point of view, than those obtained with other parameterization methods, although with higher computational efforts. In this regard, we conclude that it is better to put the computational effort in the computation of the knot vector, by making use of deterministic parameterizations; only in case that a threshold on the fitting score must be met, it is recommended to apply the proposed method to further optimize the parameterization.

Further research in this topic includes the design of new variants of the simulated annealing approach well suited for obtaining the optimal data parameterization and knot vector at once, or using a sequential schema, based in [42], where the authors constructed two schemes for data fitting with rational Bézier curves. We are also interested to analyze the extension of the current method to the case of B-spline surfaces. This problem becomes more difficult, not only because we have now to deal with duplicated sets of free variables (data parameterization for two independent variables, two knot vectors, control points arranged in matrices instead of arrays) but also because they are still related to each other in a nonlinear way. In addition, the cases of organized and non-organized data points must be addressed independently. Depending on the particular problem, some kind of clustering might also be required to obtain a preliminary arrangement of data points. From it, a base surface providing a coarse fitting could be obtained and, then, iteratively refined for higher accuracy. We also aim at carrying out a complete analysis about the SA parameter setup so that we could determine, for instance, how the desired acceptance ratio χ_0 does actually influence the convergence of the method and many other interesting open questions.

Acknowledgements This research has been kindly supported by the Computer Science National Program of the Spanish Ministry of Economy and Competitiveness, Project Ref. #TIN2012-30768, Toho University (Funabashi, Japan), and the University of Cantabria (Santander, Spain). The authors are particularly grateful to the Department of Information Science of Toho University for all the facilities given to carry out this work. We also thank the anonymous reviewers who helped us to improve the chapter with their constructive comments and suggestions. A special recognition is also owe to our editor, Prof. Xin-She Yang, for his kind assistance and encouraging support during the process of writing this chapter.

References

1. Farin, G.: Curves and Surfaces for CAGD, 5th edn. Morgan Kaufmann, San Francisco, CA, USA (2002)
2. Varady, T., Martin, R.R., Cox, J.: Reverse engineering of geometric models—an introduction. Comput. Aided Des. **29**(4), 255–268 (1997)
3. Jupp, D.L.B.: Approximation to data by splines with free knots. SIAM J. Numer. Anal. **15**, 328–343 (1978)
4. Kennedy, J., Eberhart, R.C., Shi, Y.: Swarm Intelligence. Morgan Kaufmann Publishers, San Francisco (2001)
5. Engelbretch, A.P.: Fundamentals of Computational Swarm Intelligence. Wiley, Chichester, England (2005)
6. Yang, X.S.: Nature-Inspired Metaheuristic Algorithms, 2nd edn. Luniver Press, Frome, UK (2010)
7. Yang, X.S.: Engineering Optimization: An Introduction with Metaheuristic Applications. Wiley, NJ (2010)
8. Gálvez, A., Iglesias, A., Cobo, A., Puig-Pey, J., Espinola, J.: Bézier curve and surface fitting of 3D point clouds through genetic algorithms, functional networks and least-squares approximation. Lect. Notes Comput. Sci. **4706**, 680–693 (2007)
9. Gálvez, A., Iglesias, A.: Firefly algorithm for polynomial Bézier surface parameterization. J. Appl. Math. Article ID 237984, 9 pages (2013)
10. Zhao, L., Jiang, J., Song, C., Bao, L., Gao, J.: Parameter optimization for Bézier curve fitting based on genetic algorithm. In: Advances Swarm Intelligence, pp. 451–458 (2013)
11. Gálvez, A., Iglesias, A.: Cuckoo search with Lévy flights for weighted Bayesian energy functional optimization in global–support curve data fitting. Sci. World J. 11 pages, Article ID 138760 (2014)
12. Loucera, C., Gálvez, A., Iglesias, A.: Simulated annealing algorithm for Bezier curve approximation. In: Proceedings of Cyberworlds. pp. 182–189, IEEE Computer Society Press, Los Alamitos, CA (2014)
13. Iglesias, A., Gálvez, A., Collantes, M.: Global–support rational curve method for data approximation with bat algorithm. In: Proceedings of International Conference Artificial Intelligence and Applications, AIAI'2015, Bayonne (France). IFIP Advances in Information and Communication Technology, vol. 458, pp. 191–205 (2015)
14. Iglesias, A., Gálvez, A.: Memetic firefly algorithm for data fitting with rational curves. In: Congress Evolutionary Computation–CEC'2015, Sendai (Japan). pp. 507–514, IEEE CS Press, CA (2015)
15. Iglesias, A., Gálvez, A., Avila, A.: Hybridizing mesh adaptive search algorithm and artificial immune systems for discrete rational Bzier curve approximation. Vis. Comput. **32**(3), 393–402 (2016)
16. Iglesias, A., Gálvez, A., Collantes, M.: Four adaptive memetic bat algorithm schemes for Bézier curve parameterization. Trans. Comput. Sci. **28**, 127–145 (2016)

17. Yoshimoto, F., Moriyama, M., Harada, T.: Automatic knot adjustment by a genetic algorithm for data fitting with a spline. In: Proceedings of Shape Modeling International'99, IEEE Computer Society Press, 162–169 (1999)
18. Yoshimoto, F., Harada, T., Yoshimoto, Y.: Data fitting with a spline using a real-coded algorithm. Comput. Aided Des. **35**, 751–760 (2003)
19. Park, H.: An error-bounded approximate method for representing planar curves in B-splines. Comput. Aided Geom. Des. **21**, 479–497 (2004)
20. Park, H., Lee, J.H.: B-spline curve fitting based on adaptive curve refinement using dominant points. Comput. Aided Des. **39**, 439–451 (2007)
21. Ulker, E., Arslan, A.: Automatic knot adjustment using an artificial immune system for B-spline curve approximation. Inf. Sci. **179**, 1483–1494 (2009)
22. Gálvez, A., Iglesias, A.: Efficient particle swarm optimization approach for data fitting with free knot B-splines. Comput. Aided Des. **43**(12), 1683–1692 (2011)
23. Gálvez, A., Iglesias, A., Puig-Pey, J.: Iterative two-step genetic-algorithm method for efficient polynomial B-spline surface reconstruction. Inf. Sci. **182**(1), 56–76 (2012)
24. Gálvez, A., Iglesias, A.: Particle swarm optimization for non-uniform rational B-spline surface reconstruction from clouds of 3D data points. Inf. Sci. **192**(1), 174–192 (2012)
25. Gálvez, A., Iglesias, A.: A new iterative mutually-coupled hybrid GA-PSO approach for curve fitting in manufacturing. Appl. Soft Comput. **13**(3), 1491–1504 (2013)
26. Ulker, E.: B-Spline curve approximation using Pareto envelope-based selection algorithm-PESA. Intl. J. Comput. Commun. Eng. **2**(1), 60–63 (2013)
27. Gálvez, A., Iglesias, A.: Firefly algorithm for explicit B-Spline curve fitting to data points. Math. Prob. Eng. Article ID 528215, 12 pages (2013)
28. Gálvez, A., Iglesias, A., Avila, A., Otero, C., Arias, R., Manchado, C.: Elitist clonal selection algorithm for optimal choice of free knots in B-spline data fitting. Appl. Soft Comput. **26**, 90–106 (2015)
29. Gálvez, A., Iglesias, A.: Elitist clonal selection algorithm for optimal choice of free knots in B-spline data fitting. Particle-based meta-model for continuous breakpoint optimization in smooth local-support curve fitting. Appl. Math. Comput. **275**, 195–212 (2016)
30. Gálvez, A., Iglesias, A.: New memetic self-adaptive firefly algorithm for continuous optimization. Intl. J. Bio-Inspired Comput. **8**(5), 300–317 (2016)
31. Iglesias, A., Gálvez, A.: Nature-Inspired Swarm Intelligence for Data Fitting in Reverse Engineering: Recent Advances and Future Trends. In: Yang, X.S. (ed.) Nature-Inspired Computation in Engineering. Studies in Computational Intelligence, Vol. 637, pp. 151–175 (2016)
32. Powell, M.J.D.: A direct search optimization method that models the objective and constraint functions by linear interpolation. In: Gómez, S., Hennart, J.P. (eds.) Advances in Optimization and Numerical Analysis, pp. 51–67. Springer, Netherlands (1994)
33. Piegl, L., Tiller, W.: The NURBS Book. Springer, Berlin Heidelberg (1997)
34. de Boor, C.A.: Practical Guide to Splines. Springer (2001)
35. Laurent-Gengoux, P., Mekhilef, M.: Optimization of a NURBS representation. Comput. Aided Des. **25**(11), 699–710 (1993)
36. Kirkpatrick, S., Gelatt, C.D., Vecchi, M.P.: Optimization by simulated annealing. Science **220**(4598), 671–680 (1983)
37. Dowsland, K.A., Thompson, J.M.: Simulated annealing. In: Rozenberg, G., Bäck, T., Kok, J.N. (eds.) Handbook of Natural Computing, pp. 1623–1655. Springer, Berlin, Heidelberg (2012)
38. Yang, X.S., Deb, S.: Cuckoo search via Lévy flights. In: Proceedings World Congress on Nature & Biologically Inspired Computing (NaBIC). pp. 210–214, IEEE (2009)
39. Iglesias, A., Gálvez, A.: Cuckoo search with Lévy flights for reconstruction of outline curves of computer fonts with rational Bézier curves. In: Congress Evolutionary Computation–CEC'2016, Vancouver (Canada). pp. 2247–2254, IEEE CS Press, CA (2016)
40. Jensi, R.: Jiji, G W.: An enhanced particle swarm optimization with Lévy flight for global optimization. Appl. Soft Comput. **43**, 248–261 (2016)
41. Yang, X.S..: Flower pollination algorithm for global optimization. In: Proc. Int. Conf. on Unconventional Computing and Natural Computation. Springer, pp. 240–249 (2012)

42. Iglesias, A., Gálvez, A., Loucera, C.: Two simulated annealing optimization schemas for rational Bézier curve fitting in the presence of noise. Math. Prob. Eng. 13 pages, Article ID 351648 (2014)
43. Vanderbilt, D., Louie, S.G.: A Monte Carlo simulated annealing approach to optimization over continuous variables. J. Comput. Phys. **56**(2), 259–271 (1984)
44. Bohachevsky, I.O., Johnson, M.E., Stein, M.L.: Generalized simulated annealing for function optimization. Technometrics **28**(3), 209–217 (1986)
45. Szu, H., Hartley, R.: Fast simulated annealing. Phys Lett. A **122**(3), 157–162 (1987)
46. Ingber, L.: Adaptive simulated annealing (ASA): lessons learned. Control cybern. **25**, 33–54 (1996)
47. Johnson, D.S., Aragon, C.R., McGeoch, L.A., Schevon, C.: Optimization by simulated annealing: An experimental evaluation; part I, graph partitioning. Oper. Res. **37**(6), 865–892 (1989)
48. Johnson, D.S., Aragon, C.R., McGeoch, L.A., Schevon, C.: Optimization by simulated annealing: An experimental evaluation; part II, graph coloring and number partitioning. Oper. Res. **39**(3), 378–406 (1991)
49. Locatelli, M.: Simulated annealing algorithms for continuous global optimization. In: Handbook of Global Optimization, pp. 179–229. Springer (2002)
50. Metropolis, N., Rosenbluth, A.W., Rosenbluth, M.N., Teller, A.H., Teller, E.: Equation of state calculations by fast computing machines. J. Chem. Phys. **21**(6), 1087 (1953)
51. Mantegna, R.N.: Fast, accurate algorithm for numerical simulation of Lévy stable stochastic processes. Phys. Rev. E **49**(5), 4677–4683 (1994)
52. Johnson, S.: The NLopt nonlinear-optimization package. http://ab-initio.mit.edu/nlopt
53. Loucera, C., Iglesias, A., Gálvez, A.: Simulated annealing and natural neighbor for rational Bézier surface reconstruction from scattered data points. In: International Conference on Harmony Search Algorithm, pp. 354–364. Springer (2017)
54. Loucera, C., Iglesias, A., Gálvez, A.: Memetic simulated annealing for data approximation with local-support curves. Procedia Comput. Sci. 1364–1373 (2017)

A Comprehensive Review of the Flower Pollination Algorithm for Solving Engineering Problems

Aylin Ece Kayabekir, Gebrail Bekdaş, Sinan Melih Nigdeli and Xin-She Yang

Abstract Engineering optimization problems are often solved by using meta-heuristic algorithms. Flower pollination algorithm (FPA) is a nature-inspired metaheuristic algorithm and FPA have been used in a variety of engineering problems. In this book chapter, the engineering applications of FPA and its variants are reviewed, and the applications include chemical engineering, civil engineering, energy and power systems, mechanical engineering, electronical and communication engineering, computer science and others. Further research topics are also outlined.

Keywords Flower pollination algorithm · Optimization · Metaheuristic methods · Civil engineering · Nonlinear optimization

1 Introduction

A good engineering design must consider all important issues such as economy, safety, performance, sustainability, manufacturability, energy efficiency, environment, utilization and architecture. The consideration of one or several of these issues is not enough. All issues must be fully considered. However, it is very challenging to consider all these issues and conventional design methods may

A.E. Kayabekir · G. Bekdaş (✉) · S.M. Nigdeli
Department of Civil Engineering, Istanbul University, 34320 Avcılar, Istanbul, Turkey
e-mail: bekdas@istanbul.edu.tr

A.E. Kayabekir
e-mail: ecekayabekir@gmail.com

S.M. Nigdeli
e-mail: melihnig@istanbul.edu.tr

X.-S. Yang
School of Science and Technology, Middlesex University, London NW4 4BT, UK
e-mail: x.yang@mdx.ac.uk

© Springer International Publishing AG 2018
X.-S. Yang (ed.), *Nature-Inspired Algorithms and Applied Optimization*,
Studies in Computational Intelligence 744, https://doi.org/10.1007/978-3-319-67669-2_8

struggle to cope. Even approximate solutions may not be easy to obtain. Nowadays, new methods and alternative solution techniques are often used.

The consideration of different issues and design requirements, design problems often becomes highly nonlinear. In addition, the initial design preference of an engineer plays a great role in the following stages of the design. Therefore, designs can be iterative, and design stages can be iteratively analysed by using numerical algorithms at least approximately.

An optimization problem with a single or multiple (N) objective functions (f_i for $i = 1, 2, ..., N$)) can be formulated as follows:

$$\text{Minimize } f_i(x), x \in \mathcal{R}^n, (i = 1, 2, \ldots N) \tag{1}$$

The objective functions are subjected to the design constraints which can be J equalities ($h_j(x)$) and/or K inequalities ($g_k(x)$) as shown in Eqs. (2) and (3).

$$h_j(x), (j = 1, 2, \ldots J), \tag{2}$$

$$g_k(x) \leq 0, (k = 1, 2, \ldots K), \tag{3}$$

where x is the set of design variables for the design problem. For a problem with n design variables, we can write them as a vector:

$$x = (x_1, x_2, \ldots x_n)^T, \quad (i = 1, 2, \ldots n) \tag{4}$$

The objective functions are generally related to the cost of the design, but safety, usability and architecture issues can be also put into the formulations. The design constraints are generally related with safety consideration according to the physics of the engineering problems and more often according to the design codes. The architectural and usability issues can be also considered as design constraints, but these issues are generally considered as the solution ranges of design variables so as to constrain the generation of possible optimum solutions.

This chapter is organized as follows. In Sect. 2, the flower pollination algorithm (FPA) is briefly explained. Then in the third section, applications of FPA are reviewed for different disciplines of engineering. The last section concludes with some suggestions.

2 Flower Pollination Algorithm

Flowering plants reproduce by pollination in nature, which is the transfer process of pollen by pollinators such as insects, birds, bats, other animals or winds. Sometimes, flower constancy exists when a specialised flower-pollinator partnership exists. Pollination has two major types. Approximately 90% of flowering plants reproduce via biotic pollination, and the rest reproduce by abiotic pollination.

By formulizing four rules of flower pollination, Yang developed a nature-inspired metaheuristic algorithm called flower pollination algorithm (FPA) [1]. FPA has also been extended by Yang et al. for multiobjective optimization problems [2].

Flower pollination characteristics and idealization:

Rule 1   Biotic and cross-pollination for global pollination.
Rule 2   Abiotic and self-pollination for local pollination.
Rule 3   Flower constancy.
Rule 4   A switch probability (p) controlling global and local pollination.

Biotic and cross-pollination generally occur at a long distance because pollinators can fly long distances. The flight of pollinators behaves as Lévy flight behaviour. Thus, the global pollination can be formulized by using a Lévy distribution to draw random step sizes (L) as Eq. (5) by using Rules 1 and 3.

$$x_i^{t+1} = x_i^t + L(x_i^t - g^*), \quad (i = 1, 2, \dots m) \tag{5}$$

In Eq. (5), for a design variable or a set of design variables (x), the solution of $(t + 1)$th iteration (new solution x_i^{t+1}) is generated by modifying the solution of tth iteration (existing solution x_i^t). The subscript i represents the solution of ith flower in a population of m flowers (population size m). The best existing solution in terms of objective function is denoted as g^*.

The local pollination is effective for the convergence of the solutions. Thus, two random flowers (jth and kth) are chosen and a linear distribution (ε) is used as seen in Eq. (6). The second and third rules are formulised in the local pollination.

$$x_i^{t+1} = x_i^t + \varepsilon(x_j^t - x_k^t), \quad (i = 1, 2, \dots m). \tag{6}$$

The fourth rule can be done as the probability of using the global and local pollination. At the start of the algorithm, the initial values are randomly chosen according to a solution range defined for the design variables. The comparison of the generated values of $(t + 1)$th iteration with (t)th iteration values is done with respect to the optimization objective. The results are only updated if the new solutions are better than the existing ones.

Dubey et al. [3] modified FPA by employing a scaling factor (F) to control the mutation of flowers and using an additional intensive exploitation phase. The local pollination is modified in order to increase the convergence of the method as seen in Eq. (7).

$$x_i^{t+1} = x_i^t + F(x_j^t - x_k^t), \quad (i = 1, 2, \dots m) \tag{7}$$

After the round of global and local pollination, an intensive exploitation of the best flower is done as seen in Eq. (8).

$$x_i^{t+1} = g^* + H\left(\varepsilon_1 - [(\varepsilon_2 - \varepsilon_3)g^*]\right) \tag{8}$$

Here, H is a control parameter which is calculated as Eq. (9). ε_1, ε_2, ε_3 and ε_4 are random numbers that are uniformly distributed between 0 and 1.

$$H = \begin{cases} 1, & if \ \varepsilon_4 < p, \\ 0, & otherwise. \end{cases} \tag{9}$$

Bibiks et al. [4] modified FPA for solving combinatorial optimization problems using discrete variables. The core concepts of FPA such as flower, global pollination, Lévy flight and local pollination are modified for resource constrained project scheduling problems. Namachivayam et al. [5] modified FPA in order to enhance the local and global searching abilities by including the local neighbourhood searching strategy and the dynamic switching probability strategy.

3 Applications of Flower Pollination Algorithm in Engineering

In this section, the engineering applications of FPA and its variant of are reviewed by grouping relevant applications as chemical engineering, civil engineering, energy and power systems, mechanical engineering, electronical and communication engineering, computer science and other engineering applications.

3.1 Chemical Engineering

Since FPA is a very recent algorithm, developed in 2010, the studies concerning chemical engineering are limited. Even so, there are quite a few studies. Merzougui et al. used FPA for parameter identification in liquid-liquid equilibrium modelling of food-related thermodynamic systems [6]. Other than the classical FPA, a modified flower pollination algorithm presented by Dubey et al. was employed [3]. According to the results of different numerical scenarios with and without the application of closure equations, modified FPA outperformed the classical algorithm and other heuristics such as Simulated Annealing, Genetic algorithm and Harmony Search.

Sheata et al. applied several metaheuristic algorithms including FPA for performing critical point calculations in multicomponent reservoir fluids in petroleum industry [7]. The optimizers have been compared by using black oil, volatile oil and condensate reservoir fluids with fifty components and they concluded that FPA is one of the effective algorithms in this field. Zainudin et al. developed a hybrid algorithm by combining FPA and Taguchi design in optimizing the shrinkage of triaxial porcelain containing palm oil fuel ash (POFA) and found that the shrinkage is dominantly dependent on the sintering temperature followed by POFA composition, moulding pressure, POFA particle size and soaking time [8]. In addition, Narong et al. used POFA as the cement filter for enhancing the electromagnetic interference absorption of cement-based composites and optimization of the electromagnetic interference shielding was done by employing FPA [9]. The prediction results obtained by using FPA shows comparable results with the experimental studies.

3.2 Civil Engineering

Civil engineering is one of the important areas of applied optimization since design problems in this area are highly nonlinear with stringent complex constraints. In additional to costs, design constraints resulting from the architectural, feasible and physical requirements often generate a complex engineering problem. Generally speaking, civil engineering problems are closely related with structural engineering. For this reason, structural optimization is also counted as a type of optimization in civil engineering. In this area, several basic structural mechanics problems have been used as benchmark examples for metaheuristic algorithms in the literature. The problems such as pin jointed plane frame optimization, a three bars truss system optimization, vertical deflection minimization problem of an I-beam, cost optimization of tubular column under compressive load and weight optimization of cantilever beams have been investigated by employing FPA [10]. Also, two hybrid FPA including the combination of FPA with simulated annealing [11] and the shuffled frog-leaping algorithm [12] have been proposed for the basic structural optimization problems.

The optimization of truss structures is the best known fundamental application of structural engineering. Bekdaş et al. [13] employed flower pollination algorithm for sizing optimization of planar and space frames by proposing a handling process for stress and displacement constraints. FPA is a competitive algorithm for truss structures according to the comparison of previously developed methods. Generally, the truss structural member are grouped in order to shorten the optimization effort and preventing to trap a local optima. Bekdaş et al. [14] investigated the sizing optimization of truss structures without grouping the members by using FPA.

The spatial 72-bar truss structure shown in Fig. 1 is a well-known benchmark exercise of the optimization algorithms. The maximum displacement constraint of joints is ±0.25 in for all coordinate directions, while the stress constraint is ±25 ksi. The range of design variables which are the cross sectional areas of elements, are between 0.1 and 3.0 in^2. The truss is subjected to the two independent loading cases (Table 1) and the design constraints are separately considered for all loading cases.

The optimum values of the corresponding group numbers for structural members are presented in Table 2. The number of flowers was taken as 30 and the switch probability was taken as 0.5. The results were investigated for 10000 iterations.

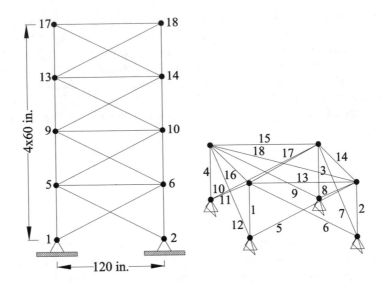

Fig. 1 The spatial 72-bar truss structure [13]

Table 1 Multiple load cases for 72-bar truss [13]	Case	Node	P_x (kips)	P_y (kips)	P_z (kips)
	1	17–20	−5.0	−5.0	−5.0
	2	17	5.0	5.0	−5.0

In Table 2, the optimum results for the methods employing ant colony optimization (ACO) [15], big bang–big crunch algorithm (BB-BC) [16], modified teaching–learning based optimization (TLBO) [17], chaotic swarming of particles (CSP) [18], colliding bodies optimization (CBO) [19], ray optimization (RO) [20] and FPA [13] are shown. It is clear that FPA outperforms the other methods in minimizing the weight of the truss structure and the number of structural analyses needed to find the optimum result.

The optimum tuning of mass dampers is an important optimization problem since the effectiveness of such mass dampers depends on the right tuning of mass dampers for structures, subject to the dynamic vibrations resulting from wind, earthquake and traffic excitations. Nigdeli et al. [21] employed FPA for optimum design of tuned mass dampers (TMDs) for seismic structures and the optimization objectives are related to time-domain solutions. In addition, a hybrid method using harmony search and FPA was developed by Nigdeli et al. [22] in order to find more effective solutions than classical algorithms for optimum TMD design. Additionally, Bekdaş et al. [23] employed FPA for TMD optimization problem by using frequency domain solutions as the objective function and good results were obtained.

For example, the objective of the optimization of TMD in the frequency (ω) domain is to minimize the top story acceleration transfer function of the structure.

Table 2 Optimization results for the 72-bar truss problem

Element group	Members	ACO	BB-BC	TLBO	CSP	CBO	RO	FPA
1	1–4	1.9480	1.9042	1.8807	1.9446	1.90280	1.8365	1.8758
2	5–12	0.5080	0.5162	0.5142	0.5026	0.51800	0.5021	0.5160
3	13–16	0.1010	0.1000	0.1000	0.1000	0.10010	0.1000	0.1000
4	17–18	0.1020	0.1000	0.1000	0.1000	0.10030	0.1004	0.1000
5	19–22	1.3030	1.2582	1.2711	1.2676	1.27870	1.2522	1.2993
6	23–30	0.5110	0.5035	0.5151	0.5099	0.50740	0.5033	0.5246
7	31–34	0.1010	0.1000	0.1000	0.1000	0.10030	0.1002	0.1001
8	35–36	0.1000	0.1000	0.1000	0.1000	0.10030	0.1002	0.1000
9	37–40	0.5610	0.5178	0.5317	0.5067	0.52400	0.5730	0.4971
10	41–48	0.4920	0.5214	0.5134	0.5165	0.51500	0.5499	0.5089
11	49–52	0.1000	0.1000	0.1000	0.1075	0.10020	0.1004	0.1000
12	53–54	0.1070	0.1007	0.1000	0.1000	0.10150	0.1001	0.1000
13	55–58	0.1560	0.1566	0.1565	0.1562	0.15640	0.1576	0.1575
14	59–66	0.5500	0.5421	0.5429	0.5402	0.54940	0.5222	0.5329
15	67–70	0.3900	0.4132	0.4081	0.4223	0.40290	0.4356	0.4089
16	71–72	0.5920	0.5756	0.5733	0.5794	0.55040	0.5972	0.5731
Best weight (lb)		380.240	379.660	379.632	379.970	379.6943	380.458	379.095
Average weight (lb)		383.160	381.850	379.759	381.560	379.8961	382.5538	379.534
Standard deviation on optimized weight (lb)		3.66	1.201	0.149	1.803	0.0791	1.2211	0.272
Number of structural analyses		18500	13200	21542	10500	15600	19084	9029

The design variables are the parameters of TMD positioned on the top of structures as seen on the top of the shear building model given as Fig. 2.

The parameters of TMD are mass, stiffness and damping coefficients shown as m_d, k_d and c_d, respectively. The design variables of the optimization are mass (m_d), period (T_d) and damping ratio (ξ_d) of TMD which are formulated as follows:

$$T_d = 2\pi \sqrt{\frac{m_d}{k_d}} \tag{10}$$

$$\xi_d = \frac{c_d}{2m_d \sqrt{\frac{k_d}{m_d}}} \tag{11}$$

The transfer function (TF) is a dimensionless value which is the ratio of Laplace transformations of the top story acceleration and the ground acceleration. In application, the peak value of the transfer function representing the resonance state is minimized. The TF formulation for all freedoms and the objective function (f) in desiBel (dB) are as follows:

Fig. 2 System model of
multi-story building structure
with single TMD [23]

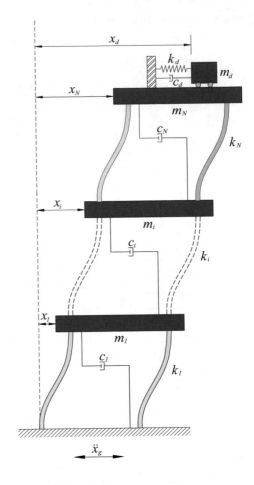

$$TF(w) = \begin{bmatrix} TF_1(\omega) \\ TF_2(\omega) \\ \vdots \\ TF_N(\omega) \\ TF_d(\omega) \end{bmatrix} = \left[-M\omega^2 + C\omega j + K \right]^{-1} M\omega^2 \mathbf{1} \qquad (12)$$

$$f = 20 Log_{10} |\max(TF_N(\omega))| \qquad (13)$$

M, C, and K (Eqs. 14–16) represent the mass, damping and stiffness matrices of a shear structure with a TMD and m_i, c_i and k_i are mass, damping coefficient, stiffness coefficients of i_{th} story, respectively. A unity vector with all entries being ones is represented with **1**.

$$M = \text{diag}[m_1, m_2 \ldots m_N, m_d] \tag{14}$$

$$C = \begin{bmatrix} (c_1+c_2) & -c_2 & & & & \\ -c_2 & (c_2+c_3) & -c_3 & & & \\ & & \cdot & \cdot & & \\ & & & \cdot & \cdot & \cdot \\ & & & & \cdot & \cdot \\ & & & -c_N & (c_N+c_d) & -c_d \\ & & & & -c_d & c_d \end{bmatrix} \tag{15}$$

$$K = \begin{bmatrix} (k_1+k_2) & -k_2 & & & & \\ -k_2 & (k_2+k_3) & -k_3 & & & \\ & & \cdot & \cdot & & \\ & & & \cdot & \cdot & \cdot \\ & & & & \cdot & \cdot \\ & & & -k_N & (k_N+k_d) & -k_d \\ & & & & -k_d & k_d \end{bmatrix} \tag{16}$$

The optimum results of the designed TMD for a 10-story structure with properties given in Table 3 are presented in Table 4. FPA, harmony search (HS) and teaching learning based optimization (TLBO) are employed and 20 independent runs are conducted. The optimum results were searched for 50000 function evaluations. The number of flowers are 25 in FPA and 25 learners in TLBO, while the switch probability is 0.5. The parameters of HS are 5, 0.5 and 0.2 for harmony memory size, harmony memory considering rate and pitch adjusting rate, respectively.

The effect of the optimally designed TMD can be clearly seen in the TF plot of the top story of the structure given as Fig. 3. As seen from the results, FPA has the best results and with least computation effort comparing to the others (see Table 4). A hybrid FPA was employed for groutability estimation of grouting process which is an efficient approach for ground improvement related to a sub-discipline of civil engineering called geotechnical engineering. By combining FPA with differential

Table 3 The properties of the example building [24]	Story	m_i (t)	k_i (MN/m)	c_i (MNs/m)
	1	179	62.47	0.81
	2	170	52.26	0.67
	3	161	56.14	0.72
	4	152	53.02	0.68
	5	143	49.91	0.64
	6	134	46.79	0.60
	7	125	43.67	0.56
	8	116	40.55	0.52
	9	107	37.43	0.48
	10	98	34.31	0.44

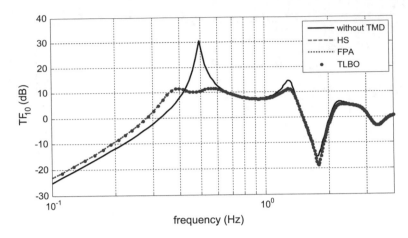

Fig. 3 TF plot of the top story of the structure [23]

Table 4 Optimum values for TMD optimization problem [23]

	HS	FPA	TLBO
m_d (t)	138.5	138.5	138.5
T_d (s)	2.2926	2.2917	2.2917
ξ_d	0.2782	0.2763	0.2763
Best TF	11.7316	11.7303	11.7303
Average TF	11.7322	11.7303	11.7303
Standard derivative	3.15×10^{-4}	1.05×10^{-6}	5.55×10^{-10}
Number of analyses to find the optimum result	16529	9125	49500

evolution, the hybrid algorithm was used to optimize the input factor selection and hyper-parameter tuning process of the support vector machine based groutability prediction model by Hoang et al. [25].

3.3 Energy and Power Systems

The solution of economic dispatch problems in modern systems is an important optimization application of power systems and biologically inspired algorithms are more effective comparing to gradient-based algorithms. Dubey et al. [3] modified FPA by using a scaling factor for the local pollination and adding an intensive exploitation phase to solve four large practical power systems test cases. Prathiba et al. [26] investigated the economic load dispatch by employing FPA in order to minimize the fuel cost and tested the method on a bus system. Lenin et al. [27]

proposed the hybridization of FPA with chaotic harmony search algorithm to solve the reactive power dispatch optimization problem. For the comparison of results of bus test system, the hybrid method is effective than the compared ones with the help of the chaotic sequences.

Abdelaziz et al. [28] employed FPA for optimal sizing and locations of capacitors in radial distribution systems such as bus radial distribution systems and the effectiveness of the method was proved in means of minimizing of loses, total cost, enhancing the voltage profile and net saving. Also, Abdelaziz et al. [29] investigated economic load dispatch and dual-objective combined economic emission dispatch considering the accumulation from emission of gaseous pollutants of fossil-fuelled power plants and employed FPA in order to solve six power system problems. The results show that FPA outperforms the other swarm intelligent algorithm even for large scale power systems. In another FPA employing study of Abdelaziz et al. [30], the most candidate buses for installing capacitors are suggested by using loss sensitivity factors.

In order to optimize the horizontal axis tidal current turbines, metaheuristic algorithm such as the second version of non-dominated sorting genetic algorithm (NSGA-II), multi objective particle swarm optimization, multiobjective cuckoo search algorithm and multiobjective FPA are combined with the blade element momentum theory by Tahani et al. [31]. According to the results, the Pareto fronts achieved by multiobjective FPA and NSGA-II have better quality than the others. Tahani et al. [32] used hybrid FPA/Simulated Annealing algorithm in optimization of Photovoltaic (PV)/Wind Battery stand-alone systems.

The influence of PV panels tilt angle was considered for the wind speed by using computational fluid dynamics simulation. Tahani et al. [33] developed a new heuristic method by combining FPA, grey wolf optimizer and elephant herding optimization in order to optimize the straight blade vertical axis of wind turbines. The proposal is effective on enlarging the average velocity magnitude around the optimized blade, momentum and power coefficient. Mahtad and Srairi [34] proposed a flexible power system planning strategy using a new adoptive partitioning flower pollination algorithm for solving the security optimal power flow considering faults at critical generating unit. Shilaja and Ravi [35] developed a new methodology based on combined emission economic dispatch for PV plants and thermal power generation units. The methodology uses the variants of FPA called Euclidean affine FPA and binary FPA. Artificial Bee Colony and FPA were combined by Ram et al. [36] to generate a new hybrid bee pollinator flower pollination algorithm for solar PV parameter estimation. The hybrid method has faster execution speed in evaluation with the compared algorithms.

In addition, Dubey et al. [37] hybridised FPA with differential evaluation and combined the hybrid algorithm with time-varying fuzzy selection mechanism for solving wind integrated multi-objective dynamic economic dispatch problem of power systems. The developed method effectively searches the best compromise solution to satisfy the three objectives related with total operating cost, emission content and power loss. Namachivanyam et al. [5] employed a modified flower pollination algorithm for reconfiguration and capacitor placement of radial distribution systems because the common methods of reducing power loss and improving voltage profile are network reconfiguration, shunt capacitor placement, distributed generation and high-voltage distribution systems. Xu and Wang [38] incorporated FPA with the Nelder-Mead simplex method and the generalized opposition-based learning mechanism for parameter estimation of PV modules. The different solar cell models such as the single diode model, the double diode model and a PV module was tested in order to show the effectiveness of the algorithm.

Furthermore, FPA based methodologies have been used for controller design of power systems. Abdelaziz and Ali [39] employed FPA for robust tuning of a static VAR compensator damping controller to reduce power system oscillations. Jagatheesan et al. [40] developed a FPA based approach for optimum tuning of Proportional-Integral Derivative (PID) controllers in load frequency control of multi-area interconnected power systems by estimating the controller parameters such as proportional gain, integral time derivative time. The FPA based method shows better performance than genetic algorithm and particle swarm optimization based method for power systems with and without non-linearity effect. Dash et al. [41] optimally tuned Proportional Integral-Proportional Derivative (PI-PD) controllers by FPA for automatic generation control of multi area power system.

On the other hand, FPA has been used with neural networks in order to predict the crude oil price in Dubai by Chiroma et al. [42] and the weights and bias of neural network were optimized.

3.4 Mechanical Engineering

Several basic mechanical engineering optimization problems such as speed reducer, gear train, tension/compression spring design and pressure vessel have been investigated by employing two hybrid FPA algorithms combined with frog leaping local search [12] and simulated annealing [11]. Abdel-Baset and Hezam [11] also investigated optimum heat exchanger design, corrugated bulk head optimum design, optimum welded beam design and PID controller tuning for step response.

Kaviranyani and Kumar [43] employed FPA for stabilization of the rotary inverted pendulum system and the method minimizes the loss occurred because of time delay.

In production and manufacturing industry, the optimization of multi-pass turning parameters were done by using an improved FPA. The improvement contains the initialization of population by using good point set and Deb's heuristic rules [44]. In addition, in production, the optimization of process of transmission laser welding of dissimilar plastics was done by using FPA and response surface methodology [45].

3.5 Electronical and Communication Engineering

Metaheuristic methods have been often employed in the wireless communication systems. Chakravarthy and Rao [46] developed a FPA based method to position nulls of circular array antennas and FPA based method needs less computation time comparing to genetic algorithm. Chakravarthy et al. [47] implemented FPA to synthesis of circular array antenna for side lobe level aperture size control. By using Cauchy based global pollination, enhanced local pollination and dynamic switch probability, FPA was improved by Singh and Salgotra [48] to solve pattern synthesis of linear antenna arrays.

Shukla and Singh Bhandari [49] employed FPA for optimization problem of spectrum sensing in cognitive radios. FPA is an effective approach for spectrum sensing since it has a good convergence rate.

Mahata et al. [50] employed the hybrid algorithm of Abdel-Raouf and Abdel-Baset [51] combining particle swarm optimization and FPA for the design of wideband infinite impulse response digital differentiators and digital integrators.

Sharawi et al. [52] used FPA in the proposal of a wireless sensor network (WSN) energy aware clustering formation model. The objective of method to obtain a global optimum for WSN lifetime. The FPA based method sustains an effective balance power utilization of sensor nodes and lifetime extension of WSN comparing to classical approach. Hajjej et al. [53] developed an FPA based method to find the best nodes deployment with maximal convergence in a wireless sensor network. The method outperforms classical forms of particle swarm optimization and genetic algorithm.

3.6 Computer Science

In image compression, FPA was employed by Kaur et al. [54] in order to decrease the search complexity of matching between range block and domain block. Ouadfel and Taleb-Ahmed [55] employed FPA for multilevel image thresholding problem. Wang et al. [56] solved a planar graph colouring problem by using a variant of FPA which uses local greedy strategies such as local swap operator and local sub-sequence reverse operator. Zhou et al. [57] employed FPA on the optimization process of shape matching problem based on atomic potential matching model and the previous approaches are outperformed. Emary et al. [58] presented an automated retinal blood vessels segmentation approach employing FPA which searches the optimum clustering of retinal image into compact clusters under some constraints. A binary FPA in which the search space is an n-dimensional Boolean lattice updating the solution across the corners of hypercube is employed by Rodrigues et al. [59] for electro encephalogram signal based person identification.

In solving Sudoku puzzles, FPA combined with chaotic harmony search outperformed the classical harmony search in means of minimum numbers of iterations [60].

Data clustering is also an active area of metaheuristic algorithms. Jensi and Jiji [61] proposed a hybrid approach combining K-Means algorithm and FPA. The hybrid method finds optimal cluster centres since the F-measure value is increased. Wang et al. [62] proposed FPA with bee pollinators for clustering in date analysis and data mining technique. Comparing to classical metaheuristic and K-Means algorithm, the results of the numerical experiments proves the effectiveness of the hybrid FPA method on accuracy and stability.

3.7 Other Engineering Applications

FPA has been modified and employed for the discrete problems of project scheduling which is constrained by resources. According to the results, FPA outperformed several classical metaheuristic algorithms [4].

Lazim et al. [63] obtained and improved polygon simplification methodology which is a type of cartographic generation. The FPA based improved method outperformed the standard simplification procedure in computing time.

An improved variant of FPA was developed by Zhou and Wang [64] in order to enhance the search ability for solving optimum path planning of unmanned undersea vehicles and the improved FPA was generated by using three strategies such as particle swarm optimization in local search, dimension evaluation and improvement strategy and dynamic switching probability strategy. Zhang et al. [65] proposed a novel model with combination of complete ensemble empirical mode decomposition adaptive noise, FPA with chaotic local search, five neural networks and no negative constraints theory for short-term wind speed forecasting. The novel method is effective in high-precision wind speed predictions.

4 Conclusions

As we have seen from the above reviews, FPA has been used in the development and solution of a wider range of engineering design problems. Generally speaking, the classical form of FPA outperforms the classical and several modifications of other metaheuristic algorithms. In order to enhance the computational capacity and preventing to local optima, FPA has been modified or combined with other algorithms in development of hybrid methods and these hybrids often obtained the best results in a variety of applications.

Further research opportunities exist in many areas. For example, discrete and combinatorial optimization is an area that requires more case studies. In addition, the application of FPA in data mining such as feature selection and classifications can be very useful. In structural and shape optimization, FPA can be applied to even larger-scale problems and in combination with more complex design evaluation tools such as finite element methods. Furthermore, wireless sensor networks and smart homes with smart sensors can be an important area for further research.

References

1. Yang, X.S.: Flower pollination algorithm for global optimization. In: International Conference on Unconventional Computing and Natural Computation, pp. 240–249. Springer, Berlin (2012)
2. Yang, X.S., Karamanoglu, M., He, X.: Flower pollination algorithm: a novel approach for multiobjective optimization. Eng. Optim. 46(9), 1222–1237 (2014)
3. Dubey, H.M., Pandit, M., Panigrahi, B.K.: A biologically inspired modified flower pollination algorithm for solving economic dispatch problems in modern power systems. Cogn. Comput. 7(5), 594–608 (2015)
4. Bibiks, K., Li, J.P., Hu, F.: Discrete flower pollination algorithm for resource constrained project scheduling problem. Int. J. Comput. Sci. Inf. Secur. 13(7), 8 (2015)
5. Namachivayam, G., Sankaralingam, C., Perumal, S.K., Devanathan, S.T.: Reconfiguration and capacitor placement of radial distribution systems by modified flower pollination algorithm. Electric Power Compon. Syst. 44(13), 1492–1502 (2016)
6. Merzougui, A., Labed, N., Hasseine, A., Bonilla-Petriciolet, A., Laiadi, D., Bacha, O.: Parameter identification in liquid-liquid equilibrium modeling of food-related thermodynamic systems using flower pollination algorithms. Open Chem. Eng. J. 10(1), 59–73 (2016)
7. Shehata, M.N., Fateen, S.E.K., Bonilla-Petriciolet, A.: Critical point calculations of multi-component reservoir fluids using nature-inspired metaheuristic algorithms. Fluid Phase Equilib. 409, 280–290 (2016)
8. Zainudin, A., Sia, C.K., Ong, P., Narong, O.L.C., Nor, N.H.M.: Taguchi design and flower pollination algorithm application to optimize the shrinkage of triaxial porcelain containing palm oil fuel ash. In: IOP Conference Series: Materials Science and Engineering, vol. 165, no. 1, p. 012036. IOP Publishing (2017)
9. Narong, L.C., Sia, C.K., Yee, S.K., Ong, P., Zainudin, A., Nor, N.H.M., & Kasim, N.A.: Optimization of the EMI shielding effectiveness of fine and ultrafine POFA powder mix with OPC powder using flower pollination algorithm. In: IOP Conference Series: Materials Science and Engineering, vol. 165, no. 1, p. 012035. IOP Publishing (2017)
10. Nigdeli, S.M., Bekdaş, G., Yang, X.-S.: Application of the flower pollination algorithm in structural engineering. In: Yang X.-S., Bekdaş G., Nigdeli S.M. (eds.) Metaheuristics and Optimization in Civil Engineering, pp. 25–43. Springer (2016)
11. Abdel-Baset, M., Hezam, I.: A hybrid flower pollination algorithm for engineering optimization problems. Int. J. Comput. Appl. 140(12), 10–23 (2016)
12. Meng, O.K., Pauline, O., Kiong, S.C., Wahab, H.A., Jafferi, N.: Application of modified flower pollination algorithm on mechanical engineering design problem. In: IOP Conference Series: Materials Science and Engineering, vol. 165, no. 1, p. 012032. IOP Publishing (2017)
13. Bekdaş, G., Nigdeli, S.M., Yang, X.S.: Sizing optimization of truss structures using flower pollination algorithm. Appl. Soft Comput. 37, 322–331 (2015)
14. Bekdas, G., Nigdeli, S.M., Yang, X.S.: Size optimization of truss structures employing flower pollination algorithm without grouping structural members. Int. J. Theor. Appl. Mech. 1, 269–273 (2017)
15. Camp, C.V., Bichon, B.J.: Design of space trusses using ant colony optimization. J. Struct. Eng. 130(5), 741–751 (2004)

16. Kaveh, A., Talatahari, S.: Size optimization of space trusses using Big Bang-Big Crunch algorithm. Comput. Struct. **87**(17), 1129–1140 (2009)
17. Camp, C.V., Farshchin, M.: Design of space trusses using modified teaching–learning based optimization. Eng. Struct. **62**, 87–97 (2014)
18. Kaveh, A., Sheikholeslami, R., Talatahari, S., Keshvari-Ilkhichi, M.: Chaotic swarming of particles: a new method for size optimization of truss structures. Adv. Eng. Softw. **67**, 136–147 (2014)
19. Kaveh, A., Mahdavi, V.R.: Colliding bodies optimization method for optimum design of truss structures with continuous variables. Adv. Eng. Softw. **70**, 1–12 (2014)
20. Kaveh, A., Khayatazad, M.: Ray optimization for size and shape optimization of truss structures. Comput. Struct. **117**, 82–94 (2013)
21. Nigdeli, S.M., Bekdas, G., Yang, X.S.: Optimum tuning of mass dampers for seismic structures using flower pollination algorithm. Int. J. Theor. Appl. Mech. **1**, 264–268 (2017)
22. Nigdeli, S.M., Bekdaş, G., Yang, X.-S.: Optimum tuning of mass dampers by using a hybrid method using harmony search and flower pollination algorithm. In: Del Ser J. (ed.) Harmony Search Algorithm. Advances in Intelligent Systems and Computing, vol. 514, pp. 222–231. Springer (2017)
23. Bekdaş, G., Nigdeli, S.M., Yang, X.-S.: Metaheuristic based optimization for tuned mass dampers using frequency domain responses. In: Del Ser J. (ed.) Harmony Search Algorithm. Advances in Intelligent Systems and Computing, vol. 514, pp. 271–279. Springer (2017)
24. Sadek, F., Mohraz, B., Taylor, A.W., Chung, R.M.: A method of estimating the parameters of tuned mass dampers for seismic applications. Earthq. Eng. Struct. Dynam. **26**(6), 617–636 (1997)
25. Hoang, N.D., Bui, D.T., Liao, K.W.: Groutability estimation of grouting processes with cement grouts using differential flower pollination optimized support vector machine. Appl. Soft Comput. **45**, 173–186 (2016)
26. Prathiba, R., Moses, M.B., Sakthivel, S.: Flower pollination algorithm applied for different economic load dispatch problems. Int. J. Eng. Technol. (IJET) **6**(2), 1009–1016 (2014)
27. Lenin, K., Ravindhranath, R.B., Surya, K.M.: Shrinkage of active power loss by hybridization of flower pollination algorithm with chaotic harmony search algorithm. Control Theory Inform. **4**, 31–38 (2014)
28. Abdelaziz, A.Y., Ali, E.S., Elazim, S.A.: Optimal sizing and locations of capacitors in radial distribution systems via flower pollination optimization algorithm and power loss index. Eng. Sci. Technol. Int. J. **19**(1), 610–618 (2016)
29. Abdelaziz, A.Y., Ali, E.S., Elazim, S.A.: Combined economic and emission dispatch solution using flower pollination algorithm. Int. J. Electr. Power Energy Syst. **80**, 264–274 (2016)
30. Abdelaziz, A.Y., Ali, E.S., Elazim, S.A.: Flower pollination algorithm and loss sensitivity factors for optimal sizing and placement of capacitors in radial distribution systems. Int. J. Electr. Power Energy Syst. **78**, 207–214 (2016)
31. Tahani, M., Babayan, N., Astaraei, F.R., Moghadam, A.: Multi objective optimization of horizontal axis tidal current turbines, using Meta heuristics algorithms. Energy Convers. Manag. **103**, 487–498 (2015)
32. Tahani, M., Babayan, N., Pouyaei, A.: Optimization of PV/wind/battery stand-alone system, using hybrid FPA/SA algorithm and CFD simulation, case study: Tehran. Energy Convers. Manag. **106**, 644–659 (2015)
33. Tahani, M., Babayan, N., Mehrnia, S., Shadmehri, M.: A novel heuristic method for optimization of straight blade vertical axis wind turbine. Energy Convers. Manag. **127**, 461–476 (2016)
34. Mahdad, B., Srairi, K.: Security constrained optimal power flow solution using new adaptive partitioning flower pollination algorithm. Appl. Soft Comput. **46**, 501–522 (2016)
35. Shilaja, C., Ravi, K.: Optimization of emission/economic dispatch using euclidean affine flower pollination algorithm (eFPA) and binary FPA (BFPA) in solar photo voltaic generation. Renew. Energy **107**, 550–566 (2017)

36. Ram, J.P., Babu, T.S., Dragicevic, T., Rajasekar, N.: A new hybrid bee pollinator flower pollination algorithm for solar PV parameter estimation. Energy Convers. Manag. **135**, 463–476 (2017)
37. Dubey, H.M., Pandit, M., Panigrahi, B.K.: Hybrid flower pollination algorithm with time-varying fuzzy selection mechanism for wind integrated multi-objective dynamic economic dispatch. Renew. Energy **83**, 188–202 (2015)
38. Xu, S., Wang, Y.: Parameter estimation of photovoltaic modules using a hybrid flower pollination algorithm. Energy Convers. Manag. **144**, 53–68 (2017)
39. Abdelaziz, A.Y., Ali, E.S.: Static VAR compensator damping controller design based on flower pollination algorithm for a multi-machine power system. Electric Power Compon. Syst. **43**(11), 1268–1277 (2015)
40. Jagatheesan, K., Anand, B., Samanta, S., Dey, N., Santhi, V., Ashour, A.S., Balas, V.E.: Application of flower pollination algorithm in load frequency control of multi-area interconnected power system with nonlinearity. Neural Comput. Appl. 1–14 (2016)
41. Dash, P., Saikia, L.C., Sinha, N.: Flower pollination algorithm optimized PI-PD cascade controller in automatic generation control of a multi-area power system. Int. J. Electr. Power Energy Syst. **82**, 19–28 (2016)
42. Chiroma, H., Abdul-kareem, S., Khan, A., Abubakar, A.I., Muaz, S.A., Gital, A.Y.U., Shuib, L.M.: Bio-inspired algorithm optimization of neural network for the prediction of Dubai crude oil price. In: Second International Conference on Advanced Data and Information Engineering (DaEng-2015), Bali, Indonesi, April 25–26
43. Kavirayani, S., Kumar, G.V.: Flower pollination for rotary inverted pendulum stabilization with delay. Telkomnika **15**(1), 245–253 (2017)
44. Xu, S., Wang, Y., Huang, F.: Optimization of multi-pass turning parameters through an improved flower pollination algorithm. Int. J. Adv. Manuf. Technol. **89**(1–4), 503–514 (2017)
45. Acherjee, B., Maity, D., Kuar, A.S.: Parameters optimisation of transmission laser welding of dissimilar plastics using RSM and flower pollination algorithm integrated approach. Int. J. Math. Model. Numer. Optim. **8**(1), 1–22 (2017)
46. Chakravarthy, V., Rao, P.M.: On the convergence characteristics of flower pollination algorithm for circular array synthesis. In: 2015 2nd International Conference on Electronics and Communication Systems (ICECS), pp. 485–489. IEEE (2015)
47. Chakrravarthy, V., Chowdary, P.S., Rao, P.M., Panda, G.: Synthesis of circular array antenna for sidelobe level and aperture size control using flower pollination algorithm. Int. J. Antennas Propag. **2015** (2015). Article ID 819712
48. Singh, U., Salgotra, R.: Pattern synthesis of linear antenna arrays using enhanced flower pollination algorithm. Int. J. Antennas Propag. **2017** (2017). Article ID 7158752
49. Shukla, S., Bhandari, A.S.: Cooperative spectrum sensing in cognitive radio using flower pollination optimization algorithm. Int. J. Eng. Trends Technol. (IJETT) **37**(3), 169–174 (2016)
50. Mahata, S., Saha, S.K., Kar, R., Mandal, D.: Optimal design of wideband digital integrators and differentiators using hybrid flower pollination algorithm. Soft Comput. 1–27 (2017)
51. Abdel-Raouf, O., Abdel-Baset, M.: A new hybrid flower pollination algorithm for solving constrained global optimization problems. Int. J. Appl. Oper. Res. Open Access J. **4**(2), 1–13 (2014)
52. Sharawi, M., Emary, E., Saroit, I.A., El-Mahdy, H.: Flower pollination optimization algorithm for wireless sensor network lifetime global optimization. Int. J. Soft Comput. Eng. **4**(3), 54–59 (2014)
53. Hajjej, F., Ejbali, R., Zaied, M.: An efficient deployment approach for improved coverage in wireless sensor networks based on flower pollination algorithm. In: Natarajan M., et al. (eds.) NETCOM, NCS, WiMoNe, GRAPH-HOC, SPM, CSEIT, pp. 117–129 (2016)
54. Kaur, G., Singh, D., Kaur, M.: Robust and efficient 'RGB' based fractal image compression: flower pollination based optimization. Int. J. Comput. Appl. **78**(10), 11–15 (2013)

55. Ouadfel, S., Taleb-Ahmed, A.: Social spiders optimization and flower pollination algorithm for multilevel image thresholding: a performance study. Expert Syst. Appl. **55**, 566–584 (2016)
56. Wang, R., Zhou, Y., Zhou, Y., Bao, Z.: Local greedy flower pollination algorithm for solving planar graph coloring problem. J. Comput. Theor. Nanosci. **12**(11), 4087–4096 (2015)
57. Zhou, Y., Zhang, S., Luo, Q., Wen, C.: Using flower pollination algorithm and atomic potential function for shape matching. Neural Comput. Appl. 1–20 doi:10.1007/s00521-016-2524-0
58. Emary, E., Zawbaa, H. M., Hassanien, A. E., Tolba, M. F., & Snášel, V. (2014). Retinal vessel segmentation based on flower pollination search algorithm. In: Proceedings of the Fifth International Conference on Innovations in Bio-Inspired Computing and Applications IBICA 2014, pp. 93–100. Springer International Publishing (2014)
59. Rodrigues, D., Silva, G.F., Papa, J.P., Marana, A.N., Yang, X.S.: EEG-based person identification through binary flower pollination algorithm. Expert Syst. Appl. **62**, 81–90 (2016)
60. Abdel-Raouf, O., El-Henawy, I., Abdel-Baset, M.: A novel hybrid flower pollination algorithm with chaotic harmony search for solving sudoku puzzles. Int. J. Modern Educ. Comput. Sci. **6**(3), 38–44 (2014)
61. Jensi, R., Jiji, G.W.: Hybrid data clustering approach using K-means and flower pollination algorithm. Adv. Comput. Intell.: Int. J. (ACII) **2**(2), 15–25 (2015)
62. Wang, R., Zhou, Y., Qiao, S., Huang, K.: Flower pollination algorithm with bee pollinator for cluster analysis. Inf. Process. Lett. **116**(1), 1–14 (2016)
63. Lazim, D., Zain, A.M., Omar, A.H.: Polygon simplification improved with flower pollination algorithm (FPA). Indian J. Sci. Technol. **9**(48), 1–5 (2016)
64. Zhou, Y., Wang, R.: An improved flower pollination algorithm for optimal unmanned undersea vehicle path planning problem. Int. J. Pattern Recogn. Artif. Intell. **30**(04), 1659010 (2016)
65. Zhang, W., Qu, Z., Zhang, K., Mao, W., Ma, Y., Fan, X.: A combined model based on CEEMDAN and modified flower pollination algorithm for wind speed forecasting. Energy Convers. Manag. **136**, 439–451 (2017)

Bat Algorithm and Directional Bat Algorithm with Case Studies

Asma Chakri, Haroun Ragueb and Xin-She Yang

Abstract In recent years, the Bat Algorithm (BA) is becoming a standard optimization tool used by scientists and engineers to solve many problems in different engineering fields. One of the most important characteristics of the bat algorithm is its easy, comprehensible structure which simplifies the computer implementation, in addition to its ability to obtain reliable results for low dimensional problems. As the problem complexity increases, several studies pointed out that premature convergence may occur when the algorithm may get trapped at a local optimum. To overcome this without losing the main BA characteristics (simplicity and reliability), the directional echolocation has been introduced to the mainframe of BA to become what is known as the directional Bat Algorithm (dBA). In this paper, we discuss the main features of the dBA and their contributions in improving the exploitation and exploration capabilities of the standard BA. We also analyze the performance of dBA in optimizing unimodal and multimodal functions in addition to a constrained engineering problem. The results are compared with those obtained by BA and also a new competitive improved BA version, namely the Novel Bat Algorithm (NBA). The ANOVA one way analysis has demonstrated the superiority of the directional bat algorithm.

Keywords Bat Algorithm · Directional bat algorithm · Echolocation · Optimization · Nature-inspired algorithm · Swarm intelligence

---

A. Chakri (✉)
Industrial Mechanics Laboratory, Department of Mechanical Engineering,
University Badji Mokhtar of Annaba (UBMA), BP12-23000 Annaba, Algeria
e-mail: chakri.as623@gmail.com

H. Ragueb
Energy and Mechanical Engineering Laboratory, Department of Mechanical Engineering,
Faculty of Engineering Sciences, University M'hamed Bougara of Boumerdes (UMBB)
Avenue of Independence, 35000 Boumerdes, Algeria

X.-S. Yang
School of Science and Technology, Middlesex University London, The Burroughs,
London NW4 4BT, UK

1 Introduction

To find their way even in a complete darkness, bats use sophisticated echolocation to map their surrounding environment. By emitting a short pulse of sound waves and then listening to their echoes, they can distingue prey from objects and dangerous predators. Based on this behavior, Xin-She Yang [1] developed a new optimization algorithm, called the Bat Algorithm (BA). BA falls into the same category of algorithms, called swarm intelligence, such as Particle Swarm Optimization (PSO) [2], and Ant Colony Optimization (ACO) [3]. BA uses a population of bats for the search of the global optimum. Soon after its appearance in the literature, BA starts to attract the attention of several researchers around the world due to its two major characteristics. The first one, BA is highly efficient and reliable in the search of the global optimum for low dimensional problems. The second is its easy structure. BA is so easy to implement that it can be programmed using any computer languages under a few dozen lines of codes.

The bat algorithm has been used to solve several engineering problems [4]. It was used to optimize the brushless DC wheel motor [5], sizing battery for energy storage [6], power system stabilizer [7, 8] and power dispatch [9]. Moreover, researchers also found various applications of BA in many disciplines such as the path planning of uninhabited combat air vehicle (UCAV) [10], structural damage detection [11], fault diagnosis [12], image processing [13, 14], and others such as flow shop scheduling [15] or simply planning sports training sessions [16]. However, as the problems' complexity increases, the algorithm's performance may show some premature convergence [17, 18]. This premature convergence of the algorithm may be due to the lack in the exploration ability. To overcome this deficiency, some researchers have proposed several improvements with the aim to enhance the standard BA's performance for general optimization use [19–22], while others modified BA to fit for certain specific tasks such as the traveling salesman problem [23], large-scale truss structures [24], structural reliability [25], heart attack detection [26], micro-grid management [27] and others [28–30]. These improved variants of BA were built using different techniques like hybridization, adaptation, bio-inspiration and others that will be discussed later in this chapter.

One of the prominent variants of BA is the directional Bat Algorithm (dBA) proposed by Chakri et al. [31]. The key idea of this algorithm is to use the directional echolocation with other modifications to improve the exploration and the exploitation capabilities of the bat algorithm. The newly proposed algorithm was tested on several complex benchmarks and the results were compared with those of 20 other standard and sophisticated algorithms including 6 improved variants of BA. The non-parametric statistical tests showed the superiority of the directional bat algorithm. This algorithm was successfully applied to solve probabilistic constrained problems, typical to structural reliability based design optimization field [32]. In this chapter, we will explore more the working system of dBA, and how the proposed improvements can enhance the exploration and the exploitation ability and how to adjust their proportion during the optimization process.

In the next section, a brief description of the standard bat algorithm followed by a detailed review of different BA variants in the current literature. We will present and discuss the dBA properties in Sect. 4. After that, we conduct a series of tests on benchmark problems and comparisons with standard BA and new improved variant called the Novel Bat Algorithm (NBA). We will use the ANOVA One Way test to examine the performances of the three algorithms. Finally, we conclude in Sect. 6.

2 Description of the Standard BA

The standard Bat Algorithm is a swarm-intelligence-based algorithm, developed by Xin-She Yang [1], inspired by the echolocation process of microbats. The bats are masters of sensory, with their large ears, they can detect the bounced sound waves they have emitted, and process the echo signals to create a mental configuration of their environment in a similar way to the sonar. This behavior, called echolocation, enables the bats to fly freely and with the ability to detect food or prey. From observations, Yang [1] developed the standard BA using three major idealized rules:

- All bats use echolocation to sense distance and the location of a bat x_i is encoded as a solution vector to an optimization problem under consideration [1].
- Bats fly randomly with velocity v_i at position x_i with a varying frequency (from a minimum f_{min} to a maximum frequency f_{max}) and loudness A to search for prey. They can automatically adjust the frequencies of their emitted pulses and the rate of pulse emission r depending on the proximity of the target [1].
- Loudness varies from a large positive value A_0 to a minimum constant value A_{min} [1], while pulse emission rate r varies from a lower constant value to a higher value.

For more details on the bats behavior and characteristics during roaming and foraging, the readers can refer to the original work of Yang [1].

From the implementation point of view, a bat's motion is governed by two modes of flight. The first mode (or we can call it a global step) is the guided flight mode in which all bats are directed toward the bat with the best location (that is the solution with the best fitness value). Thus, for the i-th bat at a location, x_i, with velocity, v_i, in a d-dimensional search space, the rules for updating its location (solution) and velocity are given as in [1]:

$$f_i = f_{min} + (f_{max} - f_{min})\beta \tag{1}$$

$$v_i^{t+1} = v_i^t + (x^* - x_i^t)f_i \tag{2}$$

$$x_i^{t+1} = x_i^t + v_i^{t+1} \tag{3}$$

where $\beta \in [0,1]$ is a random vector drawn from a uniform distribution, and x^* is the current global best solution found so far (the best bat/location). As it can be seen from Eq. (2), the motion of the bats is subject to the information from the best bat. By directing the bats to the best one, it enables them to exploit more the possessed information and seek for a better solution.

The second mode of flight is what we call the local search step. A new location for each bat is generated locally using the following updating equation:

$$x_{new} = x_{old} + \varepsilon <A^{t+1}> \tag{4}$$

where $\varepsilon \in [-1,1]$ is a random number, while $<A_i^{t+1}>$ is the average loudness of all the bats at time t.

The control of the auto-switch between the first and the second mode of flight is obtained by the tuning of two parameters, namely, the loudness A_i and the rate of pulses emission r_i (or the pulse rate). These parameters are updated during the iterations process, the loudness decreases while the pulse rate increases as the bat gets closer to its prey. The equation for updating the loudness and the pulse rate are:

$$A_i^{t+1} = \alpha A_i^t \tag{5}$$

$$r_i^{t+1} = r_i^0[1 - \exp(-\gamma t)] \tag{6}$$

where $0 < \alpha < 1$ and $\gamma > 0$ are constants. As $t \to \infty$, we have $A_i^t \to 0$ and $r_i^t \to r_i^0$. Yang [1] proposed that the initial loudness A_0 can be $A_0 \in [1, 2]$, while the initial pulse rate $r_0 \in [0, 1]$. The pseudocode of the standard bat algorithm is summarized in the pseudocode as shown in Algorithm 1.

Algorithm 1.

The standard bat algorithm.

01.	Define the objective function
02.	Initialize the bat population $L_i \le x_i \le U_i$ (i=1,2,..,n) and v_i
03.	Define frequencies f_i at x_i
04.	Initialize pulse rates r_i and loudness A_i
05.	*While* ($t \le t_{max}$)
06.	Adjust frequency, Eq. (1).
07.	Update velocities, Eq. (2).
08.	Update locations/solutions, Eq. (3).
09.	*if* ($rand > r_i$)
10.	Generate a local solution around the selected best solution, Eq. (4).
11.	*end if*
12.	*if* ($rand < A_i$ & $F(x_i) < F(x^*)$)
13.	Accept the new solutions
14.	Reduce A_i, Eq. (5).
15.	Increase r_i, Eq. (6).
16.	*end if*
17.	Rank the bats and find the current best x^*
28.	*end while*
29.	Results processing

3 Survey on the BA Improvements

There are quite a lot of techniques that have been used by researchers to improve the bat algorithms. The most popular method is to hybridize BA with other metaheuristic algorithms in order to overcome the exploration deficiency with techniques borrowed from the other algorithms. Another method consists of using adaptive parameters for a better control of the balance between exploration and exploitation in the search process. Some authors suggested to increase the randomness in the bats movements by either transforming the search space into other spaces (i.e., binary space, complex space), or using chaotic sequences.

Other authors revised the bats behavior for bio-inspired improvements. For a comprehensive review of BA's variants, we have categorized the improved variants of BA into six categories: the hybrid variants, adaptive parameters variants, search space alteration, chaotic variants, bio-inspired variants and others. This classification is not definitive as some BA version can be identified in more than one group; however, we focus here on the most dominant improvement.

3.1 Hybrid Variants

Differential Evolution (DE) [33] was the first to be hybridized with BA. Fister Jr. et al. [18] proposed to replace the original local search equation by the differential operator strategy "DE/rand/1/bin", while the main framework of the bat algorithm remains unchanged. The aim of this modification is to induce additional randomness in the bats movement, thus to enhance the exploitation capability of the standard BA. In [19], the authors replaced also the local search equation by four different DE strategies with self-adaptive selection mechanism. Three of these strategies, namely, DE/randToBest/1/bin, DE/best/2/bin and DE/best/1/bin have the ability to direct the bats movement toward the current best position which improves the exploitation, while the fourth strategy (DE/rand/1/bin) improve the exploration as we mentioned before. In another paper [34], the authors used ten DE strategies for the local search part, where the candidate selection is obtained through Random Forest regression [35]. Meng et al. [36] used also the "DE/rand/1/bin strategy but they added a supplementary step in addition to the two steps of the bat algorithm. However, the candidate selection is based on the fitness value and feasibility (in this case, the algorithm was built to solve constrained problems); as a result, the number of function evaluation is multiplied by two for each iteration.

Xie et al. [37] proposed to replace the bats flights equation by the differential strategy "DE/best/2" to improve the exploitation; moreover, they added an additional step similar to the local search, using the Lévy Flights trajectory to increase the randomness in the search. Unlike the original local search steps, this step is controlled by the loudness, A, instead of the pulse rate, r. He et al. [38] hybridized the bat algorithm with Simulated Annealing (SA) [39] and Gaussian perturbation. When the initial population is generated, the authors used SA to update the best solution, followed by the standard bats movements/equations. When the new candidates are obtained, Gaussian perturbation is applied. The selection is based on the fitness value, and thus the number of function evaluation is doubled at each iteration.

Wang and Guo [40] introduced the Harmony Search algorithm (HS) [41] to the standard bat algorithm. The proposed framework consists of an additional third step that controls the bats movement using the HS equations. In addition, they considered the local search as a separate step, which, in some cases where the statement "$rand > r$" is true, we end up with three candidates. The selection between these candidates is based on the fitness value, thus the number of function evaluation can be between $2 N$ and $3 N$ for single generation.

Nguyen et al. [42] hybridized the bat algorithm with Artificial Bee Colony (ABC) [43]. Each algorithm works separately; however, for a certain set of iterations, the two algorithms communicate between them the best individuals. The poor individual of each algorithm are replaced by the communicated new ones. This strategy of communication is similar to the technique used in the parallelized bat algorithm proposed by Tsai et al. [44].

Yilmaz and Küçüksille [45] proposed to replace the local search steps with (IWO) algorithm [46]. In addition to that, two other modifications were embedded to the velocity equation. First, they added an inertia weight factor to control the contribution of the old velocity to the generated ones; second, they introduced a random selected solution to the velocity with a learning factor. The results showed good improvement on the minimization of unimodal and multimodal functions.

3.2 Adaptive Parameters

Chen et al. [47] proposed to improve the standard bat algorithm by adjusting the frequency. The main idea is to incorporate the flight direction in the frequency generating process, so that the bats can adjust properly their flight toward the best position. This strategy can improve highly the exploitation capability of BA, however, the decrease in the flight randomness can reduce the probability of discovering new solutions far away from the current best position. Wang et al. [48] suggested to link the velocity with the distance between the current bat and the best bat with a speed factor. Thereby, the longer the distance between them, the faster the speed of flight. In addition, the last part of local search equation (Eq. (4)) is replaced by a shrinking factor that starts from a fixed value and decrease exponentially with the iteration process. These adjustments increase the convergence speed, however they do not improve the exploration ability of the algorithm. With analogy to the self-adaptive DE, named jDE [49], Fister Jr. et al. [50] developed the self-adaptive BA by replacing the updating equations of the loudness and the pulse rate with randomly generating process.

The Modified Bat Algorithm (MBA) introduced by Yilmaz et al. [51] has the same frame as the standard BA. The main modification proposed by the authors is that, instead of defining a scalar value of the loudness and the pulse rate for a single bat, they assigned a vector component for A and r with same dimension as x_i. Therefore, when it comes to the local search part, a random vector is generated and only the component of x_{ij} that satisfies the condition $rand_j > r_{ij}$ is updated. This strategy is also adopted for updating A_{ij} and r_{ij}. By equalizing the pulse rate and the loudness to the problem dimension, the authors aim to enhance the exploration capability of the algorithm. Kabir and Alam [52] proposed to multiply the last part of Eq. (4) by a random number generated using either Gaussian or Cauchy distribution. The Gaussian and the Cauchy distribution are known that occasionally, they generate large numbers where the appearance probability of these numbers is higher for Cauchy distribution than the Gaussian one. Thus, the Gaussian distribution is more suited for exploitation, while the Cauchy distribution is more adapted for exploration. To achieve the balance between them, a selection strategy with a learning period is adopted to select the probability of applying each distribution.

Xue et al. [53] performed an analysis on the optimal settings of BA parameters (pulse rate, loudness, frequency, α and γ). To achieve their goal, the authors use the orthogonal experimental design methodology [54], which is a famous strategy for multi-level multi-factor experiment design. The experiments have been conducted on unimodal and multimodal functions and the obtained optimal setting are: $f \in$ [0,5], $r_0 = 0.9$, $A_0 = 0.9$, $\alpha = 0.99$ and $\gamma = 0.9$. Pérez et al. [55, 56] proposed to use fuzzy logic for the dynamic adaptation of the BA parameters to improve the optimization process. In the first paper [55], they used the Mamdani-Type fuzzy system to control the adjustment of the loudness and the frequency boundaries (f_{min} and f_{max}) during the iteration process, whereas in the second paper [56], they applied Type-1 and Type-2 fuzzy logic systems for dynamic adaptation of β and the pulse rate. The presented results in both papers [55, 56] were good; however, it was not clear how the fuzzy logic system contribute in the improvement of the algorithm capabilities.

3.3 The Search Space Alteration

To solve certain kinds optimization problems such as the knapsack problem [57], fault diagnosis [12], feature selection [58] and others, Binary Bat Algorithm (BBA) has been developed to overcome the inherent difficulties of such problems. For the case of feature selection problem, the search space is an n-dimensional Boolean lattice, thus the bats fly from corner to corner of hypercube. Since the problem consists of selecting or not a given feature, Nakamura et al. [58] proposed a binary version of the BA by assigning a binary vector that represents the bats positions, where their movements are restricted to binary values using a sigmoid function. Using the same concept, Mirjalili et al. [59] generalized the BBA to solve continuous optimization problem in real space. By converting the continuous real space search to binary space search, the authors aim to improve the exploitation and exploration capabilities of the bat algorithm. Huang et al. [60] suggested to incorporate an inertia weight and learning factor to the velocity equation with a dynamic updating mechanism to improve the BBA proposed in [59].

With some similarity to the BBA in changing the search space from real to binary, the complex valued BA proposed by Li and Zhou [22] consist of transforming the real search space to a complex one. The main idea is to convert the bat position to complex representation with real and imaginary parts. Each part evolve separately using the BA equations, then, the yielded candidates are converted to the real space where their absolute value is equal to the modulus of the complex number, and the sign is obtained using the argument. This strategy can enhance the diversity of population for a better exploration of the search space. To increase the diversity, Fister et al. [61] proposed to use a quaternion representation of the bats positions. That means for a single dimension of the bat's position is represented with a vector of four dimensions that evolve separately, and the conversion from the quaternion space to the real space is obtained by evaluating the l^2-nome. This algorithm may be similar the complex-valued BA, the main differences are in the vector's dimension associated for each single coordinate value of the bat's position and the equations used to convert this vector to a real scalar value.

3.4 Chaotic Sequences

Afrabandpey et al. [62] stated that to overcame the premature convergence problem of the bat algorithm, the random initialization process of the algorithm parameters (pulse rate, loudness and frequencies) should be well distributed in their corresponding limits. To achieve their goal, chaotic sequences have been used in the initialization process, due to the fact that, they generate well distributed numbers far from the strange attractors of the chaotic maps. An implementation technique to avoid the chaotic attractors in Gauss and Tent maps is presented, and the results show good improvement compared to the standard BA on benchmark functions. Instead of the initialization process only, Abdel-Raouf et al. [63] applied the chaotic sequence for frequency generation during the iteration process. They used the sinusoidal map and they modified the standard BA to solve integers programming problems. The results show good potential in solving NP-hard problems.

In the proposed Chaotic Lévy flight BA [64], the authors used the logistic map to generate chaotic sequences which then applied to control the frequency and as a parameter for the Lévy flight. The authors also replaced the local search equation in the original BA by the Lévy flight, and applied the proposed algorithm in the reconstruction of dynamic nonlinear biological systems. Jordehi [65] analyzed the efficiency of six variants of the chaotic BA, the previously discussed three variants [62–64] and three others versions presented within the main paper [65]. The first version of the chaotic BA proposed by [65] consists of replacing the loudness updating equation with a linearly decreasing chaotic function, the second version uses chaotic linearly increasing function to update the pulse rate, while the last one hybridizes the previous two strategies. Eleven chaotic maps have been considered, and the results showed that the first version of the chaotic BA that uses the iterative chaotic map with infinite collapses exposed the best performance among the other variants.

Gandomi and Yang [21] also analyzed the use of chaotic sequences to increase randomness in the optimization process. They proposed four strategies to use chaotic sequences with eleven chaotic maps. The first strategy consists of replacing the parameter β in Eq. (1) with chaotic sequence. The second is to integrate chaotic sequence in the computation of the velocity. The third is developed by replacing the loudness updating equation with chaotic maps, while in the fourth they replace the pulse emission rate. The results showed that replacing the pulse rate with chaotic sequence based on sinusoidal map is more effective than the others.

3.5 Bio-Inspired Improvement

The Evolved BA propose by Tsai et al. [17], has the same framework of the standard BA but with different updating strategy of the bats positions. The authors reanalyzed the bats behavior with consideration the general characteristics of the whole species of bats. According to their analysis, they redefined a new equation of movement based on the propagation of the sound's waves on the air from the source to the target and the way-back. The distance from the source to the target is measured, and then used to generate a new solution. In addition, new equation for the local search part is proposed in the form $x_i^{lR} = \beta \cdot (x_{best} - x_i^t)$, where β is a random number. The last equation has one meticulous property, that it directs the new generated solutions toward the position $x = [0,...,0]^T$ (the zero vector or the origin). Since the majority of the benchmark functions have their optimal solutions at the origin (the zero vector), such as the sphere function, Rastrigin's function, Ackley's function and others, the results obtained through this algorithm can mislead the readers on the efficiency of the proposed approach. Two others papers have been propose to improve the search capacity of the Evolved BA [66, 67], however, in both works, the proposed equation for local search part was kept. The resulted algorithms are characterized by a rapid convergence toward the zero vector with weak search capacity.

The bat algorithm with recollection proposed by Wang et al. [68], was developed after new observations on the flight behavior of bats. The first observation is that the bats can use various flight modes so that they cannot be trapped in a blind alley. The second is that they have a conditioned reflex in their flights which is regarded as a kind of recollection function. To adjust the standard BA according to these new rules, the authors inserted an inertia weight that decreases exponentially from a maximum to a minimum in the updating equation of the velocity. In addition, they embedded a dynamic control parameter in the last part of Eq. (3) to control the bats search range. The authors also considered the time-delay that exists in the hunting process of bats. They assumed that when a bat detects a prey and starts to fly toward it, in this short period of time, the prey can change its location to the surrounding space, thus, to adapt BA to this new process, they proposed a new local search equation controlled by a time-delay disturbance-factor.

In the proposed Guidable BA (GBA) by Chen et al. [69], the authors introduced the Doppler Effect to enable the guidance of bats by shifted frequency. As it is known, the Doppler Effect produces a frequency shift as the sound travel in the air between the source and the observer. In order to achieve their objective, the authors introduced a new strategy they named Next-generation Evolutionary computing (EC 2.0), based on the collective-effect and context-awareness, to accord the bats the ability to sense the environmental changes by physical laws. The guidable BA is based on six steps: initialization, guidable search, refined search, updating the current best bat, divers search and lastly updating the bat behavior. The resulted algorithm is composed of fourteen updating equations. However, if we analyze the presented results, the algorithm was tested with two benchmark functions namely, the Griewangk function and the Rastrigin function, where their optimal solution is the zero vector. The assumed population consists of 40 bats, and the maximum number of iterations was set to 1000. From their tables, we can see that for the best run, the optimal solution of the Griewangk is achieved after 15 iterations for dimension of 64 and 335 iterations for dimension 128. The optimal solution of the Rastrigin function was achieved after 12 iterations for dimension 20, and 17 iterations for dimension 30. Similar to the evolved bat algorithm [17], one can conclude that the presented framework of the guidable bat algorithm is directed toward the zero vector, which can explain the remarkable achievement of the GBA.

The Novel Bat Algorithm (NBA) proposed by Meng et al. [70], incorporates the bat's habitat selection and a strategy for adaptive compensation for the Doppler effect. The authors indicated that the foraging habitats of bats differ from one species to another. Some of them may forage in forest, some other near water source, and moreover, a study recorded bat activity in urban landscape [71]. In the standard BA, Meng et al. [70] stated the bats' motions are governed by the laws of classical mechanics which restrict their foraging space to a single habitat. To enable the bats to forage in different habitats, quantum behavior has been attributed for each bat so they can forage in a wide range of habitats. Thus, the motion of bats is controlled by the wave equation, Ψ, which allow them to appear spontaneously in a position, x, based on the probability density function $|\Psi|^2$. In another part, the bats are allowed to use classical mechanics in their flights (the standard updating equation of BA), however, a compensation strategy for the Doppler effect is integrated to the frequency generation equation and an inertia weight to the velocity. The local search part was also modified by inserting random Gaussian distribution with mean 0 and standard deviation related to the loudness. The experiments conducted by the authors on benchmark functions and engineering problems shows the efficiency of the proposed algorithm over the standard BA, with a convergence rate slightly superior to the PSO algorithm.

Cai et al. [72] stated that when the bats seek for food (prey or as they called energies), they must spend calories in the seeking process. Therefore, it is logical that the bats prefer to select a seeking strategy that cost less in energy. This strategy was called by the authors "*optimal forage strategy*". Inspired by this phenomenon, the authors proposed to replace x_{old} in Eq. (4) of the local search part, by the position of the bat that achieved the best benefit when seeking for food. This benefit is computed as a ratio between the difference of the old and the new fitness value over the distance among all the bats. In addition to that, random disturbance strategy is applied to the velocity updating equation to improve the global exploration capability of the algorithm. Their results showed significant enhancement over the standard BA.

3.6 Others

Ghanem and Jantan [73] proposed an enhanced BA with a mutation operator. The basic framework of the standard BA was kept, however, they added a third step, the mutation stage, after the local search part with the aim to improve the global exploration. The yielded algorithm was tested on 24 classical benchmark functions with dimension set to 20, 50 and 100. Fifty bats were used in the optimization process for 50 iterations. The results were compared with those of 12 other algorithms including the standard BA. A remarkable observation is that the proposed algorithm achieved the theoretical optimum of 13 functions over 24 even for dimension 100 considering the allocated effort (50 bats and 50 generation). Another remark is that the functions, where the algorithm had the best performance, all have their optimal solution located at the space origin (the zero vector). This can be explained b the fact that the proposed mutation equation has tendency to converge toward the zero vector. The mutation operator consists of two equations, $x_v^t = 0.5 \times (x_{worst}^t - x_v^t) \times rand[0, 1)$ and $x_v^t = 0.5 \times (x_v^t - x_{best}^t) \times rand[0, 1)$. Therefore, when one of these updating equations is selected, there is a high probability that x_v^t can be too close to x_{best}^t (or x_{worst}^t) or even equal, thus, the resulted new position is approximately the zero vector. Since the selection is based on the fitness value, it is evident that the new position will be accepted; consequently, in the next generation, all the bats will be directed to this location which explains the high convergence ratio of this algorithm for the 13 functions.

The bat algorithm with Gaussian walk proposed by [20], is similar to the standard BA; however, Gaussian random walk is used in the local search step to enable the algorithm to escape from local minima. In addition, the bats are allowed to update their velocity either by using the old velocity or not. Their results showed a good improvement compared to the standard BA.

Inspired from the compact PSO [74], Dao et al. [75] developed the compact bat algorithm with the aim to reduce the memory consumption so that the BA can be installed on low-price hardware. The main idea is to use a probabilistic representation instead of a population of solution, thus, a small number of parameters have to be stored. The simulation shows significant reduction in memory usage and computation time.

4 The Directional Bat Algorithm

The directional Bat Algorithm (dBA) proposed by Chakri et al. [31] has the same procedure or flowchart as the standard bat algorithm. The strategy of flight with two modes was kept, however, the directional echolocation has been embedded to the first mode of flight as a main navigation system as well as to others modifications.

The directional echolocation can be described as follows: when the bats are flying, they emit continuously short sound waves to their environment so by analyzing the echoes, they can create a 3D map of their surroundings. In addition, they can retrieve information of other bats such as their positions, and if there is food around them or not. We assume that each bat emits two pulses into two different directions before deciding in which direction that it will fly, one pulse toward the leader with best position (solution), and another pulse to the direction of a randomly selected bat. From the echo (feedback) the bat can knows the existence of food around these two bats or not. The food is represented by the fitness value, thus, around the best bat (or leader) the food is assumed to exist, but around the second randomly selected bat, it depends on it fitness value. If it has a better fitness value as the actual bat, then the food is considered to exist, otherwise there is no source of food.

As it is shown in Fig. 1, there are two flight scenarios. The first one is once the food is confirmed to exist around the two selected bats. That means, a rich source of

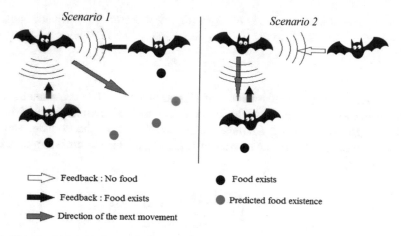

Fig. 1 Hypothetical figure of the directional echolocation

food exists in abundance at the surrounding neighborhood of the two bats, and the bat will fly to this region. If the food does not exist around the randomly selected bat as the food source became scarce, the bat will fly toward the leader to compete for food (the second scenario). The updating equations of the bat's position can be described as the following:

$$\begin{cases} x_i^{t+1} = x_i^t + (x^* - x_i^t)f_1 + (x_k^t - x_i^t)f_2 \\ x_i^{t+1} = x_i^t + (x^* - x_i^t)f_1 \; Otherwise \end{cases} \tag{7}$$

where x_k^t is the position of randomly selected bat ($k \neq i$) and x^* is the position of the leader bat (or the best solution). The $F(.)$ is fitness function, while f_1 and f_2 are the frequencies of the two pulses and defined as follows:

$$\begin{cases} f_1 = f_{min} + (f_{max} - f_{min})rand1 \\ f_2 = f_{min} + (f_{max} - f_{min})rand2 \end{cases} \tag{8}$$

Both $rand1$ and $rand2$ are two random vectors drawn from a uniform distribution between 0 and 1.

One particularity of Eq. (7) that it has the ability to diversify the flights' directions of the bat swarm. At the begging of the iteration process, the distances between bats are large which gives them the ability to explore wide ranges of the search space, thus we enhance the exploration capability and avoid premature convergence. Furthermore, as the iteration process proceeds, the bats have a tendency to gather around the leader bat which reduces the distance between them. Consequently, a strong exploitation process is engaged which enhances the speed of convergence. Equation (7) can promote different capabilities at different stages of iterations, leading to a better flight strategy and enhanced performance [31].

To improve the second mode of flight (the local search step), Chakri et al. [31] introduced a scale factor to the updating equation as the following:

$$x_i^{t+1} = x_i^t + <A^t> \varepsilon \, w_i^t \qquad (9)$$

where $<A^t>$ is the average loudness of all bats and $\varepsilon \in [-1,1]$ is a random vector. Here, w_i is used to control the scales of the search as the iterations proceed. It starts from a large value about a quarter of the space length of the domain, and then decreases to around 1% of the quarter of this length. This parameter is updated as the following:

$$w_i^t = \left(\frac{w_{i0} - w_{i\infty}}{1 - t_{max}}\right)(t - t_{max}) + w_{i\infty} \qquad (10)$$

where t is the current iteration and t_{max} is the maximum number of iterations. Here, w_{i0} and $w_{i\infty}$ are the initial and final values, respectively. The w_{i0} and $w_{i\infty}$ parameters are set as the following:

$$w_{i0} = (U_i - L_i)/4 \qquad (11)$$

$$w_{i\infty} = w_{i0}/100 \qquad (12)$$

where U_i and L_i are the upper and lower bounds, respectively.

At the initial stage, w_i starts with a large value. It allows the bats to move randomly with large steps, which gives the algorithm the ability to explore the whole search space more effectively. As the iteration proceeds, the value of w_i decreases, which focuses the search on the neighborhood around the best solution, and thus the exploitation capability of the algorithm is also enhanced.

The updating equation of the loudness and the pulse emission rate, Eqs. (5–6), proposed by Yang [1] reach their final value during the iterative process quickly. Consequently, it reduces the amount of the auto-switch between the two modes of flight due to a higher pulse rate, and the acceptance rate of a new solution (low loudness). Therefore, in the dBA structure, the following linearly descending and ascending loudness and pulse rate are used, respectively:

$$A^t = \left(\frac{A_0 - A_\infty}{1 - t_{max}}\right)(t - t_{max}) + A_\infty \qquad (13)$$

$$r^t = \left(\frac{r_0 - r_\infty}{1 - t_{max}}\right)(t - t_{max}) + r_\infty \qquad (14)$$

where the index 0 and ∞ stand for the initial and final values, respectively.

The pulse rate plays an important role in the balance between the exploration and the exploitation capability of the algorithm. Starting from a low value, r_0, it promotes the use of the second mode of flight so the algorithm can explore effectively the search space and avoid the premature convergence. However, this value should not be too low, thus allowing to a small fraction of bats to exploit the solutions of the bat with the good positions [31]. As the pulse rate value increases, the first mode of flight takes over the second which has a better exploitation capability. For the case of loudness, this parameter controls the acceptance or rejection of newly generated solutions. It allows the algorithm to avoid being trapped in local optima by rejecting some solutions. The recommended settings by [31], for the pulse rate and loudness are: $r_0 = 0.1$, $r_\infty = 0.7$, $A_0 = 0.9$ and $A_\infty = 0.6$.

In the original BA, the bats are allowed to update their positions if and only if two conditions were satisfied simultaneously. The first one is that a randomly generated number must be lower than loudness A, and for the second, the fitness value of the newly generated solution must be better than the best solution (see Algorithm 1, line 12). The last condition can reduce the diversity of the population, thus, we will allow to the bats to improve their position if the yielded fitness value is better than the old one without neglecting the fulfillment of the first condition. In addition, the best position is updated whenever the bats' motion produces a better solution. The framework of the proposed directional bat algorithm is illustrated in Algorithm 2.

Algorithm 2

The directional bat algorithm.

01.	Define the objective function
02.	Initialize the bat population $L_i \le x_i \le U_i$ $(i=1,2,..,n)$
03.	Evaluate fitness $F_i(x_i)$
04.	Initialize pulse rates r_i loudness A_i and w_i
05.	*While* $(t \le t_{max})$
06.	Select a random bat $(k \ne i)$
07.	Generate frequencies Eq. (8)
08.	Update locations/solutions Eq. (7)
09.	*if* $(rand > r_i)$
10.	Generate a local solution Eq. (9)
11.	Update w_i Eq. (10)
12.	*end if*
13.	*if* $(rand < A_i \& F(x_i^{t+1}) < F(x_i^t))$
14.	Accept the new solutions
15.	Reduce A_i Eq. (13)
16.	Increase r_i Eq. (14)
17.	*end if*
18.	*if* $(F(x_i^{t+1}) < F(x^*))$
19.	Update the best solution x^*
20	end
21.	*end while*
22.	Output results for post-processing

5 Numerical Results and Discussion

To test the performances of the directional bat algorithm, two experiments were conducted. The first one consists of conducting unconstrained optimization of the first 14 benchmark functions used in the CEC-2005 competition [76]. These functions are characterized by having their optimal solutions far and different from the zero vector, which allow as to examine the certainty the capability of the algorithm to discover the global optimum. In the second experiment, dBA is applied to solve a constrained engineering problem. For the purpose of comparison, two other algorithms have been used, namely, the standard bat algorithm and one of state-of-the-art BA variant, the novel bat algorithm (NBA) proposed by Meng et al. [70]. We have implemented the standard BA with the following parameter settings: $r_0 = 0.1$, $A_0 = 0.9$, $\alpha = \gamma = 0.9$, $f_{min} = 0$ and $f_{max} = 2$. The NBA's Matlab code has been posted on MathWorks website [77] by the lead author Xian-Bing Meng for free use. We used this code in our experiment as it is where the setting parameters were as recommended in [70].

5.1 Unconstrained Optimization

In this experiment, we consider a set of 14 functions of CEC-2005 competition on real-parameter optimization. These functions were built using on the classical benchmark functions such as Sphere, Rastrigin's, Rosenbrock's and others; however, their optimal solutions have been shifted to unknown positions. The use of these functions can provide the experimenter significant information on the performance of the tested algorithm. For more details, readers can refer to [76].

For a meaningful comparison, the three algorithms, dBA, BA and NBA, were run 25 times on each problem with the same allocations of population and number of iterations. The dimension of the considered functions was set to $D = 10$, the size of population was fixed to $N = 100$ and the maximum number of iteration was capped to $t_{max} = 1000$. The real optimal minimums of the 14 functions were provided in [76], and used to compute the error between computed optimum and the real one, thus $error = f_{comp} - f_{opt}$. Figure 2 shows the mean of the error computed at each generation of the 25 runs. As it can be seen, at the initial stage of the optimization, the convergence of dBA is slower than the other algorithm due to intensive exploration process that we deliberately favored over exploitation, so that it can explore effectively the search space. However, as the iteration proceeds, the exploitation takes over exploration, thus, the convergence speed up toward a better solution than the competitors in most cases.

Figure 3 presents the ANOVA One-Way test results [78]. The particularity of the ANOVA One-Way test is that it shows where the solutions of 25 runs are located. The square mark ($\square$) points out the mean; the mark ($\times$) locates the minimum and the maximum. The large rectangle indicates where 75% of solutions

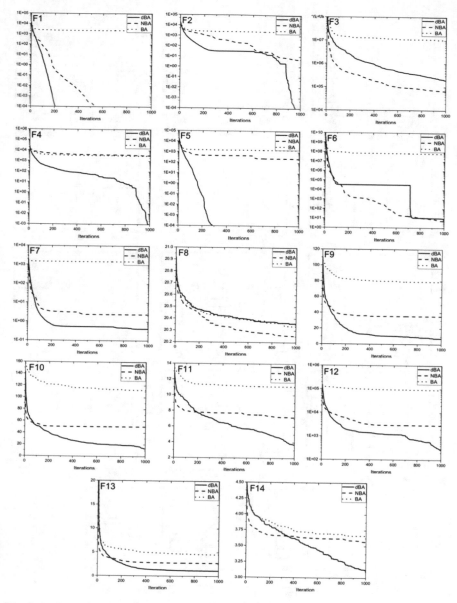

Fig. 2 Mean optimization progress of BA, NBA and dBA

are located while the horizontal mid bar specifies the median. As it can be seen, for 25 runs with random initial population, dBA achieved better results than BA and NBA in most cases; in addition, the fact the rectangle area is less in case of dBA than the other, means that the proposed algorithm is more reliable in terms of the solution's quality.

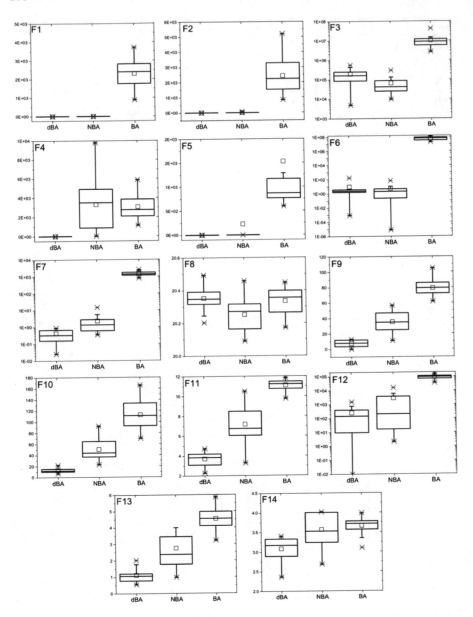

Fig. 3 Box plot of the ANOVA results

Table 1 presents the mean and the standard deviation of the errors obtained in 25 runs. As it can be seen, dBA outperformed BA in all the 13 functions (except F8), and NBA in 11 functions (except F3, F6 and F8). From the number of wins and according to the two-tailed sign test with level of significance $\alpha = 0.05$ (see Table 4 in [79]), we can say that dBA had significantly outperformed the BA and NBA.

Table 1 Mean and standard deviation of 25 runs

	F1		F2		F3	
	Mean	StD	Mean	StD	Mean	StD
dBA	**0.000E + 00**	0.000E + 00	**9.924E-07**	4.779E-06	1.922E + 05	1.356E + 05
NBA	1.054E-11	5.039E-11	4.986E + 00	1.577E + 01	**6.581E + 04**	6.762E + 04
BA	2.289E + 03	7.562E + 02	2.428E + 03	1.096E + 03	1.042E + 07	7.907E + 06

	F4		F5		F6	
	Mean	StD	Mean	StD	Mean	StD
dBA	**6.495E-04**	2.871E-03	**0.000E + 00**	0.000E + 00	9.104E + 00	3.240E + 01
NBA	3.298E + 03	2.632E + 03	2.273E + 02	1.108E + 03	**5.364E + 00**	1.253E + 01
BA	3.063E + 03	1.361E + 03	1.526E + 03	1.645E + 03	6.418E + 07	3.030E + 07

	F7		F8		F9	
	Mean	StD	Mean	StD	Mean	StD
dBA	**4.181E-01**	2.687E-01	2.035E + 01	6.478E-02	**7.442E + 00**	3.921E + 00
NBA	2.288E + 00	2.904E + 00	**2.025E + 01**	9.496E-02	3.526E + 01	1.176E + 01
BA	1.474E + 03	4.603E + 02	2.033E + 01	8.157E-02	7.908E + 01	1.005E + 01

	F10		F11		F12	
	Mean	StD	Mean	StD	Mean	StD
dBA	**1.289E + 01**	3.378E + 00	**3.711E + 00**	6.391E-01	**2.565E + 02**	3.488E + 02
NBA	5.037E + 01	1.870E + 01	7.174E + 00	1.615E + 00	2.894E + 03	4.439E + 03
BA	1.123E + 02	2.474E + 01	1.109E + 01	4.934E-01	9.125E + 04	2.597E + 04

	F13		F14	
	Mean	StD	Mean	StD
dBA	**1.114E + 00**	2.606E-01	**3.082E + 00**	4.027E-01
NBA	2.746E + 00	4.039E-01	3.567E + 00	1.437E + 00
BA	4.536E + 00	1.878E-01	3.662E + 00	6.385E-01

StD: Standard deviation

5.2   Constrained Engineering Problem

In this second experiment, we consider the geometrical optimization of five stage cantilever beam, under a concentric load on tip as it is shown in Fig. 4. The constrained optimization problem consists of 10 geometrical variables that have to be optimized subject to 11 constraints. The mathematical description of the problem is as defined by [80]:

Minimize the beam volume

$$V = l(x_1 x_2 + x_3 x_4 + x_5 x_6 + x_7 x_8 + x_9 x_{10}) \tag{15}$$

subject to five nonlinear constraints

$$g_1 = \frac{6Pl}{x_9 x_{10}^2} - \sigma_{\max} \leq 0 \tag{16}$$

$$g_2 = \frac{6P(2l)}{x_7 x_8^2} - \sigma_{\max} \leq 0 \tag{17}$$

$$g_3 = \frac{6P(3l)}{x_5 x_6^2} - \sigma_{\max} \leq 0 \tag{18}$$

$$g_4 = \frac{6P(4l)}{x_3 x_4^2} - \sigma_{\max} \leq 0 \tag{19}$$

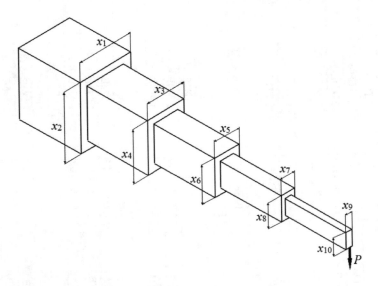

Fig. 4 Geometry of the cantilever beam

$$g_5 = \frac{6P(5l)}{x_1 x_2^2} - \sigma_{max} \leq 0 \tag{20}$$

one stiffness constraint

$$g_6 = \frac{Pl^3}{E} \left(\frac{244}{x_1 x_2^3} + \frac{148}{x_3 x_4^3} + \frac{76}{x_5 x_6^3} + \frac{28}{x_7 x_8^3} + \frac{4}{x_9 x_{10}^3} \right) - \delta_{max} \leq 0 \tag{21}$$

and five geometrical constraints

$$g_7 = (x_2/x_1) - 20 \leq 0 \tag{22}$$

$$g_8 = (x_4/x_3) - 20 \leq 0 \tag{23}$$

$$g_9 = (x_6/x_5) - 20 \leq 0 \tag{24}$$

$$g_{10} = (x_8/x_7) - 20 \leq 0 \tag{25}$$

$$g_{11} = (x_{10}/x_9) - 20 \leq 0 \tag{30}$$

The variables x_i are assumed to be continuous and bounded as follow:

$$1 \leq x_1 \leq 5$$
$$30 \leq x_2 \leq 65$$
$$2.4 \leq x_3, x_5 \leq 3.1$$
$$45 \leq x_4, x_6 \leq 60$$
$$1 \leq x_7, x_9 \leq 5$$
$$30 \leq x_8, x_{10} \leq 65$$

The total beam length, $L = 500$ cm, and the individual section length, $l = 100$ cm. the beam must support a load of $P = 50000$ N, and maximum deflection $\delta_{max} = 2.7$ cm. The allowable stress in each section is $\sigma_{max} = 14000$ N/cm^2, and the Young's modulus $E = 2 \times 10^7$ N/cm^2.

The resolution of the yielded constrained problem is obtained through the use of the static penalty method. The main idea of this method is to convert a constrained problem to an equivalent unconstrained problem where the feasibility of the solution is controlled by the penalty coefficients. If the solution is infeasible, the fitness function is heavily penalized; if it is feasible, there no contribution of the constraint in the fitness function. The equivalent unconstrained problem can be defined as follows:

Fig. 5 Mean fitness progress of 25 runs (constrained problem)

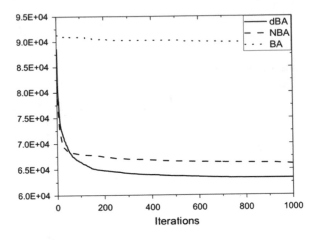

Table 2 Statistical results of 25 runs (constrained problem)

	dBA	NBA	BA
Min	63113.61	63631.55	74125.97
Median	63213.30	65288.98	90499.58
Max	64245.48	71604.09	98229.06
Mean	63282.98	65920.38	89655.69
StD	265.0299	1995.299	4856.954

$$\begin{cases} MinF(x) = V(x) + \sum_{j=1}^{11} \xi_j \left(\max\left(g_j(x), 0\right) \right) \\ L_i \leq x_i \leq U_i \quad (i = 1, \ldots, 10) \end{cases} \tag{31}$$

The setting of the penalty parameters are subject to trials and errors, thus, we found that using following settings: $\xi_j = 10^2$ for $j = 1$ to 5, $\xi_6 = 10^5$ and $\xi_j = 10^3$ for $j = 7$ to 11, one can obtain a good feasible result.

As is in the previous experiment, the parameter setting of the three algorithms including the population size and the number of iterations, are the same as the used before. Each algorithm was runs 25 times, and the mean of the fitness value at every single iteration for the three algorithms are shown in Fig. 5. One can observe that the dBA have outperformed significantly the BA and NBA.

The statistical results of the 25 runs are summarized in Table 2. As it can be seen, the worst solution obtained by dBA (the max), is better than the mean and the median of the 25 runs for both BA and NBA. The low value of the standard deviation of dBA entails the robustness and reliability of the proposed algorithm in achieving high quality solutions. For a single run, dBA has a higher probability to obtain better solution than the competitors. This conclusion is backed up with the ANOVA One Way results shown in Fig. (6). The best solution in 25 runs obtained by the three algorithm and their respective constraint values are presented in Table 3.

Fig. 6 Box plot of the ANOVA results (constrained problem)

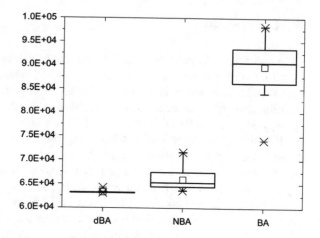

Table 3 Best solution obtained in 25 runs and their respective constraint values

Solution	dBA	NBA	BA
$F(x)$	**63113.61**	63631.55	74125.97
x_1	3.046682	3.103206	3.435927
x_2	60.93361	60.29433	56.98366
x_3	2.819806	2.795826	3.018969
x_4	56.39611	55.87571	54.81524
x_5	2.529725	2.563917	2.791017
x_6	50.59009	51.26774	51.81262
x_7	2.205159	2.247257	4.553382
x_8	44.10232	44.13850	36.22565
x_9	1.749758	1.791717	1.717729
x_{10}	34.99514	34.80139	35.32845
g_1	−3.28E-03	−1.75E + 02	−6.79E + 00
g_2	−1.09E + 01	−2.95E + 02	−3.96E + 03
g_3	−9.92E + 01	−6.45E + 02	−1.99E + 03
g_4	−6.20E + 02	−2.52E + 02	−7.71E + 02
g_5	−7.40E + 02	−7.04E + 02	−5.55E + 02
g_6	−2.21E-07	−0.00E + 00	−5.16E-02
g_7	−3.35E-05	−1.77E + 00	−1.17E + 01
g_8	−1.36E-05	−4.08E-02	−5.56E + 00
g_9	−4.42E-03	−1.06E-02	−4.01E + 00
g_{10}	−8.64E-04	−8.07E-01	−5.48E + 01
g_{11}	−1.94E-05	−1.03E + 00	+9.74E-01

6 Conclusions

In this chapter, we have presented a new improved variant of the standard BA, called the direction bat algorithm (dBA). This algorithm uses a new procedure for directional echolocation as the main foraging strategy for solutions. In addition to the other modifications that have been embedded in BA, the main advantage of this strategy is that it offers a better control of the balance between exploration and exploitation by favoring exploration during the initial stage, and then enhancing exploitation at the final stage. These modifications yield a powerful algorithm that outperforms several BA variants and can perform as much as or more of some sophisticated algorithms that can be found in [31].

During the development of dBA, two other aims were set in addition to the performance improvement. The first was keeping up with the BA framework, and the second is developing an algorithm with a structure as much as easy we can. dBA has a simple structure similar to BA with better performance. The ease structure of dBA has several advantages such as: it can be implemented in any computer language without the need of sophisticated coding, and it can be integrated easily in electronic boards such as microchips, FPGA and others.

For further works, it will be interesting to extend dBA and analyze dBA performance on multi-objective problems, and the investigation of the possibility to use multiple directional echolocations. Furthermore, it is worth realization using the electronic integration of dBA for real industrial applications.

References

1. Yang, X.-S.: A new metaheuristic bat-inspired algorithm. In: González, J.R., Pelta, D.A., Cruz, C., Terrazas, G., Krasnogor, N. (eds.) Nature Inspired Cooperative Strategies for Optimization (NICSO 2010), pp. 65–74. Springer, Heidelberg (2010)
2. Kennedy, J., Eberhart, R.: Particle swarm optimization. In: Proceedings of IEEE International Conference on Neural Networks, 1995. vol. 1944, pp. 1942–1948 (1995)
3. Dorigo, M., Birattari, M., Stutzle, T.: Ant colony optimization. IEEE Comput. Intell. Mag. **1** (4), 28–39 (2006)
4. Gandomi, A.H., Yang, X.-S., Alavi, A.H., Talatahari, S.: Bat algorithm for constrained optimization tasks. Neural Comput. Appl. **22**(6), 1239–1255 (2013)
5. Bora, T.C., Coelho, L., Lebensztajn, L.: Bat-inspired optimization approach for the brushless DC wheel motor problem. IEEE Trans. Magneti. **48**(2), 947–950 (2012)
6. Bahmani-Firouzi, B., Azizipanah-Abarghooee, R.: Optimal sizing of battery energy storage for micro-grid operation management using a new improved bat algorithm. Int. J. Electr. Power Energy Syst. **56**, 42–54 (2014)
7. Ali, E.S.: Optimization of power system stabilizers using BAT search algorithm. Int. J. Electr. Power Energy Syst. **61**, 683–690 (2014)
8. Sambariya, D.K., Prasad, R.: Robust tuning of power system stabilizer for small signal stability enhancement using metaheuristic bat algorithm. Int. J. Electri. Power Energy Syst. **61**, 229–238 (2014)

9. Biswal, S., Barisal, A.K., Behera, A., Prakash, T.: Optimal power dispatch using BAT algorithm. In: 2013 International Conference on Energy Efficient Technologies for Sustainability, 10–12 April 2013, pp. 1018–1023 (2013)
10. Wang, G., Guo, L., Duan, H., Liu, L., Wang, H.: A bat algorithm with mutation for UCAV path planning. Sci. World J. **2012**, 15 (2012)
11. Khatir, S., Belaidi, I., Serra, R., Abdel Wahab, M., Khatir, T.: Numerical study for single and multiple damage detection and localization in beam-like structures using BAT algorithm. J. VibroEng. **18**(1), 202–213 (2016)
12. Kang, M., Kim, J., Kim, J.-M.: Reliable fault diagnosis for incipient low-speed bearings using fault feature analysis based on a binary bat algorithm. Inf. Sci. **294**, 423–438 (2015)
13. Zhang, J.W., Wang, G.G.: Image matching using a bat algorithm with mutation. In: Applied Mechanics and Materials. Trans. Tech. Publ, pp. 88–93 (2012)
14. Karri, C., Jena, U.: Fast vector quantization using a Bat algorithm for image compression. Eng. Sci. Technol. Int. J. **19**(2), 769–781 (2016)
15. Marichelvam, M.K., Prabaharan, T., Yang, X.-S., Geetha, M.: Solving hybrid flow shop scheduling problems using bat algorithm. Int. J. Logist. Econo. Globalisation **5**(1), 15–29 (2013)
16. Fister, I., Rauter, S., Yang, X.-S., Ljubič, K., Fister, Jr. I.: Planning the sports training sessions with the bat algorithm. Neurocomputing **149**(Part B), 993–1002 (2015)
17. Tsai, P.W., Pan, J.S., Liao, B.Y., Tsai, M.J., Istanda, V.: Bat algorithm inspired algorithm for solving numerical optimization problems. Appl. Mech. Materi. Trans. Tech. Publ., 134–137 (2012)
18. Fister Jr., I., Fister, D., Yang, X.-S.: A hybrid bat algorithm. Elektrotehni˘Ski Vestnik **80**(1–2), 1–7 (2013)
19. Fister, I., Fong, S., Brest, J.: A novel hybrid self-adaptive bat algorithm. Sci. World J. **2014**, 12 (2014)
20. Cai, X., Wang, L., Kang, Q., Wu, Q.: Bat algorithm with Gaussian walk. Int. J. Bio-Inspired Comput. **6**(3), 166–174 (2014)
21. Gandomi, A.H., Yang, X.-S.: Chaotic bat algorithm. J. Computat. Sci. **5**(2), 224–232 (2014)
22. Li, L., Zhou, Y.: A novel complex-valued bat algorithm. Neural Comput. Appl. **25**(6), 1369–1381 (2014)
23. Osaba, E., Yang, X.-S., Diaz, F., Lopez-Garcia, P., Carballedo, R.: An improved discrete bat algorithm for symmetric and asymmetric Traveling Salesman Problems. Eng. Appl. Artif. Intell. **48**, 59–71 (2016)
24. Talatahariand, S., Kaveh, A.: Improved bat algorithm for optimum design of large-scale truss structures. Iran Univ. Sci. Technol. **5**(2), 241–254 (2015)
25. Chakri, A., Khelif, R., Benouaret, M.: Improved bat algorithm for structural reliability assessment: application and challenges. Multidiscip. Model. Mater. Struct. **12**(2), 218–253 (2016)
26. Kora, P., Kalva, S.R.: Improved Bat algorithm for the detection of myocardial infarction. SpringerPlus **4**(1), 666 (2015)
27. Li, P., Zhou, Z., Shi, R.: Probabilistic optimal operation management of microgrid using point estimate method and improved bat algorithm. In: 2014 IEEE PES General Meeting Conference and Exposition, 27–31 July 2014, pp. 1–5 (2014)
28. Enache, A.C., Sgârciu, V.: Anomaly intrusions detection based on support vector machines with an improved bat algorithm. In: 2015 20th International Conference on Control Systems and Computer Science, 27–29 May 2015, pp. 317–321 (2015)
29. Kavousi-Fard, A., Niknam, T., Fotuhi-Firuzabad, M.: A novel stochastic framework based on cloud theory and θ-modified bat algorithm to solve the distribution feeder reconfiguration. IEEE Trans. Smart Grid **7**(2), 740–750 (2016)
30. Pérez, J., Valdez, F., Castillo, O.: A new bat algorithm with fuzzy logic for dynamical parameter adaptation and its applicability to fuzzy control design. In: Castillo, O., Melin, P. (eds.) Fuzzy Logic Augmentation of Nature-Inspired Optimization Metaheuristics: Theory and Applications, pp. 65–79. Springer International Publishing, Cham (2015)

31. Chakri, A., Khelif, R., Benouaret, M., Yang, X.-S.: New directional bat algorithm for continuous optimization problems. Expert Syst. Appl. **69**, 159–175 (2017)
32. Chakri, A., Yang, X.-S., Khelif, R., Benouaret, M.: Reliability-based design optimization using the directional bat algorithm. Neural Comput. Appl. (2017)
33. Feoktistov, V.: Differential evolution: in search of solutions, vol 5. Springer Science & Business Media (2007)
34. Iztok, F.J., Fister, D., Fister, I.: Differential evolution strategies with random forest regression in the bat algorithm. In: Paper presented at the Proceedings of the 15th Annual Conference Companion on Genetic and Evolutionary Computation, Amsterdam, The Netherlands (2013)
35. Breiman, L.: Random Forests. Mach. Learn. **45**(1), 5–32 (2001)
36. Meng, X., Gao, X., Liu, Y.: A novel hybrid bat algorithm with differential evolution strategy for constrained optimization. Int. J. Hybrid Inf. Technol. **8**(1), 383–396 (2015)
37. Xie, J., Zhou, Y., Chen, H.: a novel bat algorithm based on differential operator and lévy flights trajectory. Comput. Intell. Neurosci. **2013**, 13 (2013)
38. X-s, H., Ding, W.-J., Yang, X.-S.: Bat algorithm based on simulated annealing and Gaussian perturbations. Neural Comput. Appl. **25**(2), 459–468 (2014)
39. Bertsimas, D., Tsitsiklis, J.: Simulated annealing. Statist. Sci. **8**(1), 10–15 (1993)
40. Wang, G., Guo, L.: A novel hybrid bat algorithm with harmony search for global numerical optimization. J. Appl. Math. **2013**, 21 (2013)
41. Geem, Z.W., Kim, J.H., Loganathan, G.V.: A new heuristic optimization algorithm: harmony search. Simulation **76**(2), 60–68 (2001)
42. Nguyen, T.-T., Pan, J.-S., Dao, T.-K., Kuo, M.-Y., Horng, M.-F.: Hybrid bat algorithm with artificial bee colony. In: Pan, J.-S., Snasel, V., Corchado, E.S., Abraham, A., Wang, S.-L. (Eds.), Intelligent Data analysis and its Applications, Volume II: Proceeding of the First Euro-China Conference on Intelligent Data Analysis and Applications, June 13–15, 2014, Shenzhen, China, pp. 45–55. Springer International Publishing, Cham (2014)
43. Karaboga, D., Basturk, B.: A powerful and efficient algorithm for numerical function optimization: artificial bee colony (ABC) algorithm. J. Global Optim. **39**(3), 459–471 (2007)
44. Tsai, C.-F., Dao, T.-K., Yang, W.-J., Nguyen, T.-T., Pan, T.-S.: Parallelized Bat algorithm with a communication strategy. In: Ali, M., Pan, J.-S., Chen, S.-M., Horng, M.-F. (Eds.), Modern Advances in Applied Intelligence: 27th International Conference on Industrial Engineering and Other Applications of Applied Intelligent Systems, IEA/AIE 2014, Kaohsiung, Taiwan, June 3–6, 2014, Proceedings, Part I, pp. 87–95. Springer International Publishing, Cham (2014)
45. Yılmaz, S., Küçüksille, E.U.: A new modification approach on bat algorithm for solving optimization problems. Appl. Soft Comput. **28**, 259–275 (2015)
46. Mehrabian, A.R., Lucas, C.: A novel numerical optimization algorithm inspired from weed colonization. Ecol. Inf. **1**(4), 355–366 (2006)
47. Chen, Y.T., Liao, B.Y., Lee, C.F., Tsay, W.D., Lai, M.C.: An adjustable frequency bat algorithm based on flight direction to improve solution accuracy for optimization problems. In: 2013 Second International Conference on Robot, Vision and Signal Processing, 10–12 Dec. 2013, pp. 172–177 (2013)
48. Wang, X., Wang, W., Wang, Y.: An adaptive bat algorithm. In: Huang, D.-S., Jo, K.-H., Zhou, Y.-Q., Han, K. (Eds.), Intelligent Computing Theories and Technology: 9th International Conference, ICIC 2013, Nanning, China, July 28–31, 2013. Proceedings, pp. 216–223. Springer, Heidelberg (2013)
49. Brest, J., Greiner, S., Boskovic, B., Mernik, M., Zumer, V.: Self-adapting control parameters in differential evolution: a comparative study on numerical benchmark problems. IEEE Trans. Evolut. Comput. **10**(6), 646–657 (2006)
50. Fister, Jr I., Fong, S., Brest, J., Fister, I.: Towards the self-adaptation of the bat algorithm. In: Proceedings of the 13th IASTED International Conference on Artificial Intelligence and Applications (AIA 2014) Innsbruck the 13th IASTED International Conference on Artificial Intelligence and Applications (AIA 2014). IASTED, Feb 2014, pp. 400–406 (2014)

51. Yılmaz, S., Kucuksille, E.U., Cengiz, Y.: Modified bat algorithm. Elektronika ir. Elektrotechnika **20**(2), 71–78 (2014)
52. Kabir, M.W.U., Alam, M.S.: Bat algorithm with self-adaptive mutation: a comparative study on numerical optimization problems. Int. J. Comput. Appl. **100**(10), 7–13 (2014)
53. Xue, F., Cai, Y., Cao, Y., Cui, Z., Li, F.: Optimal parameter settings for bat algorithm. Int. J. Bio-Inspired Comput. **7**(2), 125–128 (2015)
54. Ross, P.J.: Taguchi techniques for quality engineering loss function, orthogonal experiments. Parameter and Tolerance Design (1996)
55. Pérez, J., Valdez, F.: Castillo O Modification of the Bat Algorithm using fuzzy logic for dynamical parameter adaptation. In: 2015 IEEE Congress on Evolutionary Computation (CEC), 25–28 May 2015, pp. 464–47 (2015)
56. Pérez, J., Valdez, F., Castillo, O.: Modification of the bat algorithm using Type-2 fuzzy logic for dynamical parameter adaptation. In: Melin, P., Castillo, O., Kacprzyk, J. (eds.) Nature-Inspired Design of Hybrid Intelligent Systems, pp. 343–355. Springer International Publishing, Cham (2017)
57. Sabba, S., Chikhi, S.: A discrete binary version of bat algorithm for multidimensional knapsack problem. Int. J. Bio-Inspired Comput. **6**(2), 140–152 (2014)
58. Nakamura, R.Y.M., Pereira, L.A.M., Costa, K.A., Rodrigues, D., Papa, J.P., Yang, X.S.: BBA: A binary bat algorithm for feature selection. In: 2012 25th SIBGRAPI Conference on Graphics, Patterns and Images, 22–25 Aug. 2012, pp. 291–297 (2012)
59. Mirjalili, S., Mirjalili, S.M., Yang, X.-S.: Binary bat algorithm. Neural Comput. Appl. **25**(3), 663–681 (2014)
60. Huang, X., Zeng, X., Han, R.: Dynamic Inertia Weight Binary Bat Algorithm with Neighborhood Search. Comput. Intell. Neurosci. **2017**, 15 (2017)
61. Fister, I., Brest, J., Yang, X.S.: Modified bat algorithm with quaternion representation. In: 2015 IEEE Congress on Evolutionary Computation (CEC), 25–28 May 2015, pp. 491–498 (2015)
62. Afrabandpey, H., Ghaffari, M., Mirzaei, A., Safayani, M.: A novel Bat Algorithm based on chaos for optimization tasks. In: 2014 Iranian Conference on Intelligent Systems (ICIS), 4–6 Feb. 2014, pp. 1–6 (2014)
63. Abdel-Raouf, O., Abdel-Baset, M., El-Henawy, I.: An improved chaotic bat algorithm for solving integer programming problems. Int. J. Modern Educ. Comput. Sci. **6**(8), 18 (2014)
64. Lin, J.-H., Chou, C.-W., Yang, C.-H., Tsai, H.-L.: A chaotic Levy flight bat algorithm for parameter estimation in nonlinear dynamic biological systems. Comput. Inf. Technol. **2**(2), 56–63 (2012)
65. Rezaee Jordehi, A.: Chaotic bat swarm optimisation (CBSO). Appl. Soft Comput. **26**, 523–530 (2015)
66. Tsai, P.W., Zhang, J., Zhang, S., Liao, L.C., Pan, J.S., Istanda, V.: Deceleration convergence strategy for evolved bat algorithm. In: 2015 Third International Conference on Robot, Vision and Signal Processing (RVSP), 18–20 Nov. 2015, pp. 167–170 (2015)
67. Tsai, P.-W., Cai, S., Istanda, V., Liao, L.-C., Pan, J.-S.: Improving the searching capacity of evolved bat algorithm by the periodic signal. In: Zin, T.T., Lin, J.C.-W., Pan, J.-S., Tin, P., Yokota, M. (Eds.), Genetic and Evolutionary Computing: Proceedings of the Ninth International Conference on Genetic and Evolutionary Computing, August 26–28, 2015, Yangon, Myanmar – vol. 1. Springer International Publishing, Cham, pp. 3–9 (2016)
68. Wang, W., Wang, Y., Wang, X.: Bat Algorithm with recollection. In: Huang, D.-S., Jo, K.-H., Zhou, Y.-Q., Han, K. (eds.), Proceedings of Intelligent Computing Theories and Technology: 9th International Conference, ICIC 2013, Nanning, China, July 28–31, 2013, pp. 207–215. Springer, Heidelberg (2013)
69. Chen, Y.-T., Shieh, C.-S., Horng, M.-F., Liao, B.-Y., Pan, J.-S., Tsai, M.-T.: A guidable bat algorithm based on doppler effect to improve solving efficiency for optimization problems. In: Hwang, D., Jung, J.J., Nguyen, N.-T. (eds.), Proceedings of Computational Collective Intelligence. Technologies and Applications: 6th International Conference, ICCCI 2014,

Seoul, Korea, September 24–26, 2014. Springer International Publishing, Cham, pp. 373–383 (2014)

70. Meng, X.-B., Gao, X.Z., Liu, Y., Zhang, H.: A novel bat algorithm with habitat selection and Doppler effect in echoes for optimization. Expert Syst. Appl. **42**(17), 6350–6364 (2015)

71. Gehrt, S.D., Chelsvig, J.E.: Bat activity in an urban landscape: patterns at the landscape and micohabitat scale. Ecol. Appl. **13**(4), 939–950 (2003)

72. Cai, X., X-z, G., Xue, Y.: Improved bat algorithm with optimal forage strategy and random disturbance strategy. Int. J. Bio-Inspired Comput. **8**(4), 205–214 (2016)

73. Wahm, G., Jantan, A.: An enhanced Bat algorithm with mutation operator for numerical optimization problems. Neural Comput. Appl., 1–35 (2017)

74. Neri, F., Mininno, E., Iacca, G.: Compact particle swarm optimization. Inf. Sci. **239**, 96–121 (2013)

75. Dao, T.-K., Pan, J.-S., Nguyen, T.-T., Chu, S.-C., Shieh, C.-S.: Compact bat algorithm. In: Pan, J.-S., Snasel, V., Corchado, E.S., Abraham, A., Wang, S.-L. (eds.), Intelligent Data analysis and its Applications, Volume II: Proceeding of the First Euro-China Conference on Intelligent Data Analysis and Applications, June 13–15, 2014, Shenzhen, China. Springer International Publishing, Cham, pp. 57–68 (2014)

76. Suganthan, P.N., Hansen, N., Liang, J.J., Deb, K., Chen, Y.-P., Auger, A., Tiwari, S.: Problem definitions and evaluation criteria for the CEC 2005 special session on real-parameter optimization. KanGAL report 2005005 (2005)

77. Meng, X.-B.: Novel Bat Algorithm (NBA). MathWorks. https://www.mathworks.com/matlabcentral/fileexchange/51258-novel-bat-algorithm–nba-?s_tid=srchtitle. Accessed May, 5th, 2017

78. Tabachnick BG, Fidell LS, Osterlind SJ (2001) Using multivariate statistics

79. Derrac, J., García, S., Molina, D., Herrera, F.: A practical tutorial on the use of nonparametric statistical tests as a methodology for comparing evolutionary and swarm intelligence algorithms. Swarm Evolut. Comput. **1**(1), 3–18 (2011)

80. Thanedar, P.B., Vanderplaats, G.N.: Survey of discrete variable optimization for structural design. J. Struct. Eng. **121**(2), 301–306 (1995)

Applications of Flower Pollination Algorithm in Feature Selection and Knapsack Problems

Hossam M. Zawbaa and E. Emary

Abstract This chapter presents one of the recently proposed bio-inspired optimization methods, namely, flower pollination algorithm (FPA). FPA for its capability to adaptively search a large search space with maybe many local optima has been employed to solve many real problems. FPA is used to handle the feature selection problem in wrapper-based approach where it is used to search the space of feature for an optimal feature set maximizing a given criteria. The used feature selection methodology was applied in classification and regression data sets and was found to be successful. Moreover, FPA was applied to handle the knapsack problem where different data sets with different dimensions were adopted to assess FPA performance. On all the mentioned problems FPA was benchmarked against bat algorithm (BA), genetic algorithm (GA), particle swarm optimization (PSO) and is found to be very competitive.

Keywords Flower pollination algorithm · Bio-inspired optimization · Evolutionary computation · Feature selection · Knapsack problem

1 Introduction

This chapter presents the importance of flower pollination algorithm (FPA) for feature selection for regression and classification data and knapsack. In the current applications of machine learning and pattern recognition techniques, there are thousands of such features. The vast amounts of data generated today in biology offer more detailed and useful information on one hand; on the contrary, it makes the data analyzing process more difficult because not all the information is relevant. Selecting the important features of a given dataset is a complex problem. *Feature selection* is a technique for solving classification and regression problems, and it identifies

H.M. Zawbaa (✉)
Faculty of Computers and Information, Beni-Suef University, Beni Suef, Egypt
e-mail: hossam.zawbaa@gmail.com

E. Emary
Faculty of Computers and Information, Cairo University, Giza, Egypt

© Springer International Publishing AG 2018
X.-S. Yang (ed.), *Nature-Inspired Algorithms and Applied Optimization*,
Studies in Computational Intelligence 744, https://doi.org/10.1007/978-3-319-67669-2_10

the significant feature subset and removes the unnecessary ones. This mechanism is particularly useful when the size of feature subset is large, and not all of them are required for describing the data features in experiments [1]. Hence, the use of feature selection method is crucial to reduce the enormous number of features. Feature selection helps in understanding data, decreasing the computation time, reducing the effect of the curse of dimensionality and enhancing the performance of prediction model [2]. Furthermore, the feature selection process enhances the visualization and the comprehensibility of the selected feature subset [3].

In real-world applications, due to different reasons not discussed here, many features introduce noise, while others can be totally irrelevant or even misleading, affecting prediction performance. In these cases, feature selection is a must [4]. Two main criteria are employed to differentiate between the feature selection algorithms as follows:

1. *Search strategy*: the method employed to generate feature subsets or feature combinations.
2. *Subset quality (fitness)*: the criteria used to judge the quality of a feature subset.

There are two major approaches of feature selection methods: wrapper-based approach (applying machine learning algorithms) and filter-based approach (using statistical methods) [5]. The *wrapper-based approach* employs a machine learning technique as part of the assessment operation that helps to obtain better results than the filter-based [6], but it has a risk of over-fitting the model and can be computationally costly, and hence, a brilliant search method is required to minimize the computational time [7]. In contrast, the *filter-based approach* explores for a feature subset that optimizes a given data-dependent criterion rather than using classification-dependent criteria as in the wrapper methods [8].

In general, the feature selection is expressed as *multi-objective* with these two goals: (1) *minimize* the selected feature subset and (2) *maximize* the classification precision (*minimize* the prediction error in the regression problems). Commonly, these two goals are contradictory, and the optimal solution is a trade-off between them. Several search methods have been employed, based mainly on greedy search; however, these techniques have at least two drawbacks: stagnation in local optima and big computational time [9]. Evolutionary computing (EC) and population-based algorithms adaptively search the feature space by using a set of search agents that interact in a social manner to reach the optimal solution [10]. EC methods are inspired by the animal social and biological behavior in nature like (wolves, antlions, dragonflies, spiders, and so on) in a group [11].

Most of the recent optimization techniques are *nature-inspired*, i.e. they have been inspired from nature [12].

2   Related Work

Feature selection methods are composed of two elements: the search strategy and the evaluation technique (subset goodness). In the *wrapper-based* approach (alternative to the filter-based approach), the term wrapper refers to the assessment method. Learning boolean is a filter feature selection method that exhaustively explores all potential feature combinations and chooses the minimum feature subset [6].

Various heuristic techniques mimic the biological and physical conducts in nature, and they have been introduced as robust techniques for the global optimization. GA was the earliest evolutionary based technique proposed in the literature, later enhanced relying on the evolution operator during the reproduction [13]. GA feature selection method using a fuzzy set as the fitness function has been introduced in [14]. Wrapper-filter based feature selection methods combine GA with local search methods [15].

In particle swarm optimization (PSO) methods, a solution is represented by a particle with specific properties like position, fitness, and speed [16]. A binary version of PSO (BPSO) modifies the native PSO algorithm to deal with the binary optimization problems [17]. Moreover, an expanded version of BPSO is implemented to deal with feature selection [18]. The binary variant of bat algorithm (BBA) is employed to feature selection, where the search area is described as an n-cube [19].

Ant colony optimization (ACO) uses Fisher discrimination rate to adopt the heuristic information and rough set approach employed for feature selection [20]. Artificial fish swarm (AFS) algorithm mimics the stimulant reaction by controlling the tail and fin [21]. Artificial bee colony (ABC) relies on the natural conduct of honeybees that randomly produced employer bees are moved in the elite bee direction [22]. The elite bee represents the optimal (near to optimal) solution [23]. Antlion optimization algorithm (ALO) is a comparatively recent EC method, which simulates the antlions hunting in nature [24].

Artificial neural networks (ANN) particularly single hidden layer feed-forward neural networks (SLFN) are viewed as a standout amongst the most conventional machine learning models used in regression and classification domains [25]. The learning algorithm is considered the cornerstone of any neural network. Classical gradient-based learning algorithms are suffering from over-fitting, local minima, and they consume a long time to learn [26]. The back-propagation artificial neural network (BP-ANN) has average learning velocity and is likely to get caught in the local minima, leading to miserable performance and efficiency. The revised back propagation artificial neural network (RBP-ANN) is applied to defeat the constraints of BP-ANN and RBP-ANN [27].

In extreme learning machine (ELM) techniques, the output connections are tuned by solving an optimization problem, i.e. finding the minimum of the cost function by linearization [28]. Huang [29] introduced ELM in order to avoid some of the difficulties observed in gradient-based learning methods. ELM is used as a supervised learning method for SLFN neural networks [30, 31]. ELM is choosing the weights of the input and hidden layers randomly rather than completely adapting all the internal parameters. Moreover, ELM could analytically define the output layer weights [32].

3 Flower Pollination Algorithm (FPA) with Selected Applications

FPA is metaheuristic optimization technique relying on the pollination operation of flowering plants that introduced by Yang in 2012 [33]. Pollination is carried out in two modes *self pollination* (local search) and *cross pollination* (global search). Detailed information about the two ways of pollination as follow [34]:

1. *Cross pollination* happens from the pollen of a flower of a different plant at long distance via pollinators that can fly a big distance (global pollination) [34]. In the cross pollination, the pollinators convey the flower pollens and can fly long distance to assure the pollination and proliferation of the optimal solution g_*. The initial rule may be formulated as in Eq. (1):

$$X_i^{t+1} = X_i^t + L(X_i^t - g_*), \qquad (1)$$

where X_i^t represents the vector of a i solution at t iteration, g_* demonstrates the present best solution, and L describes the pollination strength that randomly pulled from the Lèvy distribution.

2. *Self pollination* is implantation of one flower from the pollen of identical flower or different flowers of the identical plant that usually happens when there is no pollinator possible. The local pollination and flower constancy is expressed as in the Eq. (2):

$$X_i^{t+1} = X_i^t + \varepsilon(X_j^t - X_k^t), \qquad (2)$$

where X_j^t and X_k^t demonstrate two random solutions, and ε drawn from the uniform distribution.

Because of local pollination may have substantial fraction (p) in the aggregate pollination actions (in our experiments, we used $p = 0.5$). A switching probability $p\varepsilon[0, 1]$ manages the local and global pollination. FPA search methodology can be outlined as in the algorithm (1).

1: **Inputs**: N Total flower agents,
   IterMax Total iterations number,
   p Switch probability,
2: **Outputs**: The best solution (g_*) and its fitness value.
3: Initialize the N flowers population randomly.
4: Choose the best solution (g_*).
5: **while** Stopping criteria do not meet **do**
6:   **for all** Flower i in the solution set **do**
7:     **if** *rand* $< p$ **then**
8:       Design the L d-dimensional vector based on Lèvy distribution.
9:       Employ the global pollination on i solution as in the equation (1).
10:     **else**
11:       Pull ϵ from the uniform distribution.
12:       Select the j and k solutions randomly.
13:       Execute the local pollination on the i solution by employing the j and k solutions as in the equation (2).
14:     **end if**
15:     Assess the new solution.
16:     **if** the new solution is better than the current one **then**
17:       Substitute the current solution i by the new solution.
18:     **end if**
19:   **end for**
20:   Upgrade the optimal solution g_*.
21: **end while**
22: Select the optimal solution and its fitness.

Algorithm 1: Flower pollination algorithm (FPA)

3.1 FPA Applied for Feature Selection

FPA is adopted here for exploiting the capabilities of filter and wrapper approaches for feature selection. The *filter approach* can be described as data-oriented methods that not directly related to classification performance. The *wrapper-based* approach is more related to prediction performance, but it does not face redundancy and dependency among the selected feature set.

We are seeking to find similarities and differences based on some evaluation criteria that may help in finding weak and strength features of each. All swarm intelligence methods regularly share the data between their multiple agents. Therefore, at every iteration, all/some agents upgrade/modify their position relied on the data of their own position and the other positions.

FPA is applied for feature selection in both classification and regression problems. For a vector with N features, the various feature selection would be 2^N that is the vast space of features to be searched *exhaustively*. Therefore, intelligent optimization is applied to explore the search area adaptively for best feature subset. The optimal feature subset is the one with *least prediction error* and a *less number of selected features* as a common objective in literature. In classification problems, the general fitness function for the proposed optimization algorithms is to maximize the classification accuracy over the validation set given the training set, as shown in Eq. (3) while keeping the minimum number of features selected:

$$\downarrow Fitness = \alpha(1 - P) + \beta \frac{|R|}{|C|}, \tag{3}$$

where R indicates the size of chosen feature set, C demonstrates the total number of features in the dataset, α and β depict the significance of classification performance and the chosen feature set length, $\alpha \in [0, 1]$ and $\beta = 1 - \alpha$, P is the classification performance measured as in Eq. (4):

$$P = \frac{N_c}{N}, \tag{4}$$

where N_c indicates the number of correctly classified instances, and N is the total number of instances.

In the case of regression problems, the general fitness function for the proposed optimization algorithms is to minimize the prediction error over the validation set given the training set as in Eq. (5) while keeping a minimum number of features selected.

$$\downarrow Fitness = \alpha * E + \beta \frac{|R|}{|C|}, \tag{5}$$

where E indicates the prediction error, α and β show the importance of prediction error and selected feature subset respectively. E is defined as:

$$E = \sum_{i=1}^{M} |a_i - t_i|, \tag{6}$$

where a_i and t_i are the actual model prediction value and target value for point i in the validation set.

The used features are the same as the number of features in a given dataset. All features are limited in the range [0, 1], where the feature value approaches to 1; its corresponding feature is a candidate to be selected in classification. In individual fitness calculation, the feature is a threshold to decide whether a feature will be selected at the evaluation stage. Therefore, a static threshold of 0.5 is used as in the Eq. (7):

$$y_{ij} = \begin{cases} 0 & \text{if } x_{ij} < 0.5 \\ 1 & \text{Otherwise,} \end{cases} \tag{7}$$

where x_{ij} is a D—dimensional point in the search space of features and y_{ij} is the binary value $\in 0, 1$ corresponding to selecting/unselecting feature j in solution i from the solution set.

3.2 FPA Applied for Knapsack Problem

Given a set of n elements with each element has a profit p_j and a weight w_j and a Knapsack of capacity C the objective is to find the most profitable solution without violating knapsack weight capacity [35]. A vector describing whether an element is selected or not can be represented in binary form with an n-dimensional vector with individual elements $x_i \in 0, 1$. So, the problem can be mathematically formulated as:

$$Maximimize \sum_{j=1}^{n} p_j x_j, \tag{8}$$

subject to

$$\sum_{j=1}^{n} w_j x_j \leqslant C. \tag{9}$$

The knapsack problem is an NP-hard problem which requires a very intelligent optimization to search the huge search space of possibilities. FPA is adopted in this work to solve a set of Knapsack problems with variant dimensions to prove the searching capability of the FPA. Death penalty [36] is adopted to handle the constraint of the knapsack while the total fitness is calculated as in Eq. (8) but with using negative sign to standardize the maximization into minimization.

4 Experimental Results and Discussion

The global and optimizer-specific parameter setting is outlined in Table 1. All the parameters are set either according to domain-specific knowledge as the α and β parameters of the used fitness function, or based on trial and error on small simulations and common in literature such as the rest of parameters.

In this study, the wrapper approach is used to find a feature subset supervised by the prediction performance. Hence, an intelligent search method is necessary for searching the feature space. In the case of classification datasets, the used classifier in the fitness function as given in Eq. (3) is KNN [37]. KNN is utilized in the experiments based on trial and error basis where the best choice of K is selected ($K = 5$) as the best performing on all the datasets.

4.1 Assessment Indicators

Each algorithm has been applied $K * M$ times with random positioning of the search agents except for the full features selected solution that was compelled to be a posi-

Table 1 The parameter setting for experiments

Parameter	Value(s)
K for cross validation	10
M total number of runs	20
Number of search agents	8
Number of iterations (dimension < 100)	100
Number of iterations (dimension ≥ 100)	200
Problem dimension	Number of features in the dataset
Search range in binary methods	{0, 1}
Search range in continuous methods	[0, 1]
α in the fitness function	0.99
β in the fitness function	0.01

tion for one of the search agents. Compelling the full features solution ensures that all consequent feature subsets; if selected as the global best solution, are fitter than it. Repeated runs of the optimization algorithms were applied to test their convergence capability. We have applied two types of indicators (measures) to compare the various algorithms.

1. Firstly, this group of indicators is applied directly to the fitness function obtained based on the validation set and used to characterize the algorithm performance as follows:

   - *Mean fitness*: is an average value of all the solutions in the final sets obtained by an optimizer in a number of individual runs [38].
   - *Median fitness*: is used to assess the average performance tolerating noise performance of the optimizer over all the M runs [38].
   - *Best fitness*: is the minimum value of the fitness function that acquired by the optimizer in M independent applications [38].
   - *Worst fitness*: is the maximum fitness function value (or worst obtained fitness value) acquired by an optimization method in M independent applications [38].
   - *Statistical standard deviation (std)*: is a representation of the variation of the obtained best solutions found for running a stochastic optimizer for M different runs. *Std* is used as an indicator for the optimizer capability to converge to same/similar optimal solution [38].

2. The second group of indicators is applied to assess the performance of the entire prediction model as follows:

   - *Average classification error*: depicts how precise the classifier of the chosen feature subset, as shown in the Eq. (10):

$$Perf = \frac{1}{M} \sum_{j=1}^{M} \frac{1}{N} \sum_{i=1}^{N} Unmatch(C_i, L_i), \tag{10}$$

where M represents the total number of runs for the optimization method, N describes the total instances in the test subset; C_i depicts the classifier output label of the i data instance. L_i denotes the source class label of the i data instance, and *Unmatch* specifies the function that yields 0 if the two labels are equivalent and yields 1 otherwise.

- *Mean square error (MSE)*: is measuring the mean square error of the difference between actual output and the predicted one as given in Eq. (11):

$$MSE = \frac{\sum_{i=1}^{n}(pred_i - obs_i)^2}{n}, \tag{11}$$

- *Root mean square error (RMSE)*: is measuring the difference among actual output and the predicted ones as given in Eq. (12):

$$RMSE = \sqrt{\frac{\sum_{i=1}^{n}(obs_i - pred_i)^2}{n}}, \tag{12}$$

where obs_i and $pred_i$ are the observed and predicted values respectively. μ represents the mean of the noticed values, n demonstrates the total of examples, and i depicts the example number in a given dataset.

- *Average selection size*: demonstrates the average size of the chosen feature subset to the aggregate amount of features as in the Eq. (13):

$$Selection\_Size = \frac{1}{M} \sum_{i=1}^{M} \frac{size(g_*^i)}{N_t}, \tag{13}$$

where N_t represents the total number of features in a given dataset.

- *Average feature reduction*: demonstrates the mean size of the reduced features to the aggregate amount of features as in the Eq. (14):

$$Reduction = 1 - \frac{1}{M} \sum_{i=1}^{M} \frac{size(g_*^i)}{N_t}, \tag{14}$$

- *Average Fisher score (F-score)*: assesses the feature subset that has large distances between the data samples in various classes, while the distances among data instances in the same class are as minimum as possible [39]. *F-score* is computed for individual features given the class labels and for M independent applications of an algorithm; as shown in Eq. (15):

$$F_j = \frac{\sum_{k=1}^{c} n_k (\mu_k^j - \mu^j)^2}{(\sigma^j)^2},\tag{15}$$

where F_j is the Fisher score for feature j, μ^j is the mean of the entire dataset. $(\sigma^j)^2$ is the standard deviation of the whole dataset, n_k denotes the size of the k class, and μ_k^j indicates the mean of k class.

- *Wilcoxon*: introduced by Wilcoxon [40] as a non-parametric test. The test allocates rank to all the scores considered as one group and afterward sums the ranks of every group. The null hypothesis originates from the same population, so any difference in the two rank sums come only from the testing error. The rank sum test is regularly depicted as the non-parametric version of the *T-test* for two independent groups.
- *T-test*: is a statistical significance that decides whether or not the difference between two classes' averages most likely reflects a real difference in the population from which the groups were sampled; as in the Eq. (16) [41].

$$t = \frac{\bar{x} - \mu_0}{\frac{s}{\sqrt{n}}}\tag{16}$$

where μ_0 is the average of the t-distribution and $\frac{s}{\sqrt{n}}$ is its standard deviation.

- *Average computational time*: is the run time for a given optimization algorithm in millisecond that calculated over the different runs as given in Eq. (17):

$$T_o = \frac{1}{M} \sum_{i=1}^{M} RunTime_{o,i},\tag{17}$$

where M demonstrates the total number of runs for the optimizer O, and $RunTime_{o,i}$ is the computational time in millisecond for optimizer o at run number i.

4.2 Datasets

All datasets were collected to have a variety of *features* and *instances* as delegates of various problem types, which the introduced methods will be examined on. Besides, we selected a set of respectively high dimensional data to ensure the performance of optimization algorithms in huge search spaces. Each dataset is split by cross-validation [42] mode for evaluation, which $K-1$ folds are employed for the training, validation, and testing sets. Each set is repeated M times, hence, each optimizer is estimated $K * M$ times for individual dataset. Each dataset is equally sized into training, validation, and testing. *Training* part is used to train the used classifier through optimization and at the final evaluation. *Validation* part used to assess the performance of the classifier at the optimization time. *Testing* part is employed to determine the finally selected features given the trained classifier. The classification and regression models are used to ensure the quality of the selected features and are assessed on the validation set inside the fitness function during the optimization process [6]. In the case of regression datasets, the regression model used in the fitness function as in Eq. (5) is extreme learning machine (ELM) with a different number of hidden layers and sigmoid basis function. ELM used for regression purposes and is adopted to evaluate the fitness function. ELM has seven nodes in input layer representation and one hundred hidden nodes (based on trial and error basis); because ELM needs more hidden nodes than the classical gradient training algorithms [28].

Table 2 outlines twenty-one datasets used in classification problems. The datasets are acquired from the UCI machine learning repository [43, 44]. Table 3 displays the ten datasets applied in the regression experiments. The used datasets are picked from the UCI machine learning repository [43].

4.3 FPA for Feature Selection Using Classification Data

In classification data category, the classifier used in fitness function as in Eq. (3) is KNN [37]. KNN is applied in the experiments based on trial and error basis where the best choice of K is selected ($K = 5$) as the best performing on all the datasets. The aggregate purpose of this part is to declare the bio-inspired optimization methods for feature selection approaches that minimize the selected feature set and maximize the classification performance from applying the whole features and conventional feature selection methods in the classification problem.

Table 4 outlines the average statistical mean fitness of FPA [45], BA [46], GA, and PSO optimization algorithms for all 21 classification datasets that calculated over the 20 runs. We can observe that all used optimization methods outperform the full features selected that proves the capability of *wrapper-based* method in feature selection problem. We can also highlight that the CS performs in general better than the other optimizers that demonstrate the ability of CS adaptively to explore the area for the optimal feature combination. For evaluating the stability of the stochastic algorithms in the study and converge to the same optimal solution. We measure the standard deviation, and the results are depicted in the Table 5. We can see that, although the FPA depends on Lèvy distribution that has infinite variance it still keeps comparable std measure.

Table 2 List of datasets used in classification data

DS	Name	No. of features	No. of samples
1	Breastcancer	9	699
2	BreastEW	30	569
3	Clean1	166	476
4	Clean2	166	6598
5	CongressEW	16	435
6	Exactly	13	1000
7	Exactly2	13	1000
8	HeartEW	13	270
9	IonosphereEW	34	351
10	KrvskpEW	36	3196
11	Lymphography	18	148
12	M-of-n	13	1000
13	PenglungEW	325	73
14	Semeion	265	1593
15	SonarEW	60	208
16	SpectEW	22	267
17	Tic-tac-toe	9	958
18	Vote	16	300
19	WaveformEW	40	5000
20	WineEW	13	178
21	Zoo	16	101

Table 3 List of datasets used in regression data

DS	Name	No. of features	No. of samples
1	CASP	9	45730
2	CBM	17	11934
3	CCPP	4	47840
4	ENB2012_Y1	8	768
5	ENB2012_Y2	8	768
6	ForestFire	12	517
7	Housing	13	506
8	RelationNetwork	22	53413
9	Slump_test	10	103
10	Yacht_hydrodynamics	6	308

Table 4 Mean fitness of 20 runs

DS	Full	BA	FPA	GA	PSO
1	0.026	0.022	**0.021**	0.022	0.027
2	0.053	0.024	**0.022**	0.025	0.030
3	0.214	0.140	**0.136**	0.150	0.148
4	0.048	0.038	**0.037**	0.038	0.038
5	0.090	0.036	**0.033**	0.043	0.048
6	0.336	0.161	**0.072**	0.219	0.296
7	0.284	0.237	**0.234**	0.240	0.242
8	0.196	0.132	**0.123**	0.138	0.133
9	0.160	0.109	**0.105**	0.111	0.127
10	0.091	0.037	**0.031**	0.036	0.050
11	0.281	0.136	**0.116**	0.161	0.165
12	0.155	0.037	**0.025**	0.081	0.114
13	0.203	0.175	**0.152**	0.193	0.180
14	0.044	0.030	0.030	0.034	**0.029**
15	0.338	**0.128**	0.132	0.136	0.164
16	0.161	0.136	**0.126**	0.141	0.136
17	0.259	0.222	**0.219**	0.224	0.229
18	0.087	0.033	**0.029**	0.034	0.041
19	0.231	0.202	**0.200**	0.206	0.223
20	0.067	0.015	**0.007**	0.015	0.019
21	0.265	0.102	**0.076**	0.132	0.125
Avg.	0.171	0.102	**0.092**	0.113	0.122

Table 6 outlines the average classification error of the selected feature subset from the optimization methods of test set averaged over the 20 runs. From the table, FPA obtains the best results on average, thus demonstrating the capability of FPA to find optimal feature combinations ensuring proper test performance. Regarding the size of selected features on the original size, Table 7 outlines the kept feature ratio to the total number of features. We can notice that FFA gets the best selection feature subset results in general. The performance over the test data is to some extent compatible with the results from the F-score calculated over the selected features by the different optimizers; as shown in the Table 8. GA has obtained the best F-score values overall. Table 9 outlines the average computational time of different optimization algorithms. From the table, FPA has the best computational time in comparison to all other algorithms.

Table 5 Std of fitness values for 20 runs

DS	Full	BA	FPA	GA	PSO
1	0.013	0.008	0.009	0.009	**0.007**
2	0.018	**0.006**	0.008	0.008	0.010
3	0.047	**0.027**	**0.027**	0.029	0.046
4	0.003	0.003	0.004	**0.002**	0.003
5	0.028	0.016	**0.009**	0.014	0.016
6	0.033	0.118	0.034	0.103	**0.029**
7	0.040	0.017	0.016	**0.012**	0.013
8	0.039	0.024	**0.015**	0.019	0.020
9	**0.005**	0.027	0.025	0.027	0.025
10	0.012	0.008	**0.005**	0.009	0.007
11	0.067	0.044	0.035	**0.028**	0.045
12	**0.019**	0.036	0.035	0.046	0.054
13	**0.005**	0.126	0.102	0.112	0.108
14	0.008	0.003	0.006	0.006	**0.005**
15	0.041	**0.030**	0.041	0.036	0.048
16	0.045	0.027	0.027	0.037	**0.024**
17	0.030	**0.011**	0.012	0.017	0.021
18	0.031	**0.013**	0.015	0.015	0.015
19	0.013	0.013	0.011	**0.008**	0.010
20	**0.000**	0.019	0.012	0.015	0.021
21	**0.029**	0.077	0.059	0.052	0.065
Avg.	0.025	0.031	**0.024**	0.029	0.028

4.4 FPA for Feature Selection Using Regression Data

In regression data, the regression model used in fitness function as in Eq. (5) is extreme learning machine (ELM). The aggregate purpose of this section is to introduce bio-inspired optimization algorithms for feature selection approach that reduce the number of selected feature subset and reduce the prediction error from applying the whole feature set and conventional feature selection techniques in regression problems.

Table 10 outlines the average statistical mean fitness of BA, CS, DA, FFA, FPA, MAKHA, GA, and PSO optimization algorithms for all ten regression datasets that calculated over the 20 runs. We can highlight that the FPA performs in general better than the other optimizers that prove the capability of FPA adaptively to explore the search area for best feature subset. For evaluating the stability of the stochastic algorithms in the study and converge to the same optimal solution. The standard

Table 6 Average classification error of 20 runs

DS	Full	BA	FPA	GA	PSO
1	0.043	0.043	0.044	**0.042**	0.045
2	**0.046**	0.056	0.059	0.063	0.067
3	0.212	0.204	0.212	0.205	**0.197**
4	0.050	0.046	**0.044**	0.048	0.048
5	0.080	0.079	**0.064**	0.068	0.074
6	0.311	0.185	**0.069**	0.253	0.314
7	0.261	0.252	**0.247**	0.253	0.252
8	**0.189**	0.212	0.214	0.210	0.220
9	0.194	0.177	0.177	**0.167**	0.170
10	0.091	0.047	**0.033**	0.041	0.058
11	0.231	0.241	0.225	0.257	**0.223**
12	0.153	0.045	**0.027**	0.092	0.124
13	0.289	0.275	**0.274**	0.311	0.297
14	0.050	0.042	**0.039**	0.042	0.042
15	0.324	0.266	0.269	**0.261**	0.282
16	0.236	0.201	0.191	**0.185**	0.187
17	0.263	0.261	0.260	**0.257**	0.275
18	0.113	0.079	0.070	0.067	**0.066**
19	0.239	0.223	**0.221**	0.224	0.236
20	0.079	**0.071**	0.077	**0.071**	0.086
21	0.286	**0.133**	0.144	0.178	0.144
Avg.	0.178	0.149	**0.141**	0.157	0.162

deviation results are depicted in the Table 11. We can see that, although the FPA depends on Lèvy distribution that has infinite variance it still keeps comparable std measure.

Table 12 describes the mean RMSE of the selected feature subset from the optimization algorithms of test data averaged over the 20 runs. From the table, FFA obtains the best results on average, thus demonstrating the capability of FFA to find optimal feature combinations ensuring proper test performance. Regarding the size of selected features on the original size, Table 13 outlines the kept feature ratio to the total number of features. We can highlight that GA obtains the best selection

Table 7 Average selection size of 20 runs

DS	BA	FPA	GA	PSO
1	**0.506**	**0.506**	0.556	**0.506**
2	**0.441**	0.448	0.456	0.493
3	**0.461**	0.482	0.483	0.468
4	**0.491**	0.515	0.502	0.516
5	0.424	**0.299**	0.396	0.403
6	0.513	**0.462**	0.556	0.556
7	0.308	0.299	0.256	**0.171**
8	0.521	0.496	**0.487**	0.556
9	0.399	**0.343**	0.395	0.422
10	**0.451**	0.454	0.488	**0.451**
11	0.401	0.481	**0.395**	0.451
12	0.513	**0.496**	0.624	0.581
13	**0.403**	0.444	0.426	0.416
14	**0.462**	0.488	0.478	0.481
15	**0.398**	0.420	0.415	0.409
16	**0.379**	0.434	0.414	0.429
17	0.654	0.605	0.605	**0.593**
18	0.340	0.299	**0.271**	0.347
19	0.542	0.533	0.583	**0.497**
20	0.393	**0.342**	0.470	0.410
21	**0.347**	0.424	0.375	0.382
Avg.	0.445	**0.441**	0.459	0.454

features size results overall. Table 14 outlines the average computational time of different optimization algorithms. From the table, DA has the best computational time in comparison to all other algorithms.

4.5 FPA for Knapsack Problem

In this section, FPA is used and benchmarked against BA, GA, and PSO on the binary Knapsack problem. A set of 20 benchmark problems were in the study having different dimensionality and capacities as in Table 15.

Table 8 Average F-score of 20 runs

DS	BA	FPA	GA	PSO
1	0.735	0.710	0.752	**0.680**
2	**0.218**	0.245	0.234	0.242
3	0.009	0.009	0.009	**0.008**
4	**0.008**	**0.008**	**0.008**	0.009
5	0.205	**0.178**	0.212	0.212
6	**0.001**	**0.001**	**0.001**	**0.001**
7	0.001	0.001	0.001	**0.000**
8	0.084	**0.078**	0.082	0.089
9	0.032	**0.027**	0.034	0.036
10	0.021	0.021	0.021	**0.020**
11	**0.132**	0.179	0.150	0.142
12	0.031	0.031	0.030	**0.028**
13	**0.310**	0.342	0.326	0.319
14	**0.009**	0.010	0.010	0.010
15	0.019	0.019	0.019	**0.018**
16	**0.021**	0.024	0.024	0.025
17	**0.005**	**0.005**	**0.005**	**0.005**
18	0.174	0.152	**0.145**	0.175
19	0.135	0.138	0.136	**0.117**
20	0.448	**0.425**	0.503	0.491
21	12.207	12.876	**10.636**	12.690
Avg.	0.705	0.737	**0.635**	0.729

Functions F1–F20 are expected to evaluate *the exploitation capability* of a given algorithm. We can see in Table 16 that the performance of the FPA optimization algorithm on the average outperforms the other methods. Such result proves the exploitation capability of the FPA algorithm. The same conclusion can be derived by remarking the median performance presented in Table 17 where the FPA still outperform the BA, GA, and PSO algorithms.

Table 18 depicts the best performance indicator for running individual optimizers over 20 runs. Such indicator targets the optimistic users. We can see from the tables that the FPA outperforms the GA and PSO. Table 19 depicts the worst fitness indicator for both simple and composite benchmark functions. Such indicator is expected to assess the worst performance of a given optimizer and hence target the pessimistic

Table 9 Average computational time (milliseconds) of 20 runs for other optimizers

DS	BA	FPA	GA	PSO
1	74.968	68.312	**48.753**	74.576
2	75.343	63.957	**55.524**	73.486
3	71.724	74.145	**39.352**	68.027
4	3081.594	2886.748	**1722.827**	3219.031
5	72.385	59.438	**32.316**	47.470
6	75.994	75.511	**57.433**	83.390
7	111.914	109.125	**59.681**	90.601
8	68.224	61.606	**47.014**	63.682
9	68.015	68.651	**61.015**	75.368
10	434.304	417.282	**395.025**	406.907
11	57.241	48.661	**33.788**	50.591
12	95.730	83.870	**61.417**	98.346
13	50.366	53.329	**23.921**	45.931
14	551.344	550.024	**305.656**	565.127
15	72.873	**71.880**	3979.521	72.161
16	69.113	68.056	**45.926**	61.559
17	103.918	99.622	**75.024**	99.064
18	70.820	71.250	**56.363**	67.409
19	914.644	885.766	**795.628**	966.532
20	41.435	33.260	**19.982**	37.951
21	48.605	41.365	**26.276**	39.603
Avg.	295.741	**280.565**	378.211	300.324

Table 10 Mean fitness of 20 runs

DS	Full	BA	FPA	GA	PSO
1	6.104	**5.491**	5.495	5.692	5.605
2	0.008	0.004	**0.003**	0.007	0.006
3	17.090	**4.621**	4.763	5.156	5.073
4	5.143	3.041	**2.792**	3.399	3.389
5	7.906	3.287	**3.173**	3.211	3.208
6	131.740	57.736	**56.612**	58.683	57.327
7	8.828	4.122	**3.906**	4.617	4.912
8	0.189	**0.049**	0.052	0.056	0.051
9	7.566	**3.262**	3.411	4.071	3.872
10	10.431	1.766	**1.964**	3.884	3.122
Avg.	19.500	8.338	**8.217**	8.878	8.656

Table 11 Std of fitness values for 20 runs

DS	Full	BA	FPA	GA	PSO
1	0.009	0.168	**0.006**	0.084	0.091
2	**0.000**	0.003	0.002	**0.000**	**0.000**
3	**0.001**	0.136	0.100	0.058	0.047
4	**1.140**	0.433	0.143	0.868	0.235
5	0.742	0.306	0.094	0.100	**0.043**
6	96.463	28.073	**27.392**	28.576	28.008
7	1.034	0.327	0.321	**0.269**	0.580
8	0.174	**0.000**	0.002	0.003	0.003
9	0.332	0.326	0.161	**0.116**	0.178
10	0.553	**0.305**	0.320	1.277	2.372
Avg.	10.045	3.008	**2.854**	3.135	3.156

Table 12 Average RMSE of 20 runs

DS	Full	BA	FPA	GA	PSO
1	6.123	5.830	5.837	**5.824**	5.834
2	0.007	0.006	**0.005**	0.006	0.007
3	12.784	7.846	8.775	**5.298**	6.302
4	4.560	3.923	**2.933**	3.419	4.110
5	4.232	3.800	**3.207**	3.325	4.432
6	92.045	178.521	203.068	**59.166**	59.853
7	7.664	5.070	**4.672**	5.086	6.176
8	0.058	**0.054**	**0.054**	0.056	0.056
9	5.847	**5.485**	5.496	5.559	5.627
10	9.442	**3.371**	3.780	4.095	4.583
Avg.	14.276	21.391	23.783	**9.183**	9.698

users' satisfaction. We can see from the table that the worst performance of the FPA still outperform the other algorithms and proves the capability of using such FPA for pessimistic applications.

Table 20 depicts the standard deviation of individual optimizer's output best solution through the 30 runs. Such indicator is expected to assess the *repeatability* of the obtained solutions and the *convergence to same/similar optima*. We can see

Table 13 Average selection size of 20 runs

DS	BA	FPA	GA	PSO
1	**0.463**	0.500	0.500	0.500
2	0.471	**0.461**	0.471	0.539
3	0.417	0.500	**0.250**	0.333
4	0.438	**0.375**	0.438	0.542
5	0.458	**0.438**	**0.438**	0.521
6	0.458	**0.403**	0.417	0.528
7	0.410	0.436	**0.359**	0.410
8	0.568	**0.553**	0.689	0.561
9	0.567	0.533	**0.467**	0.567
10	0.250	**0.222**	0.250	0.361
Avg.	0.450	0.442	**0.428**	0.486

Table 14 Average computational time (in milliseconds) of 20 runs

DS	BA	FPA	GA	PSO
1	**1516.553**	1685.297	1625.241	1600.838
2	**1385.868**	1408.185	895.429	1465.862
3	1377.098	1395.069	1330.866	**1188.551**
4	774.341	787.606	750.926	**616.100**
5	778.917	860.789	753.356	**401.869**
6	847.260	847.406	**789.993**	866.437
7	551.635	574.881	469.265	**450.445**
8	**2181.821**	2254.263	2361.871	2330.990
9	**807.589**	825.084	838.978	813.069
10	730.932	**693.650**	705.527	746.888
Avg.	1095.201	1133.223	1052.145	**1048.105**

from Table 20 that the standard deviation for the FPA outperforms the other optimizers which proves that FAP has much exploration capability it can still converge to same/similar optimal and hence can be considered as a candidate optimizer for *repeatable* results.

Table 15 Used problem sets and the corresponding dimension of each problem

Function no.	No. Dims	Function no.	No. Dims
F1	10	F11	4
F2	5	F12	13
F3	6	F13	11
F4	7	F14	18
F5	8	F15	7
F6	7	F16	16
F7	15	F17	5
F8	24	F18	14
F9	4	F19	17
F10	18	F20	20

Table 16 Mean fitness for the different used optimizers on the different problems

Function no.	BA	FPA	GA	PSO
F1	−307.750	**−309**	−288.200	−300.200
F2	−49.800	**−51**	−50	−50.600
F3	−146.900	**−150**	−138.750	−142.250
F4	−105.450	**−107**	−103.200	−105.250
F5	−895.500	**−99800**	−892.600	−896.200
F6	**−1735**	**−1735**	−1728.800	−1733.550
F7	−1457.800	**−1458**	−1442.750	−1449
F8	−13519668.200	**−13535674.350**	−13138630.650	−13406262.400
F9	−1656.400	**−1663**	**−1663**	**−1663**
F10	−5957.550	**−5959**	−5891	−5950.600
F11	−1713.950	**−1719**	**−1719**	**−1719**
F12	**−6933**	**−6933**	−6863.100	−6932.600
F13	−5479.850	**−5486**	−5401.050	−5443.450
F14	−9002.750	**−9023**	−8908	−8925.600
F15	−3332.400	**−3345**	−3335	**−3345**
F16	−9760.100	**−9773**	−9577.700	−9729.750
F17	**−3573**	**−3573**	**−3573**	**−3573**
F18	−6628	**−6636**	−6580.600	−6613.450
F19	−6696.650	**−6701**	−6448.100	−6627.600
F20	−8715.600	**−8738**	−8339.500	−8674.300

Table 17 Median fitness for the different used optimizers on the different problems

Function No.	BA	FPA	GA	PSO
F1	−309	−309	−284	−309
F2	−51	−51	−51	−51
F3	−150	−150	−150	−150
F4	−107	−107	−105	−107
F5	−900	−900	−888	−900
F6	−1735	−1735	−1735	−1735
F7	−1458	−1458	−1443	−1449.500
F8	−13520148.500	−13549094	−13109204.500	−13421603
F9	−1663	−1663	−1663	−1663
F10	−5959	−5959	−5927	−5959
F11	−1719	−1719	−1719	−1719
F12	−6933	−6933	−6929	−6933
F13	−5486	−5486	−5486	−5486
F14	−9023	−9023	−9023	−9023
F15	−3345	−3345	−3345	−3345
F16	−9773	−9773	−9688	−9773
F17	−3573	−3573	−3573	−3573
F18	−6636	−6636	−6636	−6636
F19	−6701	−6701	−6481	−6701
F20	−8738	−8738	−8328	−8738

Tables 21 and 22 depict The P-value for two of the common significance tests that are expected to assess the significance of output enhance using the proposed variants. The used significance tests are two-sided Wilcoxon test and T-test. We can see that the P-value for Wilcoxon and T-test are around 0 and hence neglecting the null hypothesis and hence proves the significance of the proposed variant that it is found to be significant using FPA rather than BA, GA, and PSO algorithms.

Table 18 Best fitness for the different used optimizers on the different problems

Function no.	BA	FPA	GA	PSO
F1	−309	−309	−309	−309
F2	−51	−51	−51	−51
F3	−150	−150	−150	−150
F4	−107	−107	−107	−107
F5	−900	−900	−900	−900
F6	−1735	−1735	−1735	−1735
F7	−1458	−1458	−1456	−1456
F8	−13549094	−13549094	−13407977	−13518963
F9	−1663	−1663	−1663	−1663
F10	−5959	−5959	−5959	−5959
F11	−1719	−1719	−1719	−1719
F12	−6933	−6933	−6933	−6933
F13	−5486	−5486	−5486	−5486
F14	−9023	−9023	−9023	−9023
F15	−3345	−3345	−3345	−3345
F16	−9773	−9773	−9773	−9773
F17	−3573	−3573	−3573	−3573
F18	−6636	−6636	−6636	−6636
F19	−6701	−6701	−6701	−6701
F20	−8738	−8738	−8738	−8738

5   Conclusions

This work assesses the performance of FPA on two application domains namely feature selection and knapsack. For feature selection, FPA can overcome the performance of BA, GA, and PSO for its capability to adaptively search the search space with many local optima avoiding premature convergence. In the domain of knapsack also FPA is found to be very competitive to PSO, GA, and BA with the tolerable difference in run time and better optimization performance.

On the basis of future performance, we have five ideas that can be investigated in addition to the work presented here:

1. The proposed FPA method will be assessed using complex datasets that have a huge number (thousands) of input features.
2. Add more statistics evaluation measures such as (sensitivity, specificity, and F-measure).

Table 19 Worst fitness for the different used optimizers on the different problems

Function no.	BA	FPA	GA	PSO
F1	−284	**−309**	−239	−247
F2	−47	**−51**	−47	−47
F3	−119	**−150**	−119	−119
F4	−93	**−107**	−91	−93
F5	−858	**−900**	−883	−888
F6	**−1735**	**−1735**	−1682	−1706
F7	−1454	**−1458**	−1427	−1441
F8	−13482886	**−13494864**	−12914151	−13125716
F9	−1531	**−1663**	**−1663**	**−1663**
F10	−5930	**−5959**	−5729	−5797
F11	−1618	**−1719**	**−1719**	**−1719**
F12	**−6933**	**−6933**	−6350	−6925
F13	−5363	**−5486**	−5054	−5058
F14	−8618	**−9023**	−8448	−8338
F15	−3093	**−3345**	−3145	**−3345**
F16	−9515	**−9773**	−8633	−9565
F17	**−3573**	**−3573**	**−3573**	**−3573**
F18	−6476	**−6636**	−6436	−6185
F19	−6641	**−6701**	−5819	−6249
F20	−8442	**−8738**	−7823	−8355

3. Employ bio-inspired optimization methods for solving the challenging problems and in different applications like big data, bioinformatics, and biomedical.
4. Use more machine learning techniques for wrapper-based fitness evaluation such as support vector machine (SVM), random forest (RF), and support vector regression (SVR).
5. Propose a multi-objective fitness function that uses bio-inspired algorithms to the find optimal feature subset.

Table 20 Standard deviation of fitness for the different used optimizers on the different problems

Function no.	BA	FPA	GA	PSO
F1	5.590	0	19.718	17.307
F2	1.881	0	1.777	1.231
F3	9.542	0	14.917	13.772
F4	3.845	0	4.753	3.810
F5	10.092	0	6.451	5.540
F6	0	0	13.513	6.485
F7	0.894	0	7.873	4.645
F8	21138.405	**16426.210**	124654.021	95227.213
F9	29.516	0	0	0
F10	6.485	0	77.911	36.165
F11	22.584	0	0	0
F12	0	0	162.916	1.789
F13	27.504	0	146.101	101.379
F14	90.561	0	169.130	191.947
F15	56.349	0	44.721	0
F16	57.691	0	284.099	57.424
F17	0	0	0	0
F18	35.777	0	72.917	100.847
F19	14.420	0	240.613	152.194
F20	72.796	0	261.054	121.372

Table 21 P-value for T-test of FPA compared to other optimizers

Optimzer_1	Optimzer_2	P-value
FPA	BA	1.835600e-02
FPA	GA	0.000
FPA	PSO	0.000

Table 22 P-value for Wilcoxon of FPA compared to other optimizers

Optimzer_1	Optimzer_2	P-value
FPA	BA	2.000000e-06
FPA	GA	0.000
FPA	PSO	0.000

References

1. Chizi, B., Rokach, L., Maimon, O.: A survey of feature selection techniques, pp. 1888–1895. IGI Global (2009)
2. Chandrashekar, G., Sahin, F.: A survey on feature selection methods. Comput. Electr. Eng. **40**(1), 16–28 (2014)
3. Huang, C.L.: ACO-based hybrid classification system with feature subset selection and model parameters optimization. Neurocomputing **73**(1–3), 438–448 (2009)
4. Chen, Y., Miao, D., Wang, R.: A rough set approach to feature selection based on ant colony optimization. Pattern Recognit. Lett. **31**(3), 226–233 (2010)
5. Kohavi, R., John, G.H.: Wrappers for feature subset selection. Artif. Intell. **97**(1), 273–324 (1997)
6. Xue, B., Zhang, M., Browne, W.N.: Particle swarm optimisation for feature selection in classification: novel initialisation and updating mechanisms. Appl. Soft Comput. **18**, 261–276 (2014)
7. Guyon, I., Elisseeff, A.: An introduction to variable and attribute selection. Mach. Learn. Res. **3**, 1157–1182 (2003)
8. Chuang, L.Y., Tsai, S.W., Yang, C.H.: Improved binary particle swarm optimization using catfish effect for feature selection. Expert Syst. Appl. **38**(10), 12699–12707 (2011)
9. Xue, B., Zhang, M., Browne, W.N.: Particle swarm optimization for feature selection in classification: a multi-objective approach. IEEE Trans. Cybern. **43**(6), 1656–1671 (2013)
10. Shoghian, S., Kouzehgar, M.: A comparison among wolf pack search and four other optimization algorithms. Comput. Electr. Autom. Control Inf. Eng. **6**(12), 1619–1624 (2012)
11. Valdez, F.: Bio-Inspired Optimization Methods. Handbook of Computational Intelligence, pp. 1533–1538. Springer (2015)
12. Jr, I.F., Yang, X.S., Fister, I., Brest, J., Fister, D.: A brief review of nature-inspired algorithms for optimization. Elektrotehniski Vestnik **80**(3), 116–122 (2013)
13. Holland, J.H.: Adaptation in natural and artificial systems. MIT Press, Cambridge, MA, USA (1992)
14. Xue, X., Yao, M., Wu, Z., Yang, J.: Genetic ensemble of extreme learning machine. Neurocomputing **129**(1), 175–184 (2014)
15. Zhu, Z.X., Ong, Y.S., Dash, M.: Wrapper-filter feature selection algorithm using a memetic framework. IEEE Trans. Syst. Man Cybern. Part B: Cybern **37**, 70–76 (2007)
16. Eberhart, R., Kennedy, J.: A new optimizer using particle swarm theory, pp. 39–43. International Symposium on Micro Machine and Human, Science (1995)
17. Kennedy, J., Eberhart, R.C.: A discrete binary version of the particle swarm algorithm. IEEE International Conference on System, Man and Cybernetics, vol. 5, pp. 4104–4108 (1997)
18. Firpi, H.A., Goodman, E.: Swarmed feature selection. In: 33rd Applied Imagery Pattern Recognition Workshop, USA, pp. 112–118 (2004)
19. Nakamura, R.Y.M., Pereira, L.A.M., Costa, K.A., Rodrigues, D., Papa, J.P., Yang, X.S.: BBA: a binary bat algorithm for feature selection. In: IEEE XXV Conference on Graphics, Patterns and Images, pp. 291–297 (2012)
20. Ming, H.: A rough set based hybrid method to feature selection. In: International Symposium on Knowledge Acquisition and Modeling, pp. 585–588 (2008)
21. Li, X.L., Shao, Z.J., Qian, J.X.: An optimizing method based on autonomous animates: Fishswarm algorithm, pp. 32–38. Methods and practices of system, engineering (2002)
22. Karaboga, D., Basturk, B.: A powerful and efficient algorithm for numerical function optimization: artificial bee colony (ABC) algorithm. J. Glob. Optim. **39**, 459–471 (2007)
23. Sundareswaran, K., Sreedevi, V.T.: Development of novel optimization procedure based on honey bee foraging behavior. In: International Conference on Systems, Man and Cybernetics, pp. 1220–1225 (2008)
24. Mirjalili, S.: The Ant Lion optimizer. Adv. Eng. Softw. **83**, 80–98 (2015)
25. Miche, Y., Sorjamaa, A., Bas, P., Simula, O., Jutten, C., Lendasse, A.: OP-ELM: optimally pruned extreme learning machine. IEEE Trans. Neural Netw. **21**(1), 158–162 (2010)

26. Han, F., Huang, D.S.: Improved extreme learning machine for function approximation by encoding a priori information. Neurocomputing **69**(1), 2369–2373 (2006)
27. Xu, H., Yu, B.: Automatic thesaurus construction for spam filtering using revised back propagation neural network. Expert Syst. Appl. **37**, 18–23 (2010)
28. Jiuwen, C., Zhiping, L.: Extreme Learning Machines on High Dimensional and Large Data Applications: A Survey. Mathematical Problems in Engineering, Hindawi Publishing Corporation, vol. 2015, no. 1, pp. 1–13 (2015)
29. Huang, G.B., Zhu, Q.Y., Siew, C.K.: Extreme learning machine: a new learning scheme of feedforward neural networks. In: International Joint Conference on Neural Networks, pp. 985–990 (2004)
30. Huang, G.B., Zhu, Q.Y., Siew, C.K.: Extreme learning machine: theory and applications. Neurocomputing **70**(1), 489–501 (2006)
31. Li, X., Xie, H., Wang, R., Cai, Y., Cao, J., Wang, F., Min, H., Deng, X.: Empirical analysis: stock market prediction via extreme learning machine. Neural Comput. Appl. **1**(3), 1–12 (2014)
32. Zhao, G.P., Hen, Z.Q., Miao, C.Y., Man, Z.H.: On improving the conditioning of extreme learning machine: a linear case. In: International Conference on Information, Communications and Signal Processing, pp. 1–5 (2009)
33. Yang, X.S.: Flower pollination algorithm for global optimization. Unconventional Computation and Natural Computation. Lecture Notes in Computer Science, vol. 7445, pp. 240–249 (2012)
34. Yang, X.S., karamanoglu, M., He, X.: Multi-objective Flower Algorithm for optimization. In: International Conference on Computational Science, Procedia Computer Science, vol. 18, pp. 861–868 (2013)
35. Ghosh, D., Goldengorin, B.: The binary knapsack problem: solutions with guaranteed quality. In: SOM-theme A Primary Processes within Firms (2001)
36. Yeniay, O.: Penalty function methods for constrained optimization with genetic algorithms. Math. Comput. Appl. **10**(1), 45–56 (2005)
37. Yang, C.S., Chuang, L.Y., Li, J.C., Yang, C.H.: Chaotic binary particle swarm optimization for feature selection using logistic map. In: IEEE Conference on Soft Computing in Industrial Applications, pp. 107–112 (2008)
38. Tilahun, S.L., Ong, H.C.: Prey-predator algorithm: a new metaheuristic algorithm for optimization problems. Inf. Technol. Decis. Mak. **14**(6), 1331–1352 (2015)
39. Duda, R.O., Hart, P.E., Stork, D.G.: Pattern Classification, 2nd edn. Wiley-Interscience (2000)
40. Wilcoxon, F.: Individual comparisons by ranking methods. Biom. Bull. **1**(6), 80–83 (1945)
41. Rice, J.A.: Mathematical Statistics and Data Analysis, 3rd edn. Duxbury Advanced (2006)
42. Hastie, T., Tibshirani, R., Friedman, J.: The Elements of Statistical Learning. Series in Statistics (2009)
43. Bache, K., Lichman, M.: UCI Machine Learning Repository, University of California, Irvine, School of Information and Computer Sciences, 2013, lastchecked on 15 May 2017. http://archive.ics.uci.edu/ml
44. Raman, B., Ioerger, T.R.: Instance-Based Filter for Feature Selection. Machine Learning Research, pp. 1–23 (2002)
45. Yang, X.S.: Nature-Inspired Metaheuristic Algorithms, 2nd edn. Luniver Press, UK (2010)
46. Yang, X.S.: A New Metaheuristic Bat-Inspired Algorithm. Nature Inspired Cooperative Strategies for Optimization, vol. 284, pp. 65–74. Springer (2010)

Why the Firefly Algorithm Works?

Xin-She Yang and Xing-Shi He

Abstract Firefly algorithm is a nature-inspired optimization algorithm and there have been significant developments since its appearance about 10 years ago. This chapter summarizes the latest developments about the firefly algorithm and its variants as well as their diverse applications. Future research directions are also highlighted.

Keywords Algorithm · Firefly algorithm · Multimodal optimization · Nature-inspired computation · Optimization · Swarm intelligence

1 Introduction

Nature-inspired computation has become a new paradigm in optimization, machine learning, data mining and computational intelligence with a diverse range of applications. The essence of nature-inspired computing is the nature-inspired algorithms such as genetic algorithm (GA) [28], particle swarm optimization (PSO) [36] and firefly algorithm (FA) [74]. Most nature-inspired algorithms use some characteristics of swarm intelligence [14], and an overview of swarm intelligence to nature-inspired computation was recently carried out by Yang [81].

Among nature-inspired algorithms, firefly algorithm (FA) was developed by Xin-She Yang in late 2007 and early 2008 [74], and it is almost 10 years since its development. Significant developments have been made in the last few years, and thus this chapter intends to provide a state-of-the-art review of FA and its variants with an emphasis on the most recent studies.

Therefore, this chapter is organized as follows. Section 2 introduces the fundamentals of the firefly algorithm, and Sect. 3 explains why FA works well in practice. Section 4 highlights the main differences between FA and PSO, and Sect. 5 summa-

X.-S. Yang (✉)
School of Science and Technology, Middlesex University, London NW4 4BT, UK
e-mail: x.yang@mdx.ac.uk; xy227@cam.ac.uk

X.-S. He
College of Science, Xi'an Polytechnic University, Xi'an, China

© Springer International Publishing AG 2018
X.-S. Yang (ed.), *Nature-Inspired Algorithms and Applied Optimization*,
Studies in Computational Intelligence 744, https://doi.org/10.1007/978-3-319-67669-2_11

rizes some of the recent variants of FA. Section 6 reviews some of the diverse applications of FA and its variants. Finally, Sect. 7 concludes with discussion of future research directions.

2   Firefly Algorithm

The bioluminescence flashes of fireflies are an amazing sight in the summer sky in tropical and temperate regions. It is estimated that there are about 2000 species of fireflies and most species produce short, rhythmic flashes. Each species can have different flashing patterns and rhythms, and one of the main functions of such flashing light acts as a signaling system to communicate with other fireflies. The rate of flashing, intensity of the flashes and the amount of time between flashes form part of the signaling system [39], and female fireflies respond to a male's unique flashing pattern. Some tropical fireflies can even synchronize their flashes, leading to self-organized behaviour.

As light intensity in the night sky decreases as the distance from the flashing source increases, the range of visibility can be typically a few hundred metres, depending on weather conditions. The attractiveness of a firefly is usually linked to the brightness of its flashes and the timing accuracy of its flashing patterns.

2.1   The Standard Firefly Algorithm

Based on the above characteristics, Xin-She Yang developed the firefly algorithm (FA) [74, 75]. Inside FA, the attractiveness of a firefly is determined by its brightness. Due to exponential decay of light absorption and inverse-square law of light variation with distance, a highly nonlinear term is used to simulate the variation of light intensity or attractiveness.

In the FA, the main algorithmic equation for the position $\mathbf{x}_i$ (as a solution vector to a problem) is

$$\mathbf{x}_i^{t+1} = \mathbf{x}_i^t + \beta_0 e^{-\gamma r_{ij}^2}(\mathbf{x}_j^t - \mathbf{x}_i^t) + \alpha\, \epsilon_i^t, \tag{1}$$

where α is a scaling factor controlling the step sizes of the random walks, while γ is a scale-dependent parameter controlling the visibility of the fireflies (and thus search modes). In addition, β_0 is the attractiveness constant when the distance between two fireflies is zero (i.e., $r_{ij} = 0$). In the above equation, the second term on the right-hand side (RHS) is the nonlinear attractiveness which varies with distance, while the third term is a randomization term and ϵ_i^t means that the random number vectors should be drawn from a Gaussian distribution at each iteration.

This system is a nonlinear system, which may lead to rich characteristics in terms of algorithmic behaviour. Loosely speaking, FA belongs to the category of swarm intelligence (SI) based algorithms, and all SI-based algorithms use some aspects of swarming intelligence [14].

It is worth pointing out that the distance r_{ij} between firefly i and firefly j can be defined as their Cartesian distance. However, for some problems such as the internet routing problems, this 'distance' can be defined as time delay. For certain combinatorial problems, it can be defined even as Hamming distance [48]. In addition, since the brightness of a firefly is associated with the objective landscape with its position as the indicator, the attractiveness of a firefly seen by others, depending on their relative positions and relative brightness. Thus, the beauty is in the eye of the beholder. Consequently, a pair comparison is needed for comparing all fireflies. The main steps of FA can be summarized as the pseudocode in Algorithm 1.

Initialize all the parameters $(\alpha, \beta, \gamma, n)$;
Initialize randomly a population of n firefies;
Evaluate the fitness of the initial population at $\mathbf{x}_i$ by $f(\mathbf{x}_i)$ for $i = 1, \ldots, n$;
while *(t < MaxGeneration)* **do**
    for *All fireflies (i = 1 : n)* **do**
        for *All other fireflies (j = 1 : n) (inner loop)* **do**
            if *Firefly j is better/brighter than i* **then**
                Move firefly i towards j according to Eq. (1);
            end
        end
        Evaluate the new solution and accept the new solution if better;
    end
    Rank and update the best solution found so far;
    Update iteration counter $t \leftarrow t + 1$;
    Reduce α (randomness strength) by a factor;
end

Algorithm 1: Firefly algorithm.

Furthermore, α is a parameter controlling the strength of the randomness or perturbations in FA. The randomness should be gradually reduced to speed up the overall convergence. Therefore, we can use

$$\alpha = \alpha_0 \theta^t, \tag{2}$$

where α_0 is the initial value and $0 < \theta < 1$ is a reduction factor. In most cases, we can use $\theta = 0.9$ to 0.99, depending on the type of problems and the desired quality of solutions.

In fact, since FA is a nonlinear system, it has the ability to automatically subdivide the whole swarm into multiple subswarms. This is because short-distance attraction is stronger than long-distance attraction, and the division of swarm is related to the mean range of attractiveness variations. After division into multi-swarms, each subswarm can potentially swarm around a local mode. Consequently, FA is naturally suitable for multimodal optimization problems. Furthermore, there is no explicit use of the best solution $\mathbf{g}^*$, thus selection is through the comparison of relative brightness according to the rule of 'beauty is in the eye of the beholder'.

2.2 Special Cases of FA

To gain more insight, let us analyze the FA system more carefully. By looking at Eq. (1) closely, we can see that γ is an important scaling parameter [74, 75].

2.2.1 Case A: $\gamma = 0$

At one extreme, we can set $\gamma = 0$, which means that there is no exponential decay and thus the visibility is very high. In this case, all fireflies can see each other in the whole domain and we have

$$\mathbf{x}_i^{t+1} = \mathbf{x}_i^t + \beta_0(\mathbf{x}_j^t - \mathbf{x}_i^t) + \alpha \epsilon_i^t. \tag{3}$$

- If $\gamma = 0$, $\alpha = 0$ and β_0 is fixed, then FA becomes a variant of differential evolution (DE) without crossover [63, 79]. In this special case, if we replace $\mathbf{x}_j$ by the best solution in the group $\mathbf{g}^*$, this reduced FA is equivalent to a special case of accelerated particle swarm optimization (APSO) [74, 79].
- If $\beta_0 = 0$, FA is equivalent to the basic simulated annealing (SA) with α as the cooling schedule [79]. In addition, if ϵ_i is further replaced by $\epsilon \mathbf{x}_i$, this special case is equivalent to the pitch adjustment of the harmony search (HS) algorithm.

Thus, it is clear that DE, APSO, SA and HS are special cases of the standard FA. In other words, FA can be considered as a good combination of APSO, HS, SA and DE enhanced in a nonlinear system. It is no surprise that FA can outperform these algorithms for many applications.

2.2.2 Case B: $\gamma \gg 1$

At the other extreme when $\gamma \gg 1$, the visibility range is very short. Fireflies are essentially flying in a dense fog and they cannot see each other clearly. Thus, each firefly flies independently and randomly. In fact, the exponential term $\exp[-\gamma r_{ij}^2]$ will decrease significantly if $\gamma r_{ij}^2 = 1$, which means that the radius R or range of influence can be defined by

$$R = \frac{1}{\sqrt{\gamma}}. \tag{4}$$

Therefore, a good value of γ should be linked to the scale or limits of the design variables so that the fireflies within a range are visible to each other.

For a given objective landscape, if the average scale of the domain is L, then γ can be estimated by

$$\gamma = \frac{1}{L^2}.$$ (5)

If there is no prior knowledge about its possible scale, we can start with $\gamma = 1$ for most problems, and then increase or decrease it when necessary. In theory, $\gamma \in [0, \infty)$, but in practice, we can use $\gamma = O(1)$, which means that we can use $\gamma = 0.001$ to 1000 for most problems we may meet.

2.3 Discrete FA

The standard FA was designed to solve continuous optimization problems. In order to solve discrete optimization problems, some discretization techniques should be used. For example, one way of converting a continuous variable x to a binary one is to use the sigmoidal function

$$S(x) = \frac{1}{1 + e^{-x}},$$ (6)

where $S \to 1$ for $x \to \infty$, while $S \to 0$ for $x \to -\infty$. However, this S-shaped function requires a large range to get a proper conversion. In practice, many researchers use an additional rule with a random threshold. A common technique is to use a random number $r \in [0,1]$. If $S > r$, then $S = 1$, otherwise, $S = 0$. Obviously, once we have $S \in \{0, 1\}$, we can use $u = 2S - 1$ to get $u \in \{+1, -1\}$ if needed.

Another way of conversion is to use random permutation. For example, a set of a uniformly distributed random number such as $r = [0.3, 0.9, \ldots, 0.7]$ can be converted to integers. On the other hand, an interesting conversion technique is to use a modulus function by

$$u = \lfloor x + k \rfloor \mod m,$$ (7)

to convert x to an integer u. Here, k and $m > 0$ are integers.

There are other methods for discretization, including random keys, random permutation, Hamming-distance based method, tanh(x), and others [55].

Many studies using FA have demonstrated how the algorithm works and the effectiveness of the algorithm. Interested readers can refer to the book by Yang [79] and reviews [15, 67]. Now let us explain in more detail why the algorithm works.

3 Why the Firefly Algorithm Works?

In the above descriptions, we have explained the main steps of the FA and how it works. We now try to summarize why it works so well in practice.

The exact reasons why FA works may require further mathematical analysis, specially for the variants to be introduced later. As it still needs a theoretical framework to explain the working mechanisms of FA and all other algorithms, we do not intend to figure out all the reasons why an algorithm works. However, from both empirical observations and the analysis of the algorithm structure, we can summarize the following four reasons why the FA works [79]:

- From the special cases discussed in the previous section, we know that APSO, SA, HS and DE are special cases of FA, and thus FA can be considered as a good combination of all these algorithms. Therefore, it is no surprise that FA can work more efficiently than these algorithms.
- Due to the nonlinear attraction mechanism in FA, the short-distance attraction is stronger than long-distance attraction; therefore, the whole swarm can automatically subdivide into multiple subswarms. Each swarm can potentially swarm around a local mode, and among all the local modes, there is always a global optimal solution. Consequently, the multiswarm nature of FA enables FA to find multiple optimal solutions simultaneously and FA is naturally suitable for solving nonlinear, multimodal optimization problems. Therefore, for a given problem with m modes, if the number of fireflies n is much higher than m (i.e., $n \gg m$), then all the optima (including the global best) can be found simultaneously.
- The influence radius or range is controlled by γ. As a small value of γ means higher influence and higher visibility, while a higher value of γ reduces its influence and visibility. Therefore, we can tune γ to control the subdivision of the swarm. If $\gamma = 0$, there is no subdivision and all fireflies belong to a single swarm. A moderate value of γ leads to multiswarms, while a much higher value of γ may lead to individual random walks without a swarm. As a result, the diversity and properties of the population are linked to γ. This nonlinearity provides much richer dynamic characteristics.
- In comparison with PSO and other algorithms, FA does not use velocities explicitly, which means that FA does not have any drawbacks associated with velocities. In addition, FA does not use $\mathbf{g}^*$ in its equation. The use of $\mathbf{g}^*$ can potentially lead to premature convergence if the initial $\mathbf{g}^*$ lies in the wrong region, which will attract all other agents towards it. Therefore, FA can avoid any disadvantage associated with $\mathbf{g}^*$.

It is worth pointing out that all these parameters have to be tuned properly. For example, α as the strength of the random walks must be reduced gradually; otherwise, the convergence may be slowed down by too much randomness. Similarly, a proper value of γ has to be tuned to allow a good set of subswarms to emerge automatically [74, 78].

4 FA Is Not PSO

Though FA and PSO are both swarm intelligence based algorithms, they thus share some similarity; however, FA is not PSO because they have some significant differences. Apart from the different inspiration from nature, we briefly summarize here the main differences between FA and PSO:

- FA is a nonlinear system due to the nonlinear attraction term $\beta_0 \exp(-\gamma r_{ij}^2)$, while PSO is a linear system because its updating equations are linear in terms of $\mathbf{x}_i$ and $\mathbf{v}_i$. The nonlinear dynamic nature of FA can lead to much richer characteristics in terms of algorithmic behaviour and population properties.
- The strong nonlinearity of FA means that FA has an ability of multi-swarming, while PSO cannot. Thus, FA can find multiple optimal solutions simultaneously and consequently deal with multimodal problems more effectively.
- PSO uses velocities, but FA does not. Thus, FA does not have the drawbacks associated with velocity initialization and instability for high velocities of particles.
- FA has some scaling control (via γ), while PSO has no scaling control. Such scaling control can give FA more flexibility.

All these differences enable FA to search the design spaces more effectively for multimodal objective landscapes.

5 Variants of FA

Since the development of FA in 2008, it has been applied to many applications [78]. A comprehensive review was done by Fister et al. in 2013 [15], covering the literature up to 2013. Yang and He provided another review from a different perspective in 2013 [76]. More recently, Tilahun et al. provided an updated review on the continuous versions of the firefly algorithm and its variants [67], and the discrete versions of the firefly algorithms were also reviewed by Tilahun and Ngnotchouye in 2017 [66].

Despite the success of the standard FA, many variants have been developed to enhance its performance in the last few years. Again many of these variants have been reviewed by Fister et al. [15] and Tilahun et al. [67], and we will not repeat their coverage. Instead, here we will focus only on the most recent variants that have just appeared in the last few years.

Though there are a diverse range of variants of FA, they can be loosely put into the following six major variants/categories:

- **Discrete FA**: The standard FA was designed to solve problems in the continuous domains. To solve discrete or combinatorial optimization problems, some modifications are needed. For example, Marichelvam et al. developed a discrete FA for solving hybrid flow shop scheduling problems [43, 44], while Osaba developed a discrete FA for solving vehicle routing problems with recycling policy [48]. In addition, Poursalehi et al. used an effective discrete FA for optimizing fuel reload design of nuclear reactors [51]. Zhang et al. used a discrete double-population for assembly sequence planning [87]. These variants can be used to solve scheduling and planning problems as well as routing problems.

- **Adaptive FA**: In the standard FA, parameters are fixed, and it may be advantageous to use adaptively varying parameter values. Baykasoglu and Ozsoydan developed an adaptive firefly algorithm with chaos to solve mechanical design problems [5], and Gálvex and Iglesias [19] developed a memetic self-adaptive FA for shape fitting [19].

- **Modified/Enhanced FA**: Researchers have designed various ways to modify and enhance the performance of FA. For example, Cheung et al. developed a non-homogeneous FA [8], while Chou and Ngo developed a modified FA for multi-dimensional structural optimization [9]. Darwish combined FA with a Bayesian classifier for solving classification problems [10], while Fister et al. used quaternion to represent the solutions of FA in higher dimensions [16].

  In addition, He and Huang used a modified FA for multilevel thresholding of color image segmentation [27]. Gupta used a modified FA for controller design [25]. Tesch and Kaczorowska used a rotational FA for arterial cannula shape optimization [65]. Verma et al. developed an opposition and dimensional based modified firefly algorithm [68]. Furthermore, Wang et al. developed a modified FA based on light intensity difference [70], while Wang et al. modified FA with neighborhood attraction [71] and Yu et al. developed a variable step size FA [82].

  Additionally, Zhou et al. used an information-fusing FA for wireless sensor placement for structural monitoring [89], and Zhou et al. combined FA with Newton's method to identify boundary conditions for transient heat conduction problems [90].

- **Chaotic FA**: Some of the parameters in the FA can be replaced by the outputs of some chaotic maps, which may be able to enhance the exploration ability of the FA. For example, Gandomi et al. developed a chaotic FA in 2013 [18], while Gokhale and Kale used a tent map for their chaotic FA [23]. Also, Zouache et al. developed a quantum-inspired FA for discrete optimization problems [91], and Dhal et al. developed a chaotic FA for enhancing image contrast [11]. Chaos-based FA variants were reviewed by Fister et al. [17].

- **Hybrid FA**: Hybridization can be a good way to create new algorithm tools by combing the advantages of each algorithm involved in the hybrid. For example, Aleshab and Abdullah developed a hybrid FA with a probabilistic neural network for solving classification problems [1], and Zhang et al. developed a hybrid by combing FA with DE and achieved improved performance and accuracy [86].

- **Multiobjective FA**: The standard FA was for single objective optimization and Yang extended the standard FA to multiobjective firefly algorithm (MOFA) for design optimization [77]. In addition, Eswari and Nickolas developed a modified multiobjective FA for task scheduling [13], while Wang et al. developed a hybrid multiobjective FA for big data optimization [72], and Zhao et al. developed a decomposition-based multiobjective FA for RFID network planning with uncertainty [88].

6 Applications of FA and Its Variants

The applications of FA and its variants are diverse, a quick Google scholar search gives more than 7000 outputs, and it is not possible to cover all these applications here. It is not our intention to review even a good fraction of the applications in the current literature. For comprehensive reviews, interested readers can refer to [15, 66, 67]. Here, our emphasis will be on the recent, new applications that can be representative in areas from engineering design to energy systems and from scheduling to image processing. For example, FA has been applied in the design of radial expanders in organic Rankine cycles [4], design optimization of steel frames [6], distributed generation system [7], beam design [12], wavelet neural network optimization [83], hysteresis model identification [84], detection of TEC seismo-ionospheric anomalies [2] and structural search in chemistry [3].

In the area of clustering and classification, Senthinath et al. compared and evaluated the performance of clustering using FA [59]. Gope et al. used FA for rescheduling of real power for congestion management concerning pumped storage hydro-units [24]. Long et al. used FA for heart disease predictions [40].

For applications in design and optimization, Mohanty applied FA for designing shell and tube heat exchangers [46], while Shukla and Singh used FA to select parameters for advanced machining processes [60]. Hung applied FA in OFDM systems [29] and Kamarian et al. used FA for thermal buckling optimization of composite plates [32], while Jafari and Akbari used FA to optimize micrometre-scale resonator modulators [31], and Othman et al. used a supervised FA to achieve optimal placement of distributed generators [49]. Also, Singh et al. combined FA with least-squares method to estimate power system harmonics [61], and Kaur and Ghosh used a fuzzy FA for network reconfiguration of unbalanced distribution networks [34].

In the area of energy engineering and energy systems, Ghorbani et al. used FA for prediction of gas flow rates from gas condensate reservoirs [22], and Massan et al. used FA to solve wind turbine applications [45]. Wang et al. used an FA-BP neural network to forecast electricity price [69] and Rastgou and Moshtagh used FA for multi-stage transmission expansion planning [54]. In addition, Satapathy et al. used a hybrid HS-FA based approach to improve the stability of PV-BESS diesel generator-based microgrid [57].

For image processing, Kanimozhi and Latha used FA for region-based image retrieval [33], and Rajinikanth and Couceiro used an FA-based approach for color image segmentation [53]. Sáchez et al. used FA to optimize modular granular neural networks for human recognition [58]. Rahebi and Hardalac used FA for optic disc detection in retinal images [52], while Gao et al. used FA for visual tracking [20], and Zhang et al. used a discrete FA for end member extraction from hyperspectral images [85].

In the area of time series and forecasting, Xiao et al. used a combined model for electrical load forecasting [73] and Ghorbani et al. used FA for capacity prediction in combination with support vector machine [21]. In addition, Ibrahim and Khatib used a hybrid model for solar radiation prediction [30].

In the area of planning and navigation, Ma et al. used FA for planning navigation paths [41], and Patle et al. used FA to optimize mobile robot navigation [50].

In deep learning and software engineering, Rosa et al. used FA for learning parameters in deep belief networks [56]. Srivatsava used an FA-based approach for generating optimal software test sequences [62], and Kaushik et al. integrated FA in artificial neural network for predicting software costs accurately [35].

Other applications include nanoscale structural optimization by Kougianos and Mohanty [37], protein complex identification by Lei et al. [38], protein structure prediction by Maher et al. [42]. Also, Nekouie and Yaghoobi used FA to carry out multimodal optimization [47], and Sundari et al. used an improved FA for programmed PWM in multilevel inverters with adjustable DC sources [64].

7 Conclusions and Future Directions

As we have seen from the above reviews and discussions, FA and its variants have been successfully applied in a wide spectrum of real-world applications. Despite its success, there are still some interesting future research directions concerning FA, and we will summarize them as follows.

1. *Theory*: Though we know FA and its variants work well, we do not have solid theoretical proof why they work and under exactly what conditions. A recent study by He et al. proved the global convergence of the flower pollination algorithm [26]. It can be expected that the same methodology can be applied to analyze the firefly algorithm and other algorithms. Therefore, more theoretical analysis is needed.
2. *Adaptivity*: All bio-inspired algorithms including FA have parameters, and tuning of these parameters can be tedious. Ideally, algorithms should be able to tune their parameters using a self-tuning framework [80] and also adapt their values to suit for a given type of problems. Future work can focus on the parameter adaptivity of FA and its variants.
3. *Hybrid*: Though there are many different variants of FA, it is no doubt that more hybrid variants will appear in the future. At the moment, hybridization is by trial and error, and it is not clear yet how to achieve a better hybrid by combining different algorithms, which needs more research and further insight.
4. *Co-evolution*: Simple hybridization can often work well; however, co-evolution can be more advantageous by co-evolving two or more algorithms together so as to allow the successful characteristics of an algorithm to enhance the co-evolutionary algorithm structure. It is not clear how to carry out co-evolution of algorithms.
5. *Applications*: In addition to the diverse range of applications reviewed in this chapter, there are more research opportunities of applying FA and its variants. Future applications can focus on the area in big data, deep learning and large-scale problems. Big data in combination with machine learning techniques such

as deep nets can be an active research area for many years to come, and the nature-inspired algorithms can expect to play an important role in this area.

As we can see that FA and its variants have been very successful in many applications, there are more opportunities for future research and applications. The authors hope that this work can inspire future research in the above mentioned directions with more real-world applications.

References

1. Alweshah, M., Abdullah, S.: Hybrizing firefly algorithms with a probabilistic neural network for solving classification problems. Appl. Soft Comput. **35**, 512–524 (2015)
2. Akhoondzadeh, M.: Firefly algorithm in detection of TEC seismo-ionospheric anomalies. Adv. Space Res. **56**(1), 10–18 (2015)
3. Avendaño-Franco, G., Romero, A.H.: Firefly algorithm for structural search. J. Chem. Theory Comput. **12**(7), 3416–3428 (2016)
4. Bahadormanesh, N., Rabat, S., Yarali, M.: Constrained multi-objective optimization of radial expanders in organic Rankine cycles by firefly algorithm. Energy Convers. Manage. **148**, 1179–1193 (2017)
5. Baykasoglu, A., Ozsoydan, F.B.: Adaptive firefly algorithm with chaos for mechanical design optimization problems. Appl. Soft Comput. **36**, 152–164 (2015)
6. Carbas, S.: Design optimization of steel frames using an enhanced firefly algorithm. Eng. Optim. **48**(12), 2007–2025 (2016)
7. Chaurasia, G.S., Singh, A.K., Agrawal, S., Sharma, N.K.: A meta-heuristic firefly algorithm based smart control strategy and analysis of a grid connected hybrid photovoltaic/wind distributed generation system. Solar Energy **150**, 265–274 (2017)
8. Cheung, N.J., Ding, X.M., Shen, H.B.: A non-homogeneous firefly algorithm and its convergence analysis. J. Optim. Theory Appl. **170**(2), 616–628 (2016)
9. Chou, J.S., Ngo, N.T.: Modified firefly algorithm for multidimensional optimization in structural design problems. Struct. Multi. Optim. **55**(6), 2013–2028 (2017)
10. Darwish, S.M.: Combining firefly algorithm and Bayesian classifier: new direction for automatic multilabel image annotation. IET Image Process. **10**(10), 763–772 (2016)
11. Dhal, K.G., Quraishi, M.I., Das, S.: Development of firefly algorithm via chaotic sequence and population diversity to enhance the image contrast. Nat. Comput. **15**(2), 307–318 (2016)
12. Erdal, F.: A firefly algorithm for optimum design of new-generation beams. Eng. Optim. **49**(6), 915–931 (2017)
13. Eswari, R., Nickolas, S.: Modified multi-objective firefly algorithm for task scheduling problem on heterogeneous systems. Int. J. Bio-Inspired Comput. **8**(6), 379–393 (2016)
14. Fisher, L.: The Perfect Swarm: The Science of Complexity in Everyday Life. Basic Books (2009)
15. Fister, I., Fister, I., Yang, X.S., Brest, J.: A comprehensive review of firefly algorithms. Swarm Evol. Comput. **13**(1), 34–46 (2013)
16. Fister, I., Yang, X.S., Brest, J., Fister, I.: Modified firefly algorithm using quaternion representation. Expert Syst. Appl. **40**(18), 7220–7230 (2013)
17. Fister, I., Perc, M., Kamal, S.M., Fister, I.: A review of chaos-based firefly algorithms: perspectives and research challenges. Appl. Math. Comput. **252**, 155–165 (2015)
18. Gandomi, A.H., Yang, X.S., Talatahari, S., Alavi, A.H.: Firefly algorithm with chaos. Commun. Nonlinear Sci. Numer. Simul. **18**(1), 89–98 (2013)
19. Gálvez, A., Iglesias, A.: New memetic self-adaptive firefly algorithm for continuous optimisation. Int. J. Bio-Inspired Comput. **8**(5), 300–317 (2016)

20. Gao, M.L., Li, L.L., Sun, X.M., Yin, L.J., Li, H.T., Luo, D.S.: Firefly algorithm (FA) based particle fiter method for visual tracking. Optik—Int. J. Light Electron Opt. **126**(18), 1705–1711 (2015)
21. Ghorbani, M.A., Shamshirband, S., Haghi, D.Z., Azani, A., Bonakdari, H., Ebtehaj, I.: Application of firefly algorithm-based support vector machines for prediction of field capacity and permanent wilting point. Soil Tillage Res. **172**, 32–38 (2017)
22. Ghorbani, H., Moghadasi, J., Wood, D.A.: Prediction of gas flow rates from gas condensate reservoirs through weelhead chokes using a firefly optimization algorithm. J. Nat. Gas Sci. Eng. **45**, 256–271 (2017)
23. Gokhale, S.S., Kale, V.S.: An application of a tent map initiated chaotic firefly algorithm for optimal overcurrent relay coodination. Int. J. Electr. Power Energy Syst. **78**, 336–342 (2016)
24. Gope, S., Goswami, A.K., Tiwari, P.K., Deb, S.: Rescheduling of real power for congestion management with integration of pumped storage hydro unit using firefly algorithm. Int. J. Electr. Power Energy Syst. **83**, 434–442 (2016)
25. Gupta, A., Padhy, P.K.: Modified firefly algorithm based controller design for integrating and unstable delay processed. Eng. Sci. Technol.: Int. J. **19**(1), 548–558 (2016)
26. He, X.S., Yang, X.S., Karamanoglu, M., Zhao, Y.X.: Global convergence analysis of the flower pollination algorithm: a discrete-time Markov chain approach. Proc. Comput. Sci. **108**(1), 1354–1363 (2017)
27. He, L.F., Huang, S.W.: Modified firefly algorithm based multilevel thresholding for color image segmenttion. Neurocomputing **240**(1), 152–174 (2017)
28. Holland, J.: Adaptation in Natural and Arficial Systems. University of Michigan Press, Ann Arbor (1975)
29. Hung, H.L.: Application firefly algorithm for peak-to-average power ratio reduction in OFDM systems. Telecommun. Syst. **65**(1), 1–8 (2017)
30. Ibrahim, I.A., Khatib, T.: A novel hybrid model for hourly global solar radiation prediction using random forest technique and firefly algorithm. Energy Convers. Manage. **138**, 413–425 (2017)
31. Jafari, O., Akbari, M.: Optimizaion and simulation of micrometre-scale ring resonator modulators based on p-i-n diodes using firefly algorithm. Optik—Int. J. Light Electron Opt. **128**, 101–102 (2017)
32. Kamarian, S., Shakeri, M., Yas, M.H.: Thermal buckling optimisation of composite plates using firefly algorithm. J. Exp. Theoret. Artif. Intell. **29**(4), 787–794 (2017)
33. Kanimozhi, T., Latha, K.: An integrated approach to region based image retrieval using firefly algorithm and support vector machine. Neurocomputing, **151**(Part 3), 1099–1111 (2015)
34. Kaur, M., Ghosh, S.: Network reconfiguration of unbalanced distribution networks using fuzzy-firefly algorithm. Appl. Soft Comput. **49**, 868–886 (2016)
35. Kaushik, A., Tayal, D.K., Yadav, K., Kaur, A.: Integrating firefly algorithm in artificial neural network models for accurate software cost predictions. J. Softw. Evol. Process **28**(8), 665–688 (2016)
36. Kennedy, J., Eberhart, R.C.: Particle swarm optimization. In: Proceedings of of IEEE International Conference on Neural Networks, Piscataway, NJ, pp. 1942–1948 (1995)
37. Kougianos, E., Mohanty, S.P.: A nature-inspired firefly algorithm based approach for nanoscale leakage optimal RTL structure. Integr. VLSI J. **51**, 46–60 (2015)
38. Lei, X.J., Wang, F., Wu, F.X., Zhang, A.D., Pedrycz, W.: Protein complex identification through Markov clustering with firefly algorithm on dynamic protein-protein interaction networks. Inf. Sci. **329**, 303–316 (2016)
39. Lewis, S.M., Cratsley, C.K.: Flash signal evolution, mate choice and predation in fireflies. Ann. Rev. Entomol. **53**(2), 293–321 (2008)
40. Long, N.C., Meesad, P., Unger, H.: A highly accurate firefly based algorithm for heart disease prediction. Expert Syst. Appl. **42**(21), 8221–8231 (2015)
41. Ma, Y., Zhao, Y.X., Wu, L.G., He, Y.X., Yang, X.S.: Navigability analysis of magnetic map with projecting puisuit-based selection method by using firefly algorihtm. Neurocomputing **159**, 288–297 (2015)

42. Maher, B., Albrecht, A.A., Loomes, M., Yang, X.S., Steinhöfel, K.: A firefly-inspired method for protein structure prediction in lattice models. Biomolecules **4**(1), 56–75 (2014)
43. Marichelvam, M.K., Prabaharan, T., Yang, X.S.: A discrete firefly algorithm for the multi-objective hybrid flowshop scheduling problems. IEEE Trans. Evol. Comput. **18**(2), 301–305 (2014)
44. Marichelvam, M.K., Geetha, M.: A hybrid discrete firefly algoirhtm to solve flow shop sheduling proboems to minimise total flow time. Int. J. Bio-Inspired Comput. **8**(5), 318–325 (2016)
45. Massan, S.R., Wagan, A.I., Shakh, M.M., Abro, R.: Wind turbine micrositing by using the firefly algorithm. Appl. Soft Comput. **27**, 450–456 (2015)
46. Mohanty, D.K.: Application of firefly algorithm for design optimization of a shell and tube heat exchanger from economic point of view. Int. J. Therm. Sci. **102**, 228–238 (2016)
47. Nekouie, N., Yaghoobi, M.: A new method in multimodal optimizatoin based on firefly algorithm. Artif. Intell. Rev. **46**(2), 267–287 (2016)
48. Osaba, E., Yang, X.S., Diaz, F., Onieva, E., Masegosa, A.D., Perallos, A.: A discrete firefly algorithm to solve a rich vehicle routing problem modelling a newspaper distribution system with recycling policy. Soft Comput. (2016). doi:10.1007/s00500-016-2114-1
49. Othman, M.M., El-Khattam, W., Hegazy, Y.G., Abdelaziz, A.Y.: Optimal placement and sizing of voltage controlled distributed generators in unbalanced distribution networks using supervised firefly algorithm. Int. J. Electr. Power Energy Syst. **82**, 105–113 (2016)
50. Patle, B.K., Parhi, D.R., Jagadeesh, A., Kashyap, S.K.: On firefly algorithm: optimization and application in mobile robot navigation. World J. Eng. **14**(1), 65–76
51. Poursalehi, N., Zolfaghari, A., Minuchehr, A.: A novel optimization method, effective discrete firefly algorithm, for fuel reload design of nuclear reactors. Ann. Nucl. Energy **81**, 263–275 (2015)
52. Rahebi, J., Hardalac, F.: A new approach to optic disc detection in human retinal images using the firefly algorithm. Med. Biol. Eng. Comput. **54**(2–3), 453–461 (2016)
53. Rajinikanth, V., Couceiro, M.S.: RGB histogram based color image segmentation using firefly algorithm. Proc. Comput. Sci. **46**, 1449–1457 (2015)
54. Rastgou, A., Moshtagh, J.: Application of firefly algorithm for multi-stage transmission expansion planning with adequacy-security considerations in deregularated environments. Appl. Soft Comput. **41**, 373–389 (2016)
55. Rodrigues, D., Pereira, L.A.M., Nakamura, R.Y.M., Costa, K.A.P., Yang, X.S., Souza, A.N., Papa, J.P.: A wrapper approach for feature selection based on the bat algorithm and optimum-path forest. Expert Syst. Appl. **41**(5), 2250–2258 (2014)
56. Rosa, G., Papa, J., Costa, K., Pereira, C., Yang, X.S.: Learning parameters in deep belief networks through firefly algorithm. In: ANNPR 2016: Artificial Neural Networks in Pattern Recognition, pp. 138–149. Springer (2016)
57. Satapathy, P., Dhar, S., Dash, P.K.: Stability improvement of PV-BESS diesel generator-based microgrid with a new modified harmony search-based hybrid firefly algorithm. IET Renew. Power Gener. **11**(5), 566–577 (2017)
58. Sánchez, D., Melin, P., Castillo, O.: Optimization of modular granular neural networks using a firefly algorithm for human recognition. Eng. Appl. Artif. Intell. **64**(1), 172–186 (2017)
59. Senthinath, J., Omkar, S.N., Mani, V.: Clustering using firefly algorithm: performance study. Swarm Evol. Comput. **1**(3), 164–171 (2011)
60. Shukla, R., Singh, D.: Selection of parameters for advanced machining processes using firefly algorithm. Eng. Sci. Technol.: Int. J. **20**(1), 212–221 (2017)
61. Singh, S.K., Sinha, N., Goswami, A.K., Sinha, N.: Optimal estimation of power system harmonics using a hybrid firefly algorithm-based least square method. Soft Comput. **21**(7), 1721–1734 (2017)
62. Srivatsava, P.R., Mallikarjun, B., Yang, X.S.: Optimal test sequence generation using firefly algorithm. Swarm Evol. Comput. **8**(1), 44–53 (2013)
63. Storn, R., Price, K.: Differential evolution: a simple and efficient heuristic for global optimization over continuous spaces. J. Global Optim. **11**(4), 341–59 (1997)

64. Sundari, M.G., Rajaram, M., Balaraman, S.: Application of improved firefly algorithm for programmed PWM in multilevel inverter with adjustable DC sources. Appl. Soft Comput. **41**, 169–179 (2016)
65. Tesch, K., Kaczorowska, K.: Arterial cannula shape optimization by means of the rotational firefly algorithm. Eng. Optim. **48**(3), 497–518 (2016)
66. Tilahun, S.L., Ngnotchouye, J.M.T.: Firefly algorithm for discrete optimization problems: A survey. KSCE J. Civ. Eng. **21**(2), 535–545 (2017)
67. Tilahun, S.L., Ngnotchouye, J.M.T., Hamadneh, N.N.: Continuous versions of firefly algorithm: a review. Artif. Intell. Rev. (2017). doi:10.1007/s10462-017-9568-0
68. Verma, O.P., Aggarwal, D., Patodi, T.: Opposition and dimensional based modified firefly algortihm. Expert Syst. Appl. **44**(1), 168–176 (2016)
69. Wang, D.Y., Luo, H.Y., Grunder, O., Lin, Y.B., Guo, H.X.: Multi-step electricity price forecasting using a hybrid model based on two-layer decomposition technique and BP neural network optimized by firefly algorithm. Appl. Energy **190**, 390–407 (2017)
70. Wang, B., Li, D.X., Jiang, J.P., Liao, Y.H.: A modified firefly algorithm based on light intensity difference. J. Comb. Optim. **31**(3), 1045–1060 (2016)
71. Wang, H., Wang, W.J., Zhou, X.Y., Sun, H., Zhao, J., Yu, X., Cui, Z.H.: Firefly algorithm with neighborhood attraction. Inf. Sci. **382–383**(1), 374–387 (2017)
72. Wang, H., Wang, W.J., Cui, L.Z., Sun, H., Zhao, J., Wang, Y., Xue, Y.: A hybrid multi-objective firefly algorithm for big data optimization. Appl. Soft Comput. (2017). (In press). doi:10.1016/j.asoc.2017.06.029
73. Xiao, L.Y., Shao, W., Liang, T.L., Wang, C.: A combined model based on multiple seasonal patterns and modified firefly algorithm for electrical load forecasting. Appl. Energy **167**, 135–153 (2016)
74. Yang, X.S.: Nature-Inspired Metaheuristic Algorithms. Luniver Press, Frome (2008)
75. Yang, X.S.: Firefly algorithm, stochastic test functions and design optimisation. Int. J. Bio-Inspired Comput. **2**(2), 78–84 (2010)
76. Yang, X.S., He, X.S.: Firefly algorithm: recent advances and applications. Int. J. Swarm Intell. **1**(1), 36–50 (2013)
77. Yang, X.S.: Multiobjective firefly algorithm for continuous optimization. Eng. Comput. **29**(2), 175–184 (2013)
78. Yang, X.S.: Cuckoo Search and Firefly Algorithm: Theory and Applications. Studies in Computational Intelligence, vol. 516. Springer (2014)
79. Yang, X.S.: Nature-Inspired Optimization Algorithms. Elsevier Insight, London (2014)
80. Yang, X.S., Deb, S., Loomes, M., Karamanoglu, M.: A framework for self-tuning optimization algorithm. Neural Comput. Appl. **23**(7–8), 2051–2057 (2013)
81. Yang, X.S., Deb, S., Fong, S., He, X.S., Zhao, Y.X.: From swarm intelligence to metaheuristics: nature-inspired optimization algorithms. Computer **49**(9), 52–59 (2016)
82. Yu, S.H., Zhu, S.L., Ma, Y., Mao, D.M.: A variable step size firefly algorithm for numerical optimization. Appl. Math. Comput. **263**, 214–220 (2015)
83. Zainuddin, Z., Ong, P.: Optimization of wavelet neural networks with the firefly algorithm for approximation problems. Neural Comput. Appl. **28**(7), 1715–1728 (2017)
84. Zaman, M.A., Sikder, U.: Bouc-Wen hysteresis model identification using modified firefly algorithm. J. Magn. Magn. Mater. **395**, 229–233 (2015)
85. Zhang, C.Y., Qin, Q.M., Zhang, T.Y., Sun, Y.H., Chen, C.: Endmember extraction from hyperspectral image based on discrete firefly algorithm (EE-DFA). ISPRS J. Photogr. Rem. Sens. **126**(1), 108–119 (2017)
86. Zhang, L.N., Liu, L.Q., Yang, X.S., Dai, Y.T.: A novel hybrid firefly algorithm for global optimization. PloS ONE, **11**(9), e0163230 (2016). doi:10.1371/journal.pone.0163230
87. Zhang, Z.F., Yuan, B.X., Zhang, Z.N.: A new discrete double-population firefly algorithm for assembly sequence planning. Proc. Inst. Mech. Eng. Part B: J. Eng. Manuf. **230**(12), 2229–2238 (2016)
88. Zhao, C.X., Wu, C.Z., Chai, J., Wang, X.Y., Yang, X.M., Lee, M., Kim, M.J.: Decomposition-based multi-objective firefly algorithm for RFID network planning with uncertainty. Appl. Soft Comput. **55**, 549–564 (2017)

89. Zhou, G.D., Yi, T.H., Xie, M.X., Li, H.N.: Wireless sensor placement for strutural monitoring using information-fusing firefly algoirthm. Smart Mater. Struct. (2017). (In press). http://iopscience.iop.org/article/10.1088/1361-665X/aa7930/pdf
90. Zhou, H.L., Zhao, X.H., Yu, B., Chen, H.L., Meng, Z.: Firefly algorithm combined with Newton method to identify boundary conditions for transient heat conduction problems. Numer. Heat Transf. Part B: Fundam. Int. J. Comput. Methodol. **71**(3), 253–269 (2017)
91. Zouache, D., Nouioua, F., Moussaoui, A.: Quantum-inspired firefly algorithm with particle swarm optimization for discrete optimization problems. Soft Comput. **20**(7), 2781–2799 (2016)

An Efficient Computational Procedure for Simultaneously Generating Alternatives to an Optimal Solution Using the Firefly Algorithm

Julian Scott Yeomans

Abstract In solving many "real world" mathematical programming applications, it is often preferable to formulate numerous quantifiably good approaches that provide distinct alternative solutions to the particular problem. This is because decision-making frequently involves complex problems possessing incompatible performance objectives and contain competing design requirements which prove very difficult—if not impossible—to capture and quantify at the time that the supporting decision models are actually formulated. There are invariably unmodelled design issues, not apparent at the time of model construction, which can greatly impact the acceptability of the model's solutions. Consequently, it can prove preferable to generate numerous alternatives providing contrasting perspectives to the problem. These alternatives should be near-optimal with respect to the known modelled objective(s), but be fundamentally dissimilar from each other in terms of their decision variables. This solution approach has been referred to as modelling to generate-alternatives (MGA). This chapter provides an efficient computational procedure for simultaneously generating multiple different alternatives to an optimal solution using the Firefly Algorithm. The efficacy and efficiency of this approach will be illustrated using a two-dimensional, multimodal optimization test problem.

Keywords Firefly Algorithm · Biologically-inspired metaheuristic · Modelling-to-generate-alternatives

1 Introduction

Typical "real world" decision-making involves complex problems that possess design requirements which are frequently very difficult to incorporate into their supporting mathematical programming formulations and tend to be riddled with

J.S. Yeomans (✉)
OMIS Area, Schulich School of Business, York University, 4700 Keele Street,
Toronto, ON M3J 1P3, Canada
e-mail: syeomans@schulich.yorku.ca

© Springer International Publishing AG 2018
X.-S. Yang (ed.), *Nature-Inspired Algorithms and Applied Optimization*,
Studies in Computational Intelligence 744, https://doi.org/10.1007/978-3-319-67669-2_12

competing performance objectives [3, 11, 13]. While optimal solutions provide provably best solutions to the mathematical constructions, they are generally not the best solutions to the underlying real problems as there are invariably unquantified issues and unmodelled objectives not apparent during the model formulation phase [3, 11, 12]. Hence, it is generally considered desirable to generate a reasonable number of very different alternatives that provide multiple, contrasting perspectives to the specified problem [16]. These alternatives should preferably all possess good (i.e. near-optimal) measures with respect to all of the modelled objective(s), but be as fundamentally different as possible from each other in terms of the system structures characterized by their decision variables. Several approaches collectively referred to as *modelling-to-generate-alternatives* (MGA) have been developed in response to this multi-solution creation requirement [2, 7–10, 12, 16].

The primary motivation behind MGA is to produce a manageably small set of alternatives that are good with respect to all known objective(s) yet are as different as possible from each other within the decision space. The resulting set of alternatives should provide diverse approaches that all perform similarly with respect to the known modelled objectives, yet very differently with respect to any unmodelled issues [7, 13]. Clearly the decision-makers must conduct subsequent evaluations to ascertain which alternatives are most applicable to their specific circumstances. Therefore, MGA methods must necessarily be regarded as decision support processes in contrast to the explicit solution determination methods of optimization.

In this chapter, it is shown how to simultaneously generate sets of maximally different alternatives by implementing a modified version of the nature-inspired Firefly Algorithm (FA) [14, 15] by extending previous concurrent MGA approaches [5–10]. For optimization, it has been demonstrated that the FA is more computationally efficient than such metaheuristics as enhanced particle swarm optimization, simulated annealing, and genetic algorithms [4, 15]. The MGA procedure extends the earlier efforts of Imanirad et al. [5–10] to now permit the simultaneous generation of the desired number of alternatives in a single computational run. This new simultaneous FA-based MGA procedure is extremely computationally efficient. This chapter illustrates the efficacy of the new FA approach for simultaneously constructing multiple, good-but-very-different solution alternatives on a 100-peak multimodal optimization test problem [12].

2 Firefly Algorithm for Optimization

While this section provides only an abridged outline of the steps involved in the FA process [4–6], more comprehensive explanations appear in [14, 15]. The FA is a biologically-inspired, population-based metaheuristic. Each firefly in the population represents one potential solution to a problem and the population of fireflies should initially be distributed uniformly and randomly throughout the solution space. The solution approach employs three idealized rules. (i) All fireflies within the population are considered "unisex", so that any one firefly could potentially be attracted to any other firefly irrespective of their sex. (ii) The brightness of a firefly is determined by the overall landscape of the objective function. Namely, for a maximization problem, the brightness is simply considered to be proportional to the value of the objective function. (iii) The relative attractiveness between any two fireflies is directly proportional to their respective brightness. This implies that for any two flashing fireflies, the less bright firefly will always be inclined to move towards the brighter one. However, attractiveness and brightness both decrease as the relative distance between the fireflies increases. If there is no brighter firefly within its visible neighborhood, then the particular firefly will move about randomly. Based upon these three rules, the basic operational steps of the FA can be summarized within the following pseudo-code [15].

Objective Function $F(X)$, $X = (x_1, x_2,\ldots x_d)$

Generate the initial population of n fireflies, X_i, $i = 1, 2,\ldots, n$

Light intensity I_i at X_i is determined by $F(X_i)$

Define the light absorption coefficient γ

while (t < MaxGeneration)

    for $i = 1: n$, all n fireflies

        for $j = 1: n$,all n fireflies (inner loop)

            if $(I_i < I_j)$, Move firefly i towards j; **end if**

            Vary attractiveness with distance r via $e^{-\gamma r}$

        end for j

    end for i

    Rank the fireflies and find the current global best solution G^*

end while

Postprocess the results

In the FA, there are two important issues to resolve: the variation of light intensity and the formulation of attractiveness. For simplicity, it can always be assumed that the attractiveness of a firefly is determined by its brightness which in turn is associated with its encoded objective function value. In the simplest case, the brightness of a firefly at a particular location X would be its calculated objective value $F(X)$. However, the attractiveness, β, between fireflies is relative and will vary with the distance r_{ij} between firefly i and firefly j. In addition, light intensity decreases with the distance from its source, and light is also absorbed in the media, so the attractiveness needs to vary with the degree of absorption. Consequently, the overall attractiveness of a firefly can be defined as

$$\beta = \beta_0 \exp(-\gamma r^2)$$

where β_0 is the attractiveness at distance $r = 0$ and γ is the fixed light absorption coefficient for the specific medium. If the distance r_{ij} between any two fireflies i and j located at X_i and X_j, respectively, is calculated using the Euclidean norm, then the movement of a firefly i that is attracted to another more attractive (i.e. brighter) firefly j is determined by

$$X_i = X_i + \beta_0 \exp\left(-\gamma(r_{ij})^2\right)(X_i - X_j) + a\varepsilon_i.$$

In this expression of movement, the second term is due to the relative attraction and the third term is a randomization component. Yang [15] indicates that α is a randomization parameter normally selected within the range [0,1] and ε_i is a vector of random numbers drawn from either a Gaussian or uniform (generally [−0.5,0.5]) distribution. It should be explicitly noted that this expression represents a random walk biased toward brighter fireflies and if $\beta_0 = 0$, it becomes a simple random walk. The parameter γ characterizes the variation of the attractiveness and its value determines the speed of the algorithm's convergence. For most applications, γ is typically set between 0.1 to 10 [4, 15]. For all computational approaches for the FA considered in this study, the variation of attractiveness parameter γ was fixed at 5 while the randomization parameter α was initially set at 0.6, but is then gradually decreased to a value of 0.1 as the procedure approaches its maximum number of iterations (see [15]).

In any given optimization problem, for a very large number of fireflies $n \gg k$, where k is the number of local optima, the initial locations of the n fireflies should be distributed relatively uniformly throughout the entire search space. As the FA proceeds, the fireflies begin to converge into all of the local optima (including the global ones). Hence, by comparing the best solutions among all these optima, the global optima can easily be determined. Yang [15] proves that the FA will approach the global optima when $n \to \infty$ and the number of iterations t, is set so that $t \gg 1$. In reality, the FA has been found to converge extremely quickly with n set in the range 20–50 [4, 14].

Two important limiting or asymptotic cases occur when $\gamma \to 0$ and when $\gamma \to \infty$. For $\gamma \to 0$, the attractiveness is constant $\beta = \beta_0$, which is equivalent to having a light intensity that does not decrease. Thus, a firefly would be visible to every other firefly anywhere within the solution domain. Hence, a single (usually global) optima can easily be reached. If the inner loop for j in the pseudo-code is removed and X_j is replaced by the current global best G^*, then this implies that the FA reverts to a special case of the accelerated particle swarm optimization (PSO) algorithm. Subsequently, the computational efficiency of this special FA case is equivalent to that of enhanced PSO. Conversely, when $\gamma \to \infty$, the attractiveness is essentially zero along the sightline of all other fireflies. This is equivalent to the case where the fireflies randomly roam throughout a very thick foggy region with no other fireflies visible and each firefly roams in a completely random fashion.

This case corresponds to a completely random search method. As the FA operates between these two asymptotic extremes, it is possible to adjust the parameters α and γ so that the FA can outperform both a random search and the enhanced PSO algorithms [4].

The computational efficiencies of the FA will be exploited in the subsequent MGA solution approach. As noted, within the two asymptotic extremes, the population in the FA can determine both the global optima as well as the local optima concurrently. This concurrency of population-based solution procedures holds huge computational and efficiency advantages for MGA purposes [16]. An additional advantage of the FA for MGA implementation is that the different fireflies essentially work independently of each other, implying that FA procedures are better than PSO and genetic algorithms for MGA because the fireflies will tend to aggregate more closely around each local optimum [4, 15]. Consequently, with a judicious selection of parameter settings, the FA will simultaneously converge extremely quickly into both local and global optima [4, 14, 15].

3 Modelling to Generate Alternatives

Most optimization methods appearing in the mathematical programming literature have focused almost entirely on the production of single optimal solutions to single-objective problem formulations or, equivalently, on the generation of non-inferior sets of solutions to multi-objective instances [2, 5, 6, 11, 13]. While such algorithms may efficiently generate solutions to the derived complex mathematical models, whether these outputs actually establish "best" approaches to the underlying real problems is debatable [2, 3, 11, 12]. In most "real world" applications, there are innumerable system requirements and objectives that are never included or apparent in the decision formulation stage [3, 13]. Furthermore, it may never be possible to explicitly incorporate all of the subjective components because there are frequently many incompatible, competing, design interpretations and, perhaps, adversarial stakeholders involved. Therefore most of the subjective aspects of a problem necessarily remain unquantified and unmodelled in the construction of the resultant decision models. This occurs frequently in situations where final decisions are constructed based not only upon clearly stated and modelled objectives, but also upon more fundamentally subjective socio-political-economic goals and stakeholder preferences [16]. Numerous "real world" examples describing these types of incongruent modelling dualities are discussed in [1, 2, 12, 17].

When unquantified objectives and unmodelled issues are suspected, then non-conventional approaches should be undertaken that not only search the feasible region for noninferior solutions, but also explore the feasible region for obviously *inferior* alternatives to the formulated problem. In particular, any search for good alternatives to problems known or suspected to contain unmodelled objectives must focus not only on the non-inferior solution set, but also necessarily on an explicit exploration of the problem's inferior decision space.

To illustrate the implications of an unmodelled objective on a decision search, assume that the optimal solution for a quantified, single-objective, maximization decision problem is X^* with corresponding objective value $Z1^*$. Now suppose that there exists a second, unmodelled, maximization objective $Z2$ that subjectively reflects some unquantifiable "political acceptability" component. Let the solution X^a, belonging to the noninferior, 2-objective set, represent a potential best compromise solution if both objectives could somehow have been simultaneously evaluated by the decision-maker. While X^a might be viewed as the best compromise solution to the real problem, it would appear inferior to the solution X^* in the quantified mathematical model, since it must be the case that $Z1^a \leq Z1^*$. Consequently, when unmodelled objectives are factored into the decision making process, mathematically inferior solutions for the modelled problem can prove optimal to the underlying real problem. Therefore, when unmodelled objectives and unquantified issues might exist, different solution approaches are needed in order to not only search the decision space for the noninferior set of solutions, but also to simultaneously explore the decision space for inferior alternative solutions to the modelled problem. Population-based solution methods such as the FA permit concurrent searches throughout a feasible region and thus prove to be particularly adept procedures for searching through a problem's decision space.

The primary motivation behind MGA is to produce a manageably small set of alternatives that are quantifiably good with respect to the known modelled objectives yet are as different as possible from each other in the decision space. The resulting alternatives are likely to provide truly different choices that all perform somewhat similarly with respect to the modelled objective(s) yet very differently with respect to any unknown unmodelled issues. By generating a set of good-but-different solutions, the decision-makers can explore desirable qualities within the alternatives that may prove to satisfactorily address the various unmodelled objectives to varying degrees of stakeholder acceptability.

In order to properly motivate an MGA search procedure, it is necessary to supply a more mathematically formal definition of the goals of the MGA process [7, 12, 16]. Suppose the optimal solution to an original mathematical model is X^* with objective value $Z^* = F(X^*)$. The following model can then be solved to generate an alternative solution, X, that is maximally different from X^*:

$$\text{Maximize} \quad \Delta(X, X^*) = \sum_i \left| X_i - X_i^* \right|$$

$$\text{Subject : to} \quad \begin{array}{l} X \in D \\ \left| F(X) - Z^* \right| \leq T \end{array} \tag{[P1]}$$

where Δ represents some difference function (for clarity, shown as an absolute difference in this instance) and T is a targeted tolerance value specified relative to the problem's original optimal objective Z^*. T is a user-supplied value that determines how much of the inferior region is to be explored in the search for acceptable alternative solutions. This difference function concept can be extended into a measure of difference between a set of alternatives by replacing X^* in the objective of [P1] and calculating the overall sum (or some other function) of the differences of the pairwise comparisons between each pair of alternatives—subject to the condition that each alternative is feasible and falls within the specified tolerance constraint.

4 FA-Based Simultaneous MGA Computational Algorithm

The MGA method to be introduced produces a pre-determined number of near-optimal, maximally different alternatives, by modifying the value of the bound T in [P1] and using an FA to solve the corresponding, maximal difference problem. Each solution within the FA's population contains one potential set of p different alternatives. By exploiting the co-evolutionary solution structure within the population of the algorithm, the Fireflies collectively evolve each solution toward sets of different local optima within the solution space. In this process, each desired solution alternative undergoes the common search procedure of the FA. However, the survival of solutions depends not only upon how well the solutions perform with respect to the modelled objective(s), but also by how far away they are from all of the other alternatives generated in the decision space.

A direct process for generating alternatives with the FA is to iteratively solve the maximum difference model [P1] by incrementally updating the target T whenever a new alternative needs to be created and then re-running the algorithm. This iterative approach parallels the seminal Hop, Skip, and Jump (HSJ) MGA algorithm [2] in which, once an initial problem formulation has been optimized, supplementary alternatives are systematically produced one-by-one via an incremental adjustment of the target constraint to force the sequential generation of suboptimal solutions. While this approach is straightforward, it requires a recurrent execution of the optimization algorithm [5, 6, 16].

To improve upon the stepwise alternative approach of the HSJ algorithm, a concurrent MGA technique was subsequently designed based upon the concept of co-evolution [5–7, 9]. In the co-evolutionary approach, pre-specified stratified subpopulation ranges within the algorithm's overall population were established that collectively evolved the search toward the creation of the stipulated number of maximally different alternatives. Each desired solution alternative was represented by each respective subpopulation and each subpopulation underwent the common processing operations of the FA. The survival of solutions in each subpopulation depended simultaneously upon how well the solutions perform with respect to the modelled objective(s) and by how far away they are from all of the other alternatives. Consequently, the evolution of solutions in each subpopulation toward local optima is directly influenced by those solutions contained in all of the other subpopulations, which forces the concurrent co-evolution of each subpopulation towards good but maximally distant regions within the decision space according to [P1] [16].

By employing this co-evolutionary concept, it became possible to implement an MGA procedure that concurrently produced alternatives possessing objective function bounds analogous to those created by the sequential, iterative HSJ-styled approach. In contrast, while each alternative produced by an HSJ procedure is maximally different only from the overall optimal solution (together with its bound on the objective value which is at least x% different from the best objective (i.e. x = 1%, 2%, etc.)), the concurrent procedure generated alternatives that are no more than x% different from the overall optimal solution but with each one of these solutions being as maximally different as possible from every other generated alternative that was produced. Co-evolution is also much more efficient than sequential HSJ in that it exploits the inherent population-based searches of FA procedures to concurrently create its entire set of maximally different solutions using only a single population [7, 9].

While a concurrent approach exploits the population-based nature of the FA's solution approach, the co-evolution process occurs within each of the stratified subpopulations. The maximal differences between solutions in different subpopulations is based upon aggregate subpopulation measures. Conversely, in the following simultaneous MGA algorithm, each solution in the population contains exactly one entire set of alternatives and the maximal difference is calculated only for that particular solution (i.e. the specific alternative set contained within that solution in the population). Hence, by the evolutionary nature of the FA search procedure, in the subsequent approach, the maximal difference is simultaneously calculated for the specific set of alternatives considered within each specific solution—and the need for concurrent subpopulation aggregation measures is circumvented.

The steps in the simultaneous co-evolutionary alternative generation algorithm are as follows:

Initialization Stage: In this preliminary step, solve the original optimization problem to determine the optimal solution, X^*. As with prior solution approaches [5]–[10] and without loss of generality, it is entirely possible to forego this step and construct the algorithm to find X^* as part of its solution processing. However, such a requirement increases the number of computational iterations of the overall procedure and the initial stages of the processing focus upon finding X^* while the other elements of each population solution remain essentially "computational overhead". Based upon the objective value $F(X^*)$, establish P target values. P represents the desired number of maximally different alternatives to be generated within prescribed target deviations from the X^*.

Note: The value for P has to have been set a priori by the decision-maker.

Step 1. Create the initial population of size K in which each solution is divided into P equally-sized partitions. The size of each partition corresponds to the number of variables for the original optimization problem. A_p represents the pth alternative, $p = 1,...,P$, in each solution.

Step 2. In each of the K solutions, evaluate each $A_p, p = 1,...,P$, with respect to the modelled objective. Alternatives meeting their target constraint and all other problem constraints are designated as *feasible*, while all other alternatives are designated as *infeasible*. A solution can only be designated as feasible if all of the alternatives contained within it are feasible.

Step 3. Apply an appropriate elitism operator to each solution to rank order the best individuals in the population. The best solution is the feasible solution containing the most distant set of alternatives in the decision space (the distance measure is defined in Step 5). Note: Because the best solution to date is always retained in the population throughout each iteration of the FA, at least one solution will always be feasible. A feasible solution for the first step can always consists of P repetitions of X^*.

This step simultaneously selects a set of alternatives that respectively satisfy different values of the target T while being as far apart as possible (i.e. maximally different as defined in [P1]) from the other solutions generated. By the co-evolutionary nature of the FA, the alternatives are *simultaneously* generated in one pass of the procedure rather than the P implementations suggested by the necessary increments to T in problem [P1].

Step 4. Stop the algorithm if the termination criteria (such as maximum number of iterations or some measure of solution convergence) are met. Otherwise, proceed to Step 5.

Step 5. For each solution $k = 1,..., K$, calculate D_k, a distance measure between all of the alternatives contained within solution k.

As an illustrative example for determining a distance measure, calculate

$$D_k = \sum_{i=1\,to\,P} \sum_{j=1\,to\,P} \Delta(A_i, A_j).$$

This represents the total distance between all of the alternatives contained within solution k. Alternatively, the distance measure could be calculated by some other appropriately defined function.

Step 6. Rank the solutions according to the distance measure D_k objective—appropriately adjusted to incorporate any constraint violation penalties for infeasible solutions. The goal of maximal difference is to force alternatives to be as far apart as possible in the decision space from the alternatives of each of the partitions within each solution. This step orders the specific solutions by those solutions which contain the set of alternatives which are most distant from each other.

Step 7. Apply appropriate FA "change operations" to the each of the solutions and return to Step 2.

5 Computational Testing of Simultaneous MGA Algorithm

As alluded to in the earlier sections, non-mathematically orientated, "real world" planners generally prefer to be able to select from a set of "near-optimal" alternatives that significantly differ from each other in terms of the system structures characterized by their decision variables. The ability of the FA MGA procedure to simultaneously produce such maximally different alternatives will be demonstrated using a non-linear optimization problem taken from [12]. The mathematical formulation for this multimodal problem is:

$$\text{Maximize } F(x,y) = \sin(19\pi x) + \frac{x}{1.7} + \sin(19\pi y) + \frac{y}{1.7} + 2$$

$$0.0 \le x \le 1.0 \quad 0.0 \le y \le 1.0$$

The non-linear, feasible region contains 100 peaks separated by valleys in which the amplitudes of both the peaks and valleys increase as the values of the decision variables increase from the (0,0) toward (1,1). For the design parameters employed in this formulation, the best solution of $F(x, y) = 5.146$ occurs at point $(x, y) =$ (0.974, 0.974) [12].

In order to create the set of different alternatives, extra target constraints that varied the value of T by up to 1.5% between successive alternatives were placed into the original formulation in order to force the generation of solutions maximally different from the initial optimal solution (i.e. the values of the bound were set at 1.5, 3, 4.5%, etc. for the respective alternatives). The MGA maximal difference algorithm described in the previous section was run to produce the optimal solution and the 10 maximally different solutions shown in Table 1 and illustrated in Fig. 1.

Table 1 Objective values and solutions for the 11 maximally different alternatives

Increment	1.5% Increment between alternatives		
	F(x,y)	x	y
Optimal	5.14	0.97	0.97
Alternative 1	5.11	0.97	0.98
Alternative 2	5.06	0.98	0.87
Alternative 3	5.01	0.87	0.76
Alternative 4	4.98	0.87	0.98
Alternative 5	4.92	0.76	0.98
Alternative 6	4.90	0.87	0.66
Alternative 7	4.77	0.45	0.87
Alternative 8	4.73	0.98	0.34
Alternative 9	4.66	0.13	0.97
Alternative 10	4.65	0.98	0.13

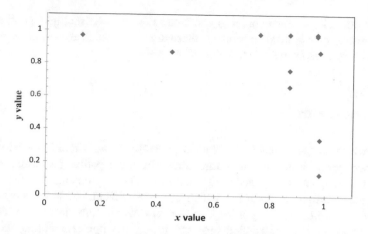

Fig. 1 Dispersion of the maximally different alternatives throughout the decision space

As described earlier, most "real world" optimization applications tend to be riddled with incongruent performance requirements that are exceedingly difficult to quantify. Consequently, it is preferable to create a set of quantifiably good alternatives that provide very different perspectives to the potentially unmodelled performance design issues during the policy formulation stage. The unique performance features captured within these dissimilar alternatives can result in very different system performance with respect to the unmodelled issues, hopefully thereby addressing some of the unmodelled issues into the actual solution process.

The example in this section underscores how a co-evolutionary MGA modelling perspective can be used to simultaneously generate multiple alternatives that satisfy known system performance criteria according to the prespecified bounds and yet remain as maximally different from each other as possible in the decision space. In addition to its alternative generating capabilities, the FA component of the MGA approach simultaneously performs extremely well with respect to its role in function optimization. It should be explicitly noted that the cost of the overall best solution produced by the MGA procedure is indistinguishable from the one determined in [12].

The computational example has demonstrated several important findings with respect to the simultaneous FA-based MGA method: (i) The co-evolutionary capabilities within the FA can be exploited to generate more good alternatives than planners would be able to create using other MGA approaches because of the evolving nature of its population-based solution searches; (ii) By the design of the MGA algorithm, the alternatives generated are good for planning purposes since all of their structures will be maximally different from each other (i.e. these differences are not just simply different from the overall optimal solution as in an HSJ-style approach to MGA); and, (iv) The approach is computationally efficient since it need only be run a single time in order to generate its entire set of multiple, good solution alternatives (i.e. to generate n solution alternatives, the MGA algorithm needs to run exactly once irrespective of the value of n).

6 Conclusions

"Real world" decision-making problems generally possess multidimensional performance specifications that are compounded by incompatible performance objectives and unquantifiable modelling features. These problems usually contain incongruent design requirements which are very difficult—if not impossible—to capture at the time that supporting decision models are formulated. Consequently, there are invariably unmodelled problem facets, not apparent during the model construction, that can greatly impact the acceptability of the model's solutions to those end users that must actually implement the solution. These uncertain and competing dimensions force decision-makers to integrate many conflicting sources into their decision process prior to final solution construction. Faced with such incongruencies, it is unlikely that any single solution could ever be constructed that simultaneously satisfies all of the ambiguous system requirements without some significant counterbalancing involving numerous tradeoffs. Therefore, any ancillary modelling techniques used to support decision formulation have to somehow simultaneously account for all of these features while being flexible enough to encapsulate the impacts from the inherent planning uncertainties.

In this chapter, an MGA procedure was presented that demonstrated how the population structures of a computationally efficient FA could be exploited to simultaneously generate multiple, maximally different, near-best alternatives. In this MGA capacity, the approach produces numerous solutions possessing the requisite structural characteristics, with each generated alternative guaranteeing a very different perspective to the problem. Since FA techniques can be modified to solve a wide variety of problem types, the practicality of this MGA approach can clearly be extended into numerous disparate planning applications. These extensions will be studied in future research.

References

1. Baugh, J.W., Caldwell, S.C., Brill, E.D.: A mathematical programming approach for generating alternatives in discrete structural optimization. Eng. Optim. **28**(1), 1–31 (1997)
2. Brill, E.D., Chang, S.Y., Hopkins, L.D.: Modelling to generate alternatives: the HSJ approach and an illustration using a problem in land use planning. Manag. Sci. **28**(3), 221–235 (1982)
3. Brugnach, M., Tagg, A., Keil, F., De Lange, W.J.: Uncertainty matters: computer models at the science-policy interface. Water Resour. Manage **21**, 1075–1090 (2007)
4. Gandomi, A.H., Yang, X.S., Alavi, A.H.: Mixed variable structural optimization using firefly algorithm. Comput. Struct. **89**(23–24), 2325–2336 (2011)

5. Imanirad, R., Yang, X.S., Yeomans, J.S.: A computationally efficient, biologically-inspired modelling-to-generate-alternatives method. J. Comput. **2**(2), 43–47 (2012)
6. Imanirad, R., Yang, X.S., Yeomans, J.S.: A Co-evolutionary, Nature-Inspired Algorithm for the Concurrent Generation of Alternatives. J. Comput. **2**(3), 101–106 (2012)
7. Imanirad, R., Yeomans, J.S.: Modelling to generate alternatives using biologically inspired algorithms. In: Yang, X.S. (ed.), Swarm Intelligence and Bio-Inspired Computation: Theory and Applications Elsevier, Amsterdam, Netherlands, pp. 313–333 (2013)
8. Imanirad, R., Yang, X.S., Yeomans, J.S.: Modelling-to-generate-alternatives via the firefly algorithm. J. Appl. Oper. Res. **5**(1), 14–21 (2013)
9. Imanirad, R., Yang, X.S., Yeomans, J.S.: A concurrent modelling to generate alternatives approach using the firefly algorithm. Int. J. Decis. Support Syst. Technol. **5**(2), 33–45 (2013)
10. Imanirad, R., Yang, X.S., Yeomans, J.S.: A biologically-inspired metaheuristic procedure for modelling-to-generate-alternatives. Int. J. Eng. Res. Appl. **3**(2), 1677–1686 (2013)
11. Janssen, J.A.E.B., Krol, M.S., Schielen, R.M.J., Hoekstra, A.Y.: The effect of modelling quantified expert knowledge and uncertainty information on model based decision making. Environ. Sci. Policy **13**(3), 229–238 (2010)
12. Loughlin, D.H., Ranjithan, S.R., Brill, E.D., Baugh, J.W.: Genetic algorithm approaches for addressing unmodeled objectives in optimization problems. Eng. Optim. **33**(5), 549–569 (2001)
13. Walker, W.E., Harremoes, P., Rotmans, J., Van der Sluis, J.P., Van Asselt, M.B.A., Janssen, P., Krayer von Krauss, M.P.: Defining uncertainty—a conceptual basis for uncertainty management in model-based decision support. Integr. Assess. **4**(1), 5–17 (2003)
14. Yang, X.S.: Firefly algorithms for multimodal optimization. Lecture Notes Comput. Sci. **5792**, 169–178 (2009)
15. Yang, X.S.: Nature-Inspired Metaheuristic Algorithms 2nd Ed. Luniver Press, Frome UK (2010)
16. Yeomans, J.S., Gunalay, Y.: Simulation-optimization techniques for modelling to generate alternatives in waste management planning. J. Appl. Oper. Res. **3**(1), 23–35 (2011)
17. Zechman, E.M., Ranjithan, S.R.: An evolutionary algorithm to generate alternatives (EAGA) for engineering optimization problems. Eng. Optim. **36**(5), 539–553 (2004)

Optimization of Relay Placement in Wireless Butterfly Networks

Quoc-Tuan Vien

Abstract As a typical model of multicast network, wireless butterfly networks (WBNs) have been studied for modelling the scenario when two source nodes wish to convey data to two destination nodes via an intermediary node namely relay node. In the context of wireless communications, when receiving two data packets from the two source nodes, the relay node can employ either physical-layer network coding or analogue network coding on the combined packet prior to forwarding to the two destination nodes. Evaluating the energy efficiency of these combination approaches, energy-delay trade-off (EDT) is worth to be investigated and the relay placement should be taken into account in the practical network design. This chapter will first investigate the EDT of network coding in the WBNs. Based on the derived EDT, algorithms that optimize the relay position will be developed to either minimize the transmission delay or minimize the energy consumption subject to constraints on power allocation and location of nodes. Furthermore, considering an extended model of the WBN, the relay placement will be studied for a general wireless multicast network with multiple source, relay and destination nodes.

Keywords Wireless butterfly network · Wireless multicast network · Network coding · Energy-delay tradeoff · Relay placement

1 Introduction

As wireless communications is growing with emerging enhanced technologies, data transmission over wireless medium turns out to be more reliable and more secured. The broadcast nature of the wireless media has been exploited to enable a variety of communication mechanisms and algorithms for enhancing the performance of the wireless communications. Apparently, there exist a number of nodes in a network and a question that can be raised is why they do not help each other in the data transmission between two end nodes. The energy waste and the unwanted interference in

Q.-T. Vien (✉)
Middlesex University, The Burroughs, London NW4 4BT, UK
e-mail: q.vien@mdx.ac.uk

© Springer International Publishing AG 2018
X.-S. Yang (ed.), *Nature-Inspired Algorithms and Applied Optimization*,
Studies in Computational Intelligence 744, https://doi.org/10.1007/978-3-319-67669-2_13

the shared media between the nodes that used to be regarded as drawbacks of the wireless communications can become a potential resource in assisting the communication between them. The attenuation in signal strength caused by severe fading of the source-destination link or even completely corrupted link could be solved with the help of intermediate nodes whose channels are independent of the channel between the source and destination nodes. The probability of successful transmission is therefore improved for a more reliable communication if these issues are satisfactorily addressed.

Cooperative communications, also known as relay communications or user cooperation in the preliminary works of Sendonaris et al. in 2003 [1, 2] and Laneman et al. in 2004 [3], has attracted an increasing interest in wireless communications aiming at throughput enhancement and quality improvement by exploiting spatial diversity gains. The relays can be used not only to improve service quality and link capacity for local users which are located near the source but also to enhance coverage and throughput for remote users. Inspired by the benefits of the relays, relay-assisted communications has been incorporated in various types of wireless systems; for instance, cellular networks in Loa et al. in 2010 [4] and Sheng et al. in 2011 [5], ad hoc networks in Sharma et al. in 2011 [6], sensor networks in Sun et al. in 2009 [7], ultra-wideband body area networks in Chen et al. in 2009 [8], and storage networks in Dimakis et al. in 2011 [9].

Conventionally, data traverses along relays in a store-and-forward manner, and thus the use of the relays does not immediately increase network throughput. In 2000, Ahlswede et al. [10] proposed the idea of network coding (NC) to increase the system throughput in lossless networks. Later in 2003, Koetter and Medard [11] developed an algebraic approach to enable the applicability of the NC. The NC has been then applied at the relays to dramatically improve the throughput of wireless relay networks, such as Zhang et al. in 2006 [12], Katti et al. in 2007 [13], and Louie et al. in 2010 [14]. By employing the NC at the relay nodes to coordinate the transmission among nodes in an efficient way, the optimality of the bandwidth could be achieved. Many NC-based protocols have also been proposed for some particular relay channel topologies such as relay-assisted bidirectional channels in Ju et al. in 2010 [15], broadcast channels in Nguyen et al. in 2009 [16], multicast channels in Chen et al. in 2010 [17], and unicast channels in Liu et al. in 2009 [18]. As a specific model of the multicast channels, butterfly networks have been investigated, e.g., Zhan et al. in 2010 [19] and Hu et al. in 2011 [20], in which the NC is applied at the relay node to help two source nodes simultaneously transmit their information to two destination nodes.

This chapter is devoted to investigating the energy efficiency for reliable communications in wireless butterfly networks (WBNs) employing various NC techniques. In particular, the relay placement (RP) problem for energy-efficient and reliable relaying in the WBNs will be discussed. In the rest of this chapter, Sect. 2 will first introduce the background of cooperative diversity starting from its foundation including the concepts of diversity and multiple-input multiple-output (MIMO). Basic principles and specific protocols for cooperative communications in wireless relay networks will be presented along with cooperative diversity techniques via distributed space-time-frequency coding and NC for a variety of relay network topologies. Section 3 will describe the system model of a typical WBN employing the NC techniques. As a common approach to improve the reliability of the wireless communications, Sect. 4 will discuss hybrid automatic repeat request with incremental redundancy (HARQ-IR) protocol and, in particular, this section will provide a detailed analysis for the energy-delay tradeoff of the HARQ-IR protocol with the NC techniques in the WBN. The RP problem in the WBN will be then formulated and optimized in Sect. 5. An extension of the RP for a general wireless multicast network will be discussed in Sect. 6. Finally, Sect. 7 will conclude this chapter with suggestion for future works.

2 Background

In this section, the basic concepts of diversity techniques will be firstly described, including temporal diversity, frequency diversity, and spatial diversity. As an approach to achieve the spatial diversity in terms of antenna diversity, MIMO systems will be presented with some well-known space-time-frequency coding schemes, based on which the motivation of cooperative diversity will be then discussed with an overview of cooperative protocols and techniques. The section will conclude by introducing NC which is regarded as a new technique to improve the throughput of wireless cooperative relay networks.

2.1 Diversity Techniques in Wireless Communications

In a communication system consisting of a sender and a receiver, the reliability of data transmission can be improved by providing more than one path between them. This technique is the main idea behind the term "diversity". In fact, by providing multiple replicas or copies of the transmitted signals over independent channels, the receiver can more reliably decode the transmitted signal by either combining all the received signal, namely a maximal ratio combiner, or selecting the best signal with the highest signal-to-noise ratio (SNR), namely a selection combiner, or choosing the signal with an SNR exceeding a threshold, namely a threshold combiner. In order to define the diversity quantitatively, Zheng and Tse in 2003 [21] formulated the relationship between the error probability, i.e., P_e, and the received SNR, i.e., γ, through a diversity gain as

$$G_d \triangleq - \lim_{\gamma \to \infty} \frac{\log P_e}{\log \gamma}. \tag{1}$$

It can be seen in Eq. (1) that the diversity gain G_d is the slope of the P_e curve in terms of γ in a log-log scale. This means that a large diversity gain is preferred to achieve a reduced P_e at a higher data rate. The problems are how to provide various copies of the transmitted signal to the receiver in an efficient way in terms of power, time, bandwidth, and complexity, and, how to take advantage of these copies at the receiver to achieve the lowest P_e. To cope with these two issues, various diversity methods, as will be shown below, can be implemented.

2.1.1 Temporal Diversity

In temporal diversity, copies of the transmitted signal are sent at different time intervals. The time interval between two transmitted replicas should be longer than the coherence time of the channel to make the fading channels uncorrelated and thus the temporal diversity can be obtained. However, the temporal diversity is bandwidth inefficient due to the delay that may be suffered at the receiver in the case of a slow fading channel, i.e., a large coherence time of the channel.

2.1.2 Frequency Diversity

Instead of using temporal separation between different replicas of the transmitted signal, the transmission of these copies can be carried out over different carrier frequencies to achieve frequency diversity. Similar to temporal diversity, frequency diversity can be achieved when there exists a necessary separation between two carrier frequencies which should be larger than the coherence bandwidth of the channel. The frequency diversity is therefore also bandwidth inefficient and the capability of frequency tuning is required at the receiver.

2.1.3 Spatial Diversity

In spatial diversity, multiple antennas are employed at the sender and/or the receiver to transmit and/or receive different copies of a signal. It is therefore also known as antenna diversity in Winters et al. in 1994 [22]. The spatial diversity does not suffer from bandwidth inefficiency which is a major drawback of the temporal and frequency diversity. However, in order to achieve the spatial diversity, a number of antennas are required at either the transmitter side or the receiver side or both sides. Also, the antennas deployed on a device are normally separated by at least half of a wavelength of the transmission frequency to guarantee that the fading channels are independent or at least low-correlated. Obviously, the condition of the antenna separation could be easily satisfied at large base stations, but may not be applicable for small handheld devices. This accordingly motivates the concept of user cooperation with cooperative diversity, which will be discussed in details in Sect. 2.3.

2.2   MIMO Systems and Space-Time-Frequency Coding

In radio communications, the negative effects of fading phenomena on quality and data rate in wireless communications can be combated with diversity in the spatial domain via the employment of multiple antennas. The concept of MIMO systems is defined for systems where multiple antennas are deployed at source and destination nodes to achieve the spatial diversity. The works of Foschini and Gans in 1998 [23] and Telatar in 1999 [24] on various MIMO techniques are regarded as the first studies of the MIMO systems. These two pioneering publications showed that a large capacity gain could be achieved with the MIMO systems compared to the traditional single-input single-output (SISO) systems. These findings have motivated a large number of research works on MIMO systems.

Also in 1998 and 1999, Tarokh et al. [25] and Guey et al. [26] derived two space-time coding (STC) design criteria based on the upper bound of pairwise error probability. One is rank criterion or diversity criterion in which an STC is said to achieve full diversity if the code difference matrix is of full rank. The other is the product criterion or determinant criterion in which the coding gain of an STC is determined by the product of eigenvalues of the code difference matrix, and thus it should be large to obtain a high coding gain.

One important means of achieving spatial diversity is by deployment of multiple antennas at the transmitter, which is known as transmit diversity. Tarokh et al. in 1998 [25] proposed space-time trellis coding (STTC) that can effectively exploit transmit diversity, but its decoding complexity increases exponentially with the transmission rate. Thus different transmit diversity schemes should be proposed to reduce the complexity of the decoding algorithm in STTC. Dealing with this issue, in the same year with the STTC, Alamouti [27] designed a new orthogonal transmit diversity scheme using two transmit antennas. This coding scheme has been widely known as the Alamouti code in honour of its inventor. The Alamouti scheme was later generalized by Tarokh et al. in 2001 [28], Ganesan and Stoica in 2001 [29], and Tirkkonen and Hottinen in 2002 [30] for more than two transmit antennas and characterized as orthogonal space-time block coding (OSTBC) for MIMO systems. Other STCs were also designed using some specific matrix structures; for instance, quasi-orthogonal space-time block code (QOSTBC) in Jafarkhani in 2001 [31], rotated QOSTBC in Su and Xia in 2004 [32], cyclic STC in Hughes in 2000 [33], unitary STC in Hochwald et al. in 2000 [34], diagonal algebraic STC in Damen et al. in 2002 [35], and group-wise STC in Du and Li in 2006 [36].

Considering wideband wireless communications when the systems are required to operate at a high data rate, the communication channels now become frequency-selective fading. The STC schemes for the narrowband communications are shown to be inappropriate and are thus required to be redesigned. Indeed, the frequency-selective or multipath fading channels cause not only severe attenuation in signal strength, but also a large amount of inter-symbol interference (ISI), which makes the signal detection unreliable. However, these multiple paths can offer multipath diversity or frequency diversity. Many studies were then dedicated to extend OST-

BCs to frequency-selective fading channels, such as Linkskog and Paulraj in 2000 [37], Al-Dhahir in 2001 [38], and Zhou and Giannakis in 2001 [39]. The newly designed codes can be viewed as a block implementation of the Alamouti code. Another approach to mitigate the frequency selectivity is orthogonal frequency-division multiplexing (OFDM) which uses multiple subcarriers to mitigate the fading effects. For wideband MIMO-OFDM systems, Agrawal et al. in 1998 [40] proposed space-frequency coding (SFC) by converting the time domain in the STC to the frequency domain. Different versions of the SFC were then developed and analyzed in Lu and Wang in 2000 [41], Blum et al. in 2001 [42], and Su et al. in 2005 [43] based on the mapping from different STCs. Adapting the SFCs to several consecutive OFDM blocks, Gong and Letaief in 2001 [44] designed space-time-frequency coding (STFC) for two transmit antennas, which was then extended by Liu et al. in 2002 [45] and Molisch et al. in 2002 [46] for multiple transmit antennas.

2.3 Cooperative Diversity Protocols and Techniques

MIMO systems are only feasible when devices employ multiple co-located antennas. However, the installation of multiple antennas may be impractical due to the inherent hardware limitation of some small devices. Instead, these devices can collaborate to form a virtual multi-antenna system. The communication between a source node and a destination node can be realized in a cooperative manner with one or multiple cooperating nodes acting as relay node(s). Drawing from user cooperation to achieve some of the benefits of MIMO systems, this form of diversity is well known as cooperative diversity or user cooperation diversity.

2.3.1 Cooperative Protocols

A very classical relay channel including three terminals was initially introduced by van der Meulen in 1971 [47] where a relay terminal simply listens to the transmitted signal from a source terminal, processes it and then sends it to a destination terminal. For this relay channel model, Cover and Gamal in 1979 [48] was the first work investigating the capacity of the relay channel and also deriving the lower and upper bounds of its capacity. The ergodic capacity of the relay channel with different coding strategies was then analysed by Kramer et al. in 2005 [49].

Motivated by the three-terminal channel model, Laneman et al. in 2004 [3] proposed low-complexity cooperative protocols for a more general system model taking into account practical aspects. These protocols were developed for time-domain division multiple access (TDMA) systems operating in half-duplex mode. Specifically, two notable cooperative protocols, namely amplify-and-forward (AF) and decode-and-forward (DF), were defined and investigated for two types of relaying techniques including fixed and adaptive relaying. While fixed relaying was shown to be easy in implementation at the cost of low bandwidth efficiency, adaptive relaying via either selective or incremental relaying methods could increase the rate at the expense of high complexity. In fact, the overall rate using the fixed relaying scheme is reduced by half for two transmissions from the source and relay, and thus results in low bandwidth efficiency. Instead of simply processing and forwarding the data to the destination, the relay in the selective relaying scheme has the capability of deciding when to transmit based on a certain threshold of quality of the received signal, i.e., SNR. If the SNR of received signal at the relay is lower than a certain threshold, i.e., the channel from the source to the relay suffers from severe fading, then the relay does not carry out any processing. A relaying scheme known as incremental relaying could improve the performance further if the source knows when to repeat the transmission and the relay knows when the destination needs help. It can be appreciated that the adaptive relaying schemes require a high-complexity processing and a feedback channel from the destination to both the source and relay is specifically required for the incremental relaying.

- *DF Protocol*: In the DF protocol, the relay tries to decode the signal from the source and then transmits the decoded signal to the destination. Since the signal detected at the relay is possibly corrupted, it may cause meaningless cooperation to the eventual decision at the destination. In order to achieve the optimal detection, the destination needs to know the error statistics of the inter-user link. This method was also mentioned by Sendonaris et al. in 2003 [1, 2] for code division multiple access (CDMA) in cellular networks. Laneman et al. showed that the diversity of the DF protocol is limited to one due to the worst link from the source to the relay and from the source to the destination [3].
- *AF Protocol*: In the AF protocol, the relay only amplifies what it receives from the source and then transmits the amplified version to the destination. The destination combines the information sent by the source and the relay, and makes a final decision on the transmitted signal. Although the noises at the relay in the AF protocol are also amplified together with the information, the destination can make a better decision with two independently faded versions of the transmitted signal. Indeed, Laneman et al. in 2004 [3] showed that the AF protocol can achieve the full diversity.
- *Other Protocols*: Besides the DF and AF protocols, compress-and-forward (CF) protocol also attracted much attention in Cover and Gamal in 1979 [48] and Kramer et al. in 2005 [49]. In the CF protocol, the relay transmits to the destination a quantized and compressed version of the signal received from the source. At the destination, the signal received from the source is used as side information to decode the information from the relay. Another cooperative protocol that was studied by Hunter and Nosratinia in 2006 [50] is coded cooperation where error-control coding is included. In the coded cooperation, the relay transmits incremental redundancy to help the destination recover the original data more reliably by combining the codewords with redundancy from both the source and the relay.

2.3.2 Cooperative Diversity via Distributed Space-Time-Frequency Coding

Relaying protocols are in fact repetition-based cooperative diversity schemes designed to achieve spatial diversity gain as a virtual MIMO system. The benefits of these cooperative protocols are achieved at the price of decreasing bandwidth efficiency with the number of cooperating terminals since each relay requires its own channel or subchannel for repetition. Inspired by the work on STCs for MIMO systems, Laneman and Wornell in 2003 [51] proposed distributed STC (DSTC) to improve the bandwidth efficiency of the cooperative communications. The basic idea of the DSTC is that each single-antenna terminal in the relay network transmits a column of the original OSTBC that was designed for multiple co-located antennas in the MIMO systems. In a distributed fashion, multiple columns of the original OSTBC can be transmitted by the source and the relays to indirectly generate coding matrices which are of the same form as that of the OSTBC, and thus it was named distributed STBC (DSTBC). Various forms of the DSTBC were then devised for flat and frequency-selective fading channels.

With regard to flat fading channels, the first DSTBC was proposed by Laneman and Wornell in 2003 [51] for DF relaying protocol. Nabar et al. in 2004 [52] analyzed different DSTBCs for AF relay networks. The original design criteria for the conventional STC in the work of Tarokh et al. in 1998 [25] was shown to be able to apply to the DSTBCs. Laneman and Wornell showed that, in order to guarantee full diversity, the number of relay nodes should be less than the number of columns in the conventional OSTBC matrix [51]. The limit on the number of relays was then solved with a new class of the DSTBC for multiple relays designed by Yiu et al. in 2006 [53]. In this scheme, the signal transmitted by an active relay node is the product of an information-carrying code matrix and a unique node signature vector to ensure that no active node transmits data using the same coding vector. This method nevertheless operates under the DF protocol and requires high-complexity processing at the relay nodes. For the AF protocol, originated from the idea of linear dispersion STC in Hassibi and Hochwald in 2002 [54], a new DSTBC for the relaying systems was constructed by Jing and Hassibi in 2006 [55] and Jing and Jafarkhani in 2007 [56]. The transmitted signal at each relay is a linear function of its received signals without any decoding but only simple processing. However, these DSTBCs, in general, cannot offer a simple decoding mechanism at the destination. To address this problem, Yi and Kim in 2007 [57] designed a new DSTBC to obtain symbol-wise decodability.

On the subject of dealing with inherent frequency-selective fading phenomena in wideband wireless communications, the DSTBCs for flat fading channels are not directly applicable. In 2005, Scutari and Barbarossa proposed a DSTBC for multi-hop transmission over frequency-selective fading channels with DF protocol [58]. For optimal detection, the error statistics at the relays must be known at the destination. However, this cannot be easily implemented in many current wireless systems. Focused on the uplink communications system with fixed wireless relay stations, Anghel and Kaveh in 2003 [59] introduced the combination of DSTBC and OFDM signaling. Another DSTBC for frequency-selective fading channels was studied in Mheidat et al. in 2007 [60] and Tran et al. in 2009 and 2012 [61, 62] where the traditional equalization techniques were extended to the DSTBC with AF protocol. Considering two-relay networks also employing the AF protocol, Vien et al. in 2009–2011 proposed other DSTBCs to obtain maximal data rate, maximal diversity gain and decoupling detection of data blocks for a low-complexity receiver structure [63–65]. Based on the QOSTBC designed by Jafarkhani in 2001 [31] for co-located antennas in MIMO systems, distributed QOSTBC (DQOSTBC) was developed for four-relay networks by Vien et al. in 2009 in [66]. Inspired by the concept of coded cooperation, Vien et al. in 2009 [67] also designed a new DSTBC combined with the hybrid automatic repeat request (HARQ) for turbo-coded relay networks to enhance the performance of the DSTBC achieving both diversity gain and coding gain. The STFC in the MIMO systems was also adapted to the cooperative communications where a distributed space-time-frequency block code (DSTFBC) was proposed by Vien et al. in 2013 and 2014 [68, 69] for non-regenerative cognitive wireless relay networks.

2.4 Network Coding Techniques

Network coding (NC) was first proposed in 2000 by Ahswede et al. [10] to increase system throughput in lossless networks. This work was regarded as a seminal publication on the NC and has motivated a vast amount of research works. The principle of the NC is that intermediate nodes are allowed to mix signals received from multiple links for subsequent transmissions.

In a typical two-way single-relay network (TWSRN) with no direct link between two end terminals, four transmissions are conventionally required to exchange the data from two terminals through a relay. By applying NC, the number of transmissions could be reduced to three, including two transmissions from two terminals to the relay and one broadcast transmission of the mixed data from the relay to both terminals. Basically, the relay in an NC-based TWSRN mixes the signals received from two terminals, and then forwards the combined signal to both terminals. An end terminal can extract from the combined signal the data sent by another terminal based on its known signal. Since the relay in the NC-based TWSRN has to avoid the collision of the two data packets to detect the data from these two terminals separately, two transmissions are required in the first phase and hence three transmissions in total.

Considering the application of NC at the physical layer, also known as physical-layer NC (PNC), the number of transmissions in a TWSRN could be reduced to two due to the fact that two terminals can transmit simultaneously. The PNC can accordingly improve the network throughput by up to 100% and 50% over the conventional relaying and the NC-based relaying, respectively. Several studies have been dedicated to investigating the application of the PNC in the TWSRN. Specifically, Zhang et al. in 2006 [12] and Katti et al. in 2007–2008 [13, 70] were the preliminary works that applied the PNC concept for the wireless environment in the TWSRN. The performance of different PNC-based protocols was summarized by Louie et al. in 2010 [14] providing a detailed analysis of the bit error rate and the achievable data rate.

Basically, the processing at the relay with PNC techniques follows DF relaying protocol, while the AF-based PNC was named analog NC (ANC) by Katti et al. in 2007 [13]. In the following, these two NC techniques will be presented along with the relevant works on their application in various system models.

2.4.1 Physical-Layer Network Coding

In one-way relay network operated under DF protocol, the relay simply decodes the signal received from the source before forwarding to the destination. However, the relay in TWSRNs receives two data sequences simultaneously from two terminals. The challenging problem is how the relay decodes this mixed signal. Dealing with this problem, Zhang et al. proposed the first strategy, namely physical-layer NC (PNC), in 2006 [12]. In the PNC, the relay decodes the sum of two signals instead of decoding each signal individually. The sum of any two signals is characterized by a point in a lattice. Based on this lattice, the relay can decode its received mixed signal and then forward it to both terminals. Using the DF protocol, the PNC technique does not suffer from noise amplification at the relay, and thus a higher data rate is expected. However, the generation of the lattice for mapping would be complicated for a general scenario where the signals transmitted from two terminals use different modulation and coding schemes. Also, this strategy requires a perfect synchronization at the relay in both time and carrier when receiving signals from two terminals.

Another approach to PNC was proposed by Rankov and Wittneben in 2007 [71] where the relay separately decodes two signals from the mixed signal received from two terminals, then combines and forwards them to both terminals. The decoding of these two signals could be implemented using multiuser detection techniques in a well-known textbook of Verdu published in 1998 [72]. Similar to the technique proposed by Zhang et al. in 2006 [12], the error amplification at the relay does not have any effects on this strategy. As an advantage of this scheme, the generation of lattice matrices for mapping is not necessary; however, the separate decoding at the relay requires a higher complexity and produces a lower data rate.

2.4.2 Analog Network Coding

With AF protocol, the operation at the relay in TWSRNs is much simpler. The relay only amplifies the mix of two signals received from two terminals and then forwards this amplified version to both terminals. Since the relay performs processing upon the analog signals received from the terminals, this AF-based technique was named analog network coding (ANC) in Katti et al. in 2007 [13]. Similar to the AF protocol for one-way relay networks, the ANC has some advantages and disadvantages. The complexity at the relay using the ANC is significantly reduced compared to PNC technique, but the performance and data rate could be nonetheless affected since the noises at the relay are also amplified and forwarded to both terminals. Moreover, in order to extract the interested signal sent by another terminal from the mixed signal, channel information has to be estimated at both terminals to remove its own signal which is regarded as an interference to the interested signal.

2.4.3 Related Works on PNC and ANC

Related works on the application of NC in TWSRNs can be found in Popovski and Yomo in 2007 [73], Zhang et al. in 2009 [74], Song et al. in 2010 [75], Louie et al. in 2010 [14], Ju and Kim in 2010 [15], Wang et al. in 2010 [76], and Vien et al. in 2010–2014 [77–80], where different PNC and ANC approaches were investigated and evaluated. For instance, the performance analysis of PNC and ANC protocols for TWSRN in Popovski and Yomo in 2007 [73], Louie et al. in 2010 [14], and Ju and Kim in 2010 [15], beamforming for ANC in Zhang et al. in 2009 [74], differential modulation for ANC in Song et al. in 2010 [75], ANC for asynchronous TWSRNs in Wang et al. in 2010 [76], and automatic repeat request (ARQ) with PNC in Vien et al. in 2010 and 2011 [77, 78], channel quality indicator (CQI) reporting with PNC in AF-based TWSRNs in Vien et al. in 2012 and 2014 [79, 80].

Many NC-based protocols have also been proposed for a variety of relay channel topologies. A summary of these protocols can be found in the Ph.D. thesis of Vien et al. in 2013 [81], such as broadcast channels in Nguyen et al. in 2009 [16]; multicast channels in Chen et al. in 2010 [17], and Vien et al. in 2011–2015 [82–89]; unicast channels in Liu et al. in 2009 [18]; and multi-relay channels in Vien et al. in 2011–2013 [90–93]. As a specific model of the multicast channels, butterfly networks have been investigated in Zhan et al. in 2010 [19], Hu et al. in 2011 [20], and Vien et al. in 2013 and 2015 [94–96], in which the NC is applied at the relay node to help two source nodes simultaneously transmit their information to two destination nodes.

3 Network Coding in Wireless Butterfly Networks

The basic system model of a WBN is shown in Fig. 1 where data transmitted from two source nodes $\mathscr{S}_1$ and $\mathscr{S}_2$ to two destination nodes $\mathscr{D}_1$ and $\mathscr{D}_2$ is assisted by one relay node $\mathscr{R}$. A half-duplex system is considered where all nodes can either transmit or receive data, but not simultaneously. In the WBN, the NC is applied at $\mathscr{R}$ to help $\mathscr{S}_1$ and $\mathscr{S}_2$ simultaneously transmit their data packets $\mathbf{s}_1$ and $\mathbf{s}_2$, respectively, to $\mathscr{D}_1$ and $\mathscr{D}_2$ in two time slots. In the first time slot, $\mathscr{S}_1$ transmits $\mathbf{s}_1$ to both $\mathscr{R}$ and $\mathscr{D}_1$ while $\mathscr{S}_2$ transmits $\mathbf{s}_2$ to both $\mathscr{R}$ and $\mathscr{D}_2$. Then, $\mathscr{R}$ performs NC on the mixed signals received from $\mathscr{S}_1$ and $\mathscr{S}_2$ and broadcasts the network coded signals to both $\mathscr{D}_1$ and $\mathscr{D}_2$ in the second time slot. Accordingly, $\mathscr{D}_1$ can extract the signal transmitted

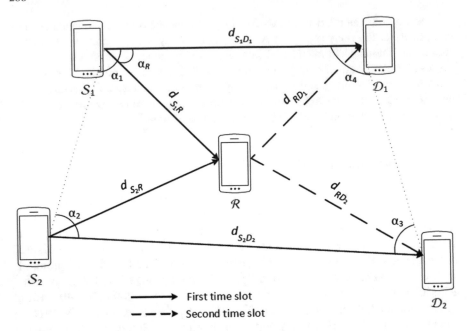

Fig. 1 System model of a wireless butterfly network

from $\mathscr{S}_2$, i.e., $\mathbf{s}_2$, and $\mathscr{D}_2$ can extract the signal transmitted from $\mathscr{S}_1$, i.e., $\mathbf{s}_1$. The data transmission in the first time slot consists of two direct (DR) transmissions ($\mathscr{S}_1 \rightarrow \mathscr{D}_1$ and $\mathscr{S}_2 \rightarrow \mathscr{D}_2$) and a multiple access (MA) transmission ($\{\mathscr{S}_1 \mathscr{S}_2\} \rightarrow \mathscr{R}$), while there is only a broadcast (BC) transmission ($\mathscr{R} \rightarrow \{\mathscr{D}_1 \mathscr{D}_2\}$) in the second time slot. Note that the DR and MA transmissions are carried out simultaneously in the first time slot due to the broadcast nature of the wireless medium. This chapter is focused on energy efficiency for a conventional butterfly network when the relay plays a role of coverage extension, facilitating message delivery of indirect links ($\mathscr{S}_1 \rightarrow \mathscr{D}_2$ and $\mathscr{S}_2 \rightarrow \mathscr{D}_1$), and thus it is assumed that there is no direct link between $\mathscr{S}_1$ and $\mathscr{D}_2$ and between $\mathscr{S}_2$ and $\mathscr{D}_1$.

For convenience, the main notation used in this chapter is listed in Table 1, unless stated otherwise.

4 Hybrid Automatic Repeat Request with Incremental Redundancy Protocol and Energy-Delay Tradeoff

In addition to the merit of NC techniques providing throughput improvement, the reliability and energy efficiency of data transmission should also be taken into consideration within communication systems. This is particularly the case in wireless environments where the communication channels often suffer from deep fading and background noise, and where the energy consumption of various communication and

Table 1 Main notation in the chapter

Notation	Meaning
d_{AB}, $\{A, B\} \in \{S_1, S_2, R, D_1, D_2\}$	Distance of link $\mathscr{A} - \mathscr{B}$
$\alpha_1, \alpha_2, \alpha_3, \alpha_4, \alpha_R$	Physical angles $\widehat{\mathscr{D}_1 \mathscr{S}_1 \mathscr{S}_2}$, $\widehat{\mathscr{S}_1 \mathscr{S}_2 \mathscr{D}_2}$, $\widehat{\mathscr{S}_2 \mathscr{D}_2 \mathscr{D}_1}$, $\widehat{\mathscr{D}_2 \mathscr{D}_1 \mathscr{S}_1}$, $\widehat{\mathscr{D}_1 \mathscr{S}_1 \mathscr{R}}$, respectively
P_i, $i = 1, 2$, P_R	Transmit powers of $\mathscr{S}_i$, $\mathscr{R}$, respectively
r_i, $i = 1, 2$, r_R	Transmission rate at $\mathscr{S}_i$, $\mathscr{R}$, respectively
h_{ii}, h_{iR}, h_{Ri}, $i = 1, 2$	Channel coefficients of links $\mathscr{S}_i \to \mathscr{D}_i$, $\mathscr{S}_i \to \mathscr{R}$, $\mathscr{R} \to \mathscr{D}_i$, respectively
$\mathbf{n}_{ii}$, $\mathbf{n}_R$, $\mathbf{n}_{Ri}$, $i = 1, 2$	Independent circularly symmetric complex Gaussian (CSCG) noise vectors of links $\mathscr{S}_i \to \mathscr{D}_i$, $\{\mathscr{S}_1, \mathscr{S}_2\} \to \mathscr{R}$, $\mathscr{R} \to \mathscr{D}_i$, respectively, with each entry having zero mean and unit variance
γ_{ii}, γ_{iR}, γ_{Ri}, $i = 1, 2$	Signal-to-noise ratio (SNR) of links $\mathscr{S}_i \to \mathscr{D}_i$, $\mathscr{S}_i \to \mathscr{R}$, $\mathscr{R} \to \mathscr{D}_i$, respectively
v	Pathloss exponent between a pair of transceiver nodes
$\kappa_{(\cdot)}$	Number of transmissions required in HARQ-IR protocol to transmit a data packet
$\delta_{(\cdot)}$	Effective delay (ED) of HARQ-IR protocol
$\varepsilon_{(\cdot)}$	Energy per bit (EB) of HARQ-IR protocol
$[a]_i$	i-th realisation of a random variable a
$\bar{a}$	Mean of a random variable a
$\log(\cdot)$	Binary logarithm function
$\ln(\cdot)$	Natural logarithm function
$E[\cdot]$	Statistical expectation function

networking devices causes an increasing carbon dioxide emission. To cope with the reliability issue, hybrid automatic repeat request (HARQ) protocols were proposed to reliably deliver information over error-prone channels such as the wireless medium. A detailed study of various error control mechanisms for digital communications is summarized in a textbook of Wicker in 1995 [97]. Specifically, Caire and Tunineti in 2001 [98] showed that the HARQ with incremental redundancy (HARQ-IR) can achieve the ergodic capacity of fading and interference channels. With respect to energy efficiency, an energy-delay tradeoff (EDT) tool was developed by Choi and To in 2012 [99] to evaluate the energy efficiency of HARQ-IR protocols for NC-based two-way relay systems.

4.1 Energy-Delay Tradeoff in Point-to-Point Wireless Links

In order to investigate HARQ-IR protocols with PNC and ANC techniques in WBNs, this section will first introduce briefly a simple HARQ-IR protocol for wireless point-to-point (P2P) communications along with EDT evaluation for this system model.

Over a P2P communication channel $\mathscr{S} \to \mathscr{D}$ employing the HARQ-IR protocol, node $\mathscr{S}$ encodes a data packet $\mathbf{d}$ into a sequence of N coded packets $\{\mathbf{c}_1, \mathbf{c}_2, \ldots, \mathbf{c}_N\}$. Then, $\mathscr{S}$ sequentially transmits $\mathbf{c}_k$, $k = 1, 2, \ldots, N$, to $\mathscr{D}$ until a positive acknowledgement (ACK) is received. The signal $\mathbf{y}_k$ received at node $\mathscr{D}$ when transmitted the k-th coded packet $\mathbf{c}_k$ from node $\mathscr{S}$ can be expressed through

$$\mathbf{y}_k = \sqrt{P} h_k \mathbf{x}_k + \mathbf{n}_k, \tag{2}$$

where P is the signal power, h_k is the channel gain of link $\mathscr{S} \to \mathscr{D}$ for the k-th packet transmission, $\mathbf{x}_k$ is the modulated signal of $\mathbf{c}_k$, and $\mathbf{n}_k$ is an independent circularly symmetric complex Gaussian (CSCG) noise vector with each entry having zero mean and unit variance.

Let κ_{P2P} denote the number of transmissions required in the HARQ-IR protocol to transmit a data packet from $\mathscr{S}$ to $\mathscr{D}$. Caire and Tuninetti in 2001 [98] expressed κ_{P2P} as

$$\kappa_{\text{P2P}} = \min \left\{ k \Big| \sum_{j=1}^{k} \log(1 + P|[h_k]_j|^2) > r_{\text{P2P}} \right\}, \tag{3}$$

where r_{P2P}, in bits/sec/Hertz (or b/s/Hz), denotes the link spectral efficiency of a capacity-achieving code in P2P communications. By using the same evaluation tool developed by Choi in 2012 [99], the EDT can be characterized by two normalized metrics including energy per bit (EB) in Joules/bit/Hertz (or J/b/Hz] and effective delay (ED) in secs/bit/Hertz (or s/b/Hz). Here, the EB and ED are normalized over the link spectral efficiency r_{P2P}. Let δ_{P2P} and ε_{P2P} denote the ED and EB, respectively, of the HARQ-IR protocol for the P2P communications. These metrics can be written as

$$\delta_{\text{P2P}} = \frac{\bar{\kappa}_{\text{P2P}}}{r_{\text{P2P}}}, \tag{4}$$

$$\varepsilon_{\text{P2P}} = \frac{P \bar{\kappa}_{\text{P2P}}}{r_{\text{P2P}}} = P \delta_{\text{P2P}}, \tag{5}$$

where $\bar{\kappa}_{\text{P2P}}$ denotes the average number of transmissions for reliable P2P communications.

4.2 Energy-Delay Tradeoff in Wireless Butterfly Networks

Basically, the signal processing at relay $\mathscr{R}$ in a WBN (cf. Fig. 1) can be carried out with either PNC or ANC protocols. This section will derive the EDTs of the HARQ-IR protocols with PNC and ANC.

4.2.1 EDT of HARQ-IR Protocol with PNC

Using the PNC scheme for HARQ-IR in a WBN, $\mathscr{R}$ performs joint decoding of two signals received from $\mathscr{S}_1$ and $\mathscr{S}_2$ in MA transmission following the approach of Zhang and Liew in 2009 [100]. The number of transmissions in the MA transmission can be determined through the MA channel capacity bound derived in a book on information theory of Cover and Thomas in 2006 [101] as follows:

$$
\begin{aligned}
\kappa_{\text{PNC,MA}} = \min \Bigg\{ & k \Big| \Big\{ \sum_{j=1}^{k} \log(1 + [\gamma_{1R}]_j) > r_1 \Big\} \\
& \cap \Big\{ \sum_{j=1}^{k} \log(1 + [\gamma_{2R}]_j) > r_2 \Big\} \\
& \cap \Big\{ \sum_{j=1}^{k} \log(1 + [\gamma_{1R}]_j + [\gamma_{2R}]_j) > r_1 + r_2 \Big\} \Bigg\},
\end{aligned}
\tag{6}
$$

where γ_{iR} and r_i, $i = 1, 2$, denote the SNR of the transmission link $\mathscr{S}_i \rightarrow \mathscr{R}$ and the transmission rate at $\mathscr{S}_i$, respectively. In parallel with the MA transmission, $\mathscr{D}_i$, $i = 1, 2$, receives the packet from $\mathscr{S}_i$ in the DR transmission. The received signal at $\mathscr{D}_i$ can be written by

$$
\mathbf{y}_{ii} = \sqrt{P_i} h_{ii} \mathbf{s}_i + \mathbf{n}_{ii},
\tag{7}
$$

where P_i, h_{ii} and $\mathbf{n}_{ii}$ denote the transmission power, channel coefficient and CSCG noise vector at $\mathscr{D}_i$ of the transmission link $\mathscr{S}_i \rightarrow \mathscr{D}_i$, respectively. Similar to the transmission over P2P channels, the number of transmissions required at $\mathscr{S}_i, i = 1, 2$, to transmit $\mathbf{s}_i$ to $\mathscr{D}_i$ in the DR transmission can be computed by

$$
\kappa_{\text{PNC,DR}_i} = \min \Bigg\{ k \Big| \sum_{j=1}^{k} \log(1 + [\gamma_{ii}]_j) > r_i \Bigg\},
\tag{8}
$$

where γ_{ii} denotes the SNR of the transmission link $\mathscr{S}_i \rightarrow \mathscr{D}_i$. With HARQ-IR protocol, the data packet is retransmitted by $\mathscr{S}_i$, $i = 1, 2$, until both $\mathscr{R}$ and $\mathscr{D}_i$ successfully decode. Thus, the number of transmissions at $\mathscr{S}_i$ and the total number of transmissions in the first time slot are given by

$$
\kappa_{\text{PNC,S}_i} = \max\{\kappa_{\text{PNC,MA}}, \kappa_{\text{PNC,DR}_i}\},
\tag{9}
$$

$$
\kappa_{\text{PNC,1}} = \max\{\kappa_{\text{PNC,MA}}, \kappa_{\text{PNC,DR}_1}, \kappa_{\text{PNC,DR}_2}\},
\tag{10}
$$

respectively. Then, $\mathscr{R}$ encodes the superimposed packet, and then broadcasts the encoded packet to both $\mathscr{S}_1$ and $\mathscr{S}_2$ in the second time slot. The number of transmissions required at $\mathscr{R}$ to transmit the mixed packet to $\mathscr{D}_i$, $i = 1, 2$, in the BC transmission is similarly determined as in P2P communications in Sect. 4.1, i.e.,

$$\kappa_{\text{PNC,BC}_i} = \min \left\{ k \Big| \sum_{j=1}^{k} \log(1 + [\gamma_{Ri}]_j) > r_{i'} \right\}, \tag{11}$$

where $i' = 1$ if $i = 2$ and $i' = 2$ if $i = 1$ (or $i' = i - (-1)^i$). Here, γ_{Ri} denotes the SNR of the transmission link $\mathscr{R} \to \mathscr{D}_i$. In order to help both $\mathscr{D}_1$ and $\mathscr{D}_2$ detect the data packets from $\mathscr{S}_2$ and $\mathscr{S}_1$, respectively, $\mathscr{R}$ retransmits the packet until both $\mathscr{D}_1$ and $\mathscr{D}_2$ successfully detect it. Thus, the number of transmissions in the second time slot is computed by

$$\kappa_{\text{PNC,2}} = \max\{\kappa_{\text{PNC,BC}_1}, \kappa_{\text{PNC,BC}_2}\}. \tag{12}$$

Overall, the resulting ED and EB of the HARQ-IR protocol with the PNC are respectively given by

$$\delta_{\text{PNC}} = \frac{\bar{\kappa}_{\text{PNC,1}} + \bar{\kappa}_{\text{PNC,2}}}{r_1 + r_2}, \tag{13}$$

$$\varepsilon_{\text{PNC}} = \frac{P_1 \bar{\kappa}_{\text{PNC,S}_1} + P_2 \bar{\kappa}_{\text{PNC,S}_2} + P_R \bar{\kappa}_{\text{PNC,2}}}{r_1 + r_2}, \tag{14}$$

where P_R denotes the transmission power at $\mathscr{R}$.

4.2.2 EDT of HARQ-IR Protocol with ANC

With the ANC protocol, in the MA transmission of the first time slot, $\mathscr{R}$ receives the data packets from both $\mathscr{S}_1$ and $\mathscr{S}_2$, which can be written by

$$\mathbf{r} = \sqrt{P_1} h_{1R} \mathbf{s}_1 + \sqrt{P_2} h_{2R} \mathbf{s}_2 + \mathbf{n}_R, \tag{15}$$

where h_{iR} and $\mathbf{n}_R$ denote the channel coefficient and CSCG noise vector at $\mathscr{R}$ of the transmission link $\mathscr{S}_i \to \mathscr{R}$, respectively. At the same time, $\mathscr{D}_i$, $i = 1, 2$, receives the data packet from $\mathscr{S}_i$ in the DR transmission. Similarly, the received signal $\mathbf{y}_{ii}$ at $\mathscr{D}_i$ is given by Eq. (7) and the number of transmissions $\kappa_{\text{ANC,DR}_i}$ is determined as $\kappa_{\text{PNC,DR}_i}$ in Eq. (8).

Prior to broadcasting the received signal to both $\mathscr{D}_1$ and $\mathscr{D}_2$, $\mathscr{R}$ normalises its received signal $\mathbf{r}$ in Eq. (15) by a factor $\lambda = 1/\sqrt{E\left[|\mathbf{r}|^2\right]} = 1/\sqrt{\gamma_{1R} + \gamma_{2R} + 1}$ to have unit average energy. Thus, in the BC transmission, the signals received at $\mathscr{D}_i$, $i = 1, 2$, can be written as

$$\mathbf{y}_{Ri} = \sqrt{P_R} h_{Ri} \lambda \mathbf{r} + \mathbf{n}_{Ri}, \tag{16}$$

where h_{Ri} and $\mathbf{n}_{Ri}$ denote the channel coefficient and CSCG noise vector at $\mathscr{D}_i$ of the transmission link $\mathscr{R} \to \mathscr{D}_i$, respectively. Then, $\mathscr{D}_i$, $i = 1, 2$, detects $\mathbf{s}_{i'}$, $i' = i - (-1)^i$, by canceling $\mathbf{s}_i$ which is detected in the DR transmission. The resulting SNR $\gamma_{i'}$ at $\mathscr{D}_i$ is expressed by

$$\gamma_{i'} = \frac{\gamma_{Ri}\gamma_{i'R}}{\gamma_{Ri} + \gamma_{i'R} + \gamma_{iR} + 1}, \tag{17}$$

where γ_{iR} and γ_{Ri} denote the SNRs of the transmission links $\mathscr{S}_i \to \mathscr{R}$ and $\mathscr{R} \to \mathscr{D}_i$, respectively. In the HARQ-IR protocol with ANC, $\mathscr{D}_1$ and $\mathscr{D}_2$ feedback to $\mathscr{S}_1$ and $\mathscr{S}_2$ over direct links to acknowledge the packets $\mathbf{s}_1$ and $\mathbf{s}_2$, respectively. Since there is no decoding process carried out at $\mathscr{R}$ in the first time slot, $\mathscr{R}$ does not perform any feedback for the links $\mathscr{S}_1 \to \mathscr{R}$ and $\mathscr{S}_2 \to \mathscr{R}$. However, $\mathscr{R}$ can help $\mathscr{D}_1$ and $\mathscr{D}_2$ forward the acknowledgement of the packets $\mathbf{s}_2$ and $\mathbf{s}_1$ to $\mathscr{S}_2$ and $\mathscr{S}_1$, respectively. Therefore, the number of transmissions required at $\mathscr{S}_i$, $i = 1, 2$, to transmit $\mathbf{s}_i$ to $\mathscr{D}_{i'}$ is determined by

$$\kappa_{\text{ANC}_i} = \min \left\{ k \,\Big|\, \sum_{j=1}^{k} \log(1 + [\gamma_i]_j) > r_i \right\}. \tag{18}$$

The total number of transmissions at $\mathscr{S}_i$, $i = 1, 2$, is accordingly given by

$$\kappa_{\text{ANC},S_i} = \max\{\kappa_{\text{ANC}_i}, \kappa_{\text{ANC,DR}_i}\}. \tag{19}$$

It is noted that, with the ANC protocol, the retransmission of the lost packets at $\mathscr{D}_1$ and $\mathscr{D}_2$ is carried out by $\mathscr{S}_1$ and $\mathscr{S}_2$. $\mathscr{R}$ only amplifies and forwards to $\mathscr{D}_1$ and $\mathscr{D}_2$ the data received from $\mathscr{S}_1$ and $\mathscr{S}_2$. This means that the number of transmissions at $\mathscr{R}$ to assist $\mathscr{S}_1$ and $\mathscr{S}_2$ is also given by κ_{ANC_1} and κ_{ANC_2}, respectively, and, $\mathscr{R}$ uses half power for each task. Therefore, the resulting ED and EB of the HARQ-IR protocol with the ANC scheme are respectively obtained as

$$\delta_{\text{ANC}} = \frac{\max\{\bar{\kappa}_{\text{ANC},S_1}, \bar{\kappa}_{\text{ANC},S_2}\} + \max\{\bar{\kappa}_{\text{ANC}_1}, \bar{\kappa}_{\text{ANC}_2}\}}{r_1 + r_2}, \tag{20}$$

$$\varepsilon_{\text{ANC}} = \frac{P_1 \bar{\kappa}_{\text{ANC},S_1} + P_2 \bar{\kappa}_{\text{ANC},S_2} + \frac{P_R}{2} \bar{\kappa}_{\text{ANC}_1} + \frac{P_R}{2} \bar{\kappa}_{\text{ANC}_2}}{r_1 + r_2}. \tag{21}$$

4.3 Analysis of EDTs in WBNs

In order to provide insights of the EDT in WBNs, this section will derive the approximations of the EDTs for various HARQ-IR protocols in WBNs in high and low power regimes. For comparison, the approximated EDTs of both relay-aided transmission, i.e., PNC and ANC, and non-relay-aided transmission, i.e., DT, are investigated. For fair comparison, both relay-aided transmission and non-relay-aided transmission require the same number of time slots to transmit data packets $\mathbf{s}_1$ and $\mathbf{s}_2$ from $\mathscr{S}_1$ and $\mathscr{S}_2$, respectively, to both $\mathscr{D}_1$ and $\mathscr{D}_2$. Specifically, in the DT scheme, in the i-th, $i = 1, 2$, time slot $\mathscr{S}_i$ transmits $\mathbf{s}_i$ to $\mathscr{D}_1$ and $\mathscr{D}_2$ over $\mathscr{S}_i$-$\mathscr{D}_1$ and $\mathscr{S}_i$-$\mathscr{D}_2$ links, respectively. In the PNC and ANC schemes, the data transmission in the first time slot consists of two DR transmissions ($\mathscr{S}_1 \to \mathscr{D}_1$ and $\mathscr{S}_2 \to \mathscr{D}_2$) and a MA transmission ($\{\mathscr{S}_1, \mathscr{S}_2\} \to \mathscr{R}$), and there is a BC transmission ($\mathscr{R} \to \{\mathscr{D}_1, \mathscr{D}_2\}$) in the second time slot. This means that all the PNC, ANC and DT schemes require 2 time slots for the data transmission.

Let P denote the total power constraint of all transmitting nodes, i.e., $P = P_1 + P_2 + P_R$. Also, denote ρ_1, ρ_2 and $(1 - \rho_1 - \rho_2)$ as the fractions of power allocated to $\mathscr{S}_1, \mathscr{S}_2$ and $\mathscr{R}$, respectively. Note that, in the DT scheme, $P_R = 0$ and $\rho_1 + \rho_2 = 1$. Accordingly, $P_1 = \rho_1 P$, $P_2 = \rho_2 P$ and $P_R = (1 - \rho_1 - \rho_2)P$. All channel links are assumed to suffer from quasi-static Rayleigh block fading with $E[|h_{11}|^2] = 1/d_{S_1 D_1}^v$, $E[|h_{22}|^2] = 1/d_{S_2 D_2}^v$, $E[|h_{iR}|^2] = 1/d_{S_i R}^v$ and $E[|h_{Rj}|^2] = 1/d_{RD_j}^v$, $i = 1, 2, j = 1, 2$.

Applying HARQ-IR protocol for DT scheme, the ED and EB can be simply derived as

$$\delta_{\mathrm{DT}} = \frac{\bar{\kappa}_{\mathrm{DT},1} + \bar{\kappa}_{\mathrm{DT},2}}{r_1 + r_2}, \tag{22}$$

$$\varepsilon_{\mathrm{DT}} = \frac{P_1 \bar{\kappa}_{\mathrm{DT},1} + P_2 \bar{\kappa}_{\mathrm{DT},2}}{r_1 + r_2}. \tag{23}$$

Here, $\kappa_{\mathrm{DT},i}$, $i = 1, 2$, denotes the total number of transmissions required at $\mathscr{S}_i$ to transmit $\mathbf{s}_i$ to both $\mathscr{D}_1$ and $\mathscr{D}_2$, which is given by

$$\kappa_{\mathrm{DT},i} = \max \left\{ \min \left\{ k \Big| \sum_{j=1}^{k} \log(1 + [\gamma_{ii}]_j) > r_i \right\}, \\ \min \left\{ k \Big| \sum_{j=1}^{k} \log(1 + [\gamma_{ii'}]_j) > r_i \right\} \right\}, \tag{24}$$

where $i' = i - (-1)^i$, $i = 1, 2$, and $\gamma_{ii'}$ denotes the SNR of the transmission link $\mathscr{S}_i \to \mathscr{D}_{i'}$.

In the high power regime, all HARQ-IR protocols for both relay-aided and non-relay-aided transmissions in a WBN require 2 time slots in total to transmit successfully 2 data packets $\mathbf{s}_1$ and $\mathbf{s}_2$ from $\mathscr{S}_1$ and $\mathscr{S}_2$ to $\mathscr{D}_1$ and $\mathscr{D}_2$. This means that all the PNC, ANC and DT schemes achieve the same EDT performance with $\{\delta_{\mathrm{PNC}}, \delta_{\mathrm{ANC}}, \delta_{\mathrm{DT}}\} \to \frac{2}{r_1+r_2}$ and $\{\varepsilon_{\mathrm{PNC}}, \varepsilon_{\mathrm{ANC}}, \varepsilon_{\mathrm{DT}}\} \to \infty$ as $P \to \infty$, and there is no advantageous scheme in the high power regime.

In the low power regime, the transmission power at all transmitting nodes is assumed to be equally allocated as $P_1 = P_2 = P_R = P/3$ in PNC and ANC schemes and $P_1 = P_2 = P/2$ in the DT scheme. Note that, although the equal power allocation is not optimal in general, it is reasonable to assume the equal power allocation at all transmitting nodes as $P \to 0$. Also, for simplicity, the data transmission from $\mathscr{S}_1$ and $\mathscr{S}_2$ to $\mathscr{D}_1$ and $\mathscr{D}_2$ is assumed to be carried out at the same data rate, i.e., $r_1 = r_2 = R$. The EDTs of the HARQ-IR protocol in the WBN with the DT, PNC and ANC schemes as P approaches to 0 can be derived as in the following theorems.

Theorem 1 *If P approaches 0, then the ED and EB of the HARQ-IR protocol with the DT scheme are approximated by $\delta_{DT,0}$ and $\varepsilon_{DT,0}$, respectively, where*

$$\delta_{DT,0} = \frac{\ln 2}{P}(\max\{d^v_{S_1 D_1}, d^v_{S_1 D_2}\} + \max\{d^v_{S_2 D_1}, d^v_{S_2 D_2}\}), \tag{25}$$

$$\varepsilon_{DT,0} = \frac{\ln 2}{2}(\max\{d^v_{S_1 D_1}, d^v_{S_1 D_2}\} + \max\{d^v_{S_2 D_1}, d^v_{S_2 D_2}\}). \tag{26}$$

Proof It is noted that when x is sufficiently small,

$$\log(1 + ax) \approx \frac{ax}{\ln 2} + O(x^2). \tag{27}$$

Thus, when $P \to 0$,

$$\log(1 + [\gamma_{ii}]_j) \approx \frac{|[h_{ii}]_j|^2 P}{2 \ln 2}, \quad \log(1 + [\gamma_{ii'}]_j) \approx \frac{|[h_{ii'}]_j|^2 P}{2 \ln 2},$$

where $i' = i - (-1)^i$, $i = 1, 2$. Since $E\{|h_{11}|^2\} = 1/d^v_{S_1 D_1}$, $E\{|h_{22}|^2\} = 1/d^v_{S_2 D_2}$, $E\{|h_{12}|^2\} = 1/d^v_{S_1 D_2}$, $E\{|h_{21}|^2\} = 1/d^v_{S_2 D_1}$ and $r_1 = r_2 = R$, it can be deduced

$$\bar{\kappa}_{DT,1} \approx \frac{2R \ln 2}{P} \max\{d^v_{S_1 D_1}, d^v_{S_1 D_2}\}, \tag{28}$$

$$\bar{\kappa}_{DT,2} \approx \frac{2R \ln 2}{P} \max\{d^v_{S_2 D_2}, d^v_{S_2 D_1}\}. \tag{29}$$

Substituting Eqs. (28) and (29) into Eqs. (22) and (23) with $r_1 = r_2 = R$ and $P_1 = P_2 = P/2$, the theorem is proved.

Theorem 2 *If P approaches 0, then the ED and EB of the HARQ-IR protocol with the PNC scheme are approximated by $\delta_{PNC,0}$ and $\varepsilon_{PNC,0}$, respectively, where*

$$\delta_{PNC,0} = \frac{3 \ln 2}{2P}(\max\{d^v_{S_1 D_1}, d^v_{S_2 D_2}\} + \max\{d^v_{RD_1}, d^v_{RD_2}\}), \tag{30}$$

$$\varepsilon_{PNC,0} = \frac{\ln 2}{2}(d^v_{S_1 D_1} + d^v_{S_2 D_2} + \max\{d^v_{RD_1}, d^v_{RD_2}\}). \tag{31}$$

Proof Consider Eqs. (13) and (14). When $P \to 0$, applying the approximation in Eq. (27) to $\kappa_{\text{PNC,MA}}, \kappa_{\text{PNC,DR}_i}$ and $\kappa_{\text{PNC,BC}_i}, i = 1, 2$, given by Eqs. (6), (8) and (11) with $r_i = R$ and $P_i = P_R = P/3$, i.e.,

$$\bar{\kappa}_{\text{PNC,MA}} \approx \frac{6R \ln 2}{P} \frac{d^v_{S_1 R} d^v_{S_2 R}}{d^v_{S_1 R} + d^v_{S_2 R}}, \tag{32}$$

$$\bar{\kappa}_{\text{PNC,DR}_1} \approx \frac{3R \ln 2}{P} d^v_{S_1 D_1}, \tag{33}$$

$$\bar{\kappa}_{\text{PNC,DR}_2} \approx \frac{3R \ln 2}{P} d^v_{S_2 D_2}, \tag{34}$$

$$\bar{\kappa}_{\text{PNC,BC}_i} \approx \frac{3R \ln 2}{P} d^v_{Ri}. \tag{35}$$

It is noted that $d_{S_1 R}$ and d_{3R} should be both less than $d_{S_1 D_1}$ and $d_{S_2 D_2}$. Thus,

$$\frac{d^v_{S_1 R} d^v_{S_2 R}}{d^v_{S_1 R} + d^v_{S_2 R}} < \frac{d^v_{S_1 D_1}}{2}, \quad \frac{d^v_{S_1 R} d^v_{S_2 R}}{d^v_{S_1 R} + d^v_{S_2 R}} < \frac{d^v_{S_2 D_2}}{2}.$$

Substitute Eqs. (32)–(35) into Eqs. (9), (10) and (12) as

$$\bar{\kappa}_{\text{PNC,S}_1} \approx \frac{3R \ln 2}{P} d^v_{S_1 D_1}, \tag{36}$$

$$\bar{\kappa}_{\text{PNC,S}_2} \approx \frac{3R \ln 2}{P} d^v_{S_2 D_2}, \tag{37}$$

$$\bar{\kappa}_{\text{PNC,1}} \approx \frac{3R \ln 2}{P} \max\{d^v_{S_1 D_1}, d^v_{S_2 D_2}\}, \tag{38}$$

$$\bar{\kappa}_{\text{PNC,2}} \approx \frac{3R \ln 2}{P} \max\{d^v_{RD_1}, d^v_{RD_2}\}. \tag{39}$$

Then, substituting Eqs. (36)–(39) into Eqs. (13) and (14) with $r_1 = r_2 = R$, the theorem is proved.

Theorem 3 *If P approaches 0, then the ED and EB of the HARQ-IR protocol with the ANC scheme are approximated by $\delta_{\text{ANC,0}}$ and $\varepsilon_{\text{ANC,0}}$, respectively, where*

$$\delta_{\text{ANC,0}} = \frac{9 \ln 2}{P^2} \max\{d^v_{S_1 R} d^v_{RD_2}, d^v_{S_2 R} d^v_{RD_1}\}, \tag{40}$$

$$\varepsilon_{\text{ANC,0}} = \frac{9 \ln 2}{4P} (d^v_{S_1 R} d^v_{RD_2} + d^v_{S_2 R} d^v_{RD_1}). \tag{41}$$

Proof Consider Eqs. (20) and (21). When $P \to 0$, applying the approximation in Eq. (27) to κ_{ANC_i}, $i = 1, 2$, given by Eq. (18) with $r_i = R$ and $P_i = P_R = P/3$, i.e.,

$$\bar{\kappa}_{\text{ANC}_i} \approx \frac{9R \ln 2}{P^2} d_{S_i R}^{\nu} d_{RD_i'}^{\nu}, \tag{42}$$

where $i' = 2$ if $i = 1$ and $i' = 1$ if $i = 2$. Substituting Eqs. (33), (34) and (42) into Eq. (19), it can be deduced

$$\bar{\kappa}_{\text{ANC},S_1} \approx \max\{\frac{9R \ln 2}{P^2} d_{S_1 R}^{\nu} d_{RD_2}^{\nu}, \frac{3R \ln 2}{P} d_{S_1 D_1}^{\nu}\}, \tag{43}$$

$$\bar{\kappa}_{\text{ANC},S_2} \approx \max\{\frac{9R \ln 2}{P^2} d_{S_2 R}^{\nu} d_{RD_1}^{\nu}, \frac{3R \ln 2}{P} d_{S_2 D_2}^{\nu}\}. \tag{44}$$

Since $P \to 0$, it can be shown that

$$\frac{9R \ln 2}{P^2} d_{S_1 R}^{\nu} d_{RD_2}^{\nu} > \frac{3R \ln 2}{P} d_{S_1 D_1}^{\nu},$$

$$\frac{9R \ln 2}{P^2} d_{S_2 R}^{\nu} d_{RD_1}^{\nu} > \frac{3R \ln 2}{P} d_{S_2 D_2}^{\nu}.$$

Thus, Eqs. (43) and (44) can be rewritten as

$$\bar{\kappa}_{\text{ANC},S_1} \approx \frac{9R \ln 2}{P^2} d_{S_1 R}^{\nu} d_{RD_2}^{\nu}, \tag{45}$$

$$\bar{\kappa}_{\text{ANC},S_2} \approx \frac{9R \ln 2}{P^2} d_{S_2 R}^{\nu} d_{RD_1}^{\nu}, \tag{46}$$

respectively. Substituting Eqs. (42), (45) and (46) into Eqs. (20) and (21) with $r_1 = r_2 = R$ and $P_1 = P_2 = P_R = P/3$, the theorem is proved.

From the above theorems, the following remarks can be noticed in the low power regime.

Remark 1 (*Energy inefficiency with ANC*) It can be seen in Eq. (41) that $\varepsilon_{\text{ANC},0}$ increases as P decreases. This means that the ANC scheme is not energy efficient when compared to the DT and PNC schemes for the HARQ-IR protocol in WBN.

Remark 2 (*Higher energy efficiency with PNC when relay node is located far from source nodes*) When $\mathscr{R}$ is far from $\mathscr{S}_1$ and $\mathscr{S}_2$, $\{d_{RD_1}^{\nu}, d_{RD_2}^{\nu}\} \ll \{d_{S_1 D_1}^{\nu}, d_{S_2 D_2}^{\nu}\}$. Thus, $d_{S_1 D_1}^{\nu} + d_{S_2 D_2}^{\nu} + \max\{d_{RD_1}^{\nu}, d_{RD_2}^{\nu}\} \approx d_{S_1 D_1}^{\nu} + d_{S_2 D_2}^{\nu}$. Accordingly, from Eqs. (26) and (31), it can be shown that $\varepsilon_{\text{PNC},0} < \varepsilon_{\text{DT},0}$, which means the HARQ-IR protocol with the PNC scheme is more energy efficient than the HARQ-IR protocol with the DT scheme.

Remark 3 (*Higher energy efficiency with DT over PNC when relay node is located nearby source nodes*) In this scenario, $\{d_{RD_1}^v, d_{RD_2}^v\} \gtrsim \{d_{S_1D_1}^v, d_{S_2D_2}^v\}$. Thus, from Eqs. (26) and (31), it can be shown that $\varepsilon_{PNC,0} > \varepsilon_{DT,0}$. This means that the DT scheme is more energy efficient than the PNC scheme for the HARQ-IR protocol in the WBN. In other words, there is no advantage of employing the relay when the relay is in the neighborhood of the sources.

For illustration, the EDT performance of the HARQ-IR protocols in a WBN is validated in two following examples, i.e., Examples 1 and 2, for different network configurations.

Example 1 A symmetric WBN is considered with $d_{S_1D_1} = d_{S_2D_2}$, $d_{S_1S_2} = d_{D_1D_2}$ and $\alpha_1 = \alpha_2 = \alpha_3 = \alpha_4 = \pi/2$. The data transmission from $\mathscr{S}_1$ and $\mathscr{S}_2$ to $\mathscr{D}_1$ and $\mathscr{D}_2$ is carried out at the same data rate with spectral efficiency of $r_1 = r_2 = R$ [b/s/Hz]. HARQ-IR protocol is employed with either DT or PNC or ANC schemes. The pathloss exponent between a pair of transceiver nodes is assumed to be $v = 3$ and all channels experience quasi-static Rayleigh block fading.

Figure 2 plots the EDT curves of three HARQ-IR protocols with different data rates at $\mathscr{S}_1$ and $\mathscr{S}_2$. The spectral efficiency, i.e., R, is assumed to vary in the ranges $\{1, 4, 16\}$ b/s/Hz. The relay is assumed to be located at the center of the network, i.e., $d_{S_1R} = d_{S_2R} = d_{RD_1} = d_{RD_2}$. The transmission powers at $\mathscr{S}_1$, $\mathscr{S}_2$ and $\mathscr{R}$ are assumed to be equally allocated. It can be seen that the PNC scheme is more energy efficient than both the ANC and DT schemes. In fact, using the HARQ-IR protocol with PNC, $\mathscr{R}$ can help $\mathscr{S}_1$ and $\mathscr{S}_2$ retransmit the corrupted combined packets to both $\mathscr{D}_1$ and $\mathscr{D}_2$. Using the HARQ-IR protocol with the ANC scheme, $\mathscr{S}_1$ and $\mathscr{S}_2$ are required to retransmit the corrupted packets to $\mathscr{R}$, then $\mathscr{R}$ combines the received packets and broadcasts the new combined packets to both $\mathscr{D}_1$ and $\mathscr{D}_2$. Using the DT scheme, there is no relay to assist $\mathscr{S}_1$ and $\mathscr{S}_2$ retransmit the corrupted combined packets to both $\mathscr{D}_1$ and $\mathscr{D}_2$. Due to the long distances from $\mathscr{S}_1$ to $\mathscr{D}_2$ and from $\mathscr{S}_2$ to $\mathscr{D}_1$, the DT scheme is shown to be less energy efficient than the PNC scheme. However, the EDT of the DT scheme is better than that of the ANC scheme since a re-combination process is required at $\mathscr{R}$ in the ANC scheme, which means more energy consumption at $\mathscr{R}$. This confirms the statements in Remarks 1 and 2 regarding a lower energy efficiency of the ANC scheme and a higher energy efficiency of the PNC scheme over the DT scheme when the relay node is located far from the source nodes. The impact of data rate on the EDT performance can also be observed in Fig. 2 where an improved EDT is achieved for all the HARQ-IR protocols as the data rate increases.

Example 2 Taking into account the practical scenario where the relay is not always located at the center of the network, Fig. 3 plots the EDT curves of various HARQ-IR protocols in the WBN with respect to different relay positions.

Three relay positions are considered, including

(i) $\mathscr{R}$ near $\{\mathscr{S}_1, \mathscr{S}_2\}$: $d_{S_1R} = 1/4$ m, $\alpha_R = \pi/4$;
(ii) $\mathscr{R}$ at the center: $d_{S_1R} = d_{S_2R} = d_{RD_1} = d_{RD_2}$;
(iii) $\mathscr{R}$ near $\{\mathscr{D}_1, \mathscr{D}_2\}$: $d_{S_1R} = 2$ m, $\alpha_R = \pi/6$.

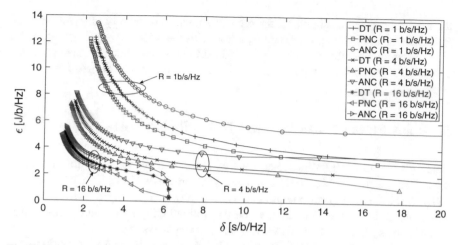

Fig. 2 EDTs of various HARQ-IR protocols in WBN with $d_{S_1D_1} = d_{S_2D_2} = 2$ m, $d_{S_1S_2} = d_{D_1D_2} = 1/2$ m and $d_{S_1R} = d_{S_2R} = d_{RD_1} = d_{RD_2}$

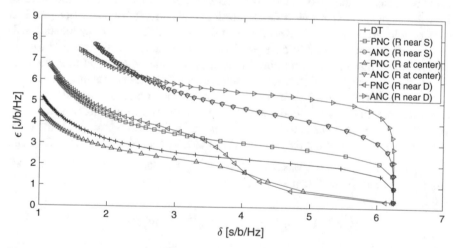

Fig. 3 EDTs of various HARQ-IR protocols in WBN with $R = 16$ b/s/Hz, $d_{S_1D_1} = d_{S_2D_2} = 2$ m, $d_{S_1S_2} = d_{D_1D_2} = 1/2$ m and various relay positions

As shown in Fig. 3, the DT scheme is the most energy efficient scheme compared to both the PNC and ANC schemes when the relay is in the neighborhood of the sources. This confirms the statement in Remark 3 in relation to the higher energy efficiency of the DT scheme when the relay is located nearby the sources. In fact, it can be intuitively observed that the relay plays the same role as the sources if the relay is located near the sources, which means the use of the relay in the PNC and ANC schemes is not as energy efficient compared to the DT scheme, though the relay can be used in this case to increase the transmit diversity order. For the scenario when the relay is near the destinations, the relay is shown to be energy efficient with the PNC scheme in the low power regime, while with the ANC scheme it is seen to be always less energy efficient. This scenario is similar to the scenario when the relay is located at the center of the network as observed in Fig. 2 of Example 1.

5 Relay Placement in Wireless Butterfly Networks

Relay placement (RP) problem has been extensively investigated in the literature, e.g., Chen et al. in 2012 [102] and Han et al. in 2013 [103]. In Chen et al. in 2012 [102], the relay position optimization was proposed to improve diversity gain of unbalanced DF relay networks, while the optimal relay placement problem was investigated by Han et al. in 2013 [103] for AF relay networks.

In WBNs employing HARQ-IR protocol with either PNC or ANC technique, the location of the relay also has a considerable impact on the energy efficiency of the network. Based on the derived EDT for HARQ-IR protocols with PNC and ANC in Sect. 3, this section will develop algorithms for solving the RP optimization problem subject to location and power constraints in the WBNs. The best relay location will be determined with respect to different HARQ-IR protocols. This is useful for the system where the mobile users play the role as the relay nodes and thus the user having the best relay location would be selected for the relay communications.

The RP problem relates to how to position the relay node in order to minimize either the total delay or the total energy consumption of all the multicast transmissions from two source nodes to two destination nodes. As shown in Fig. 1, the relay location can be determined through the distance between $\mathscr{S}_1$ and $\mathscr{R}$, i.e., d_{S_1R}, and the angle $\widehat{\mathscr{D}_1\mathscr{S}_1\mathscr{R}}$, i.e., α_R. Based on d_{S_1R} and α_R, the distances from $\mathscr{R}$ to $\mathscr{S}_2$, $\mathscr{D}_1$ and $\mathscr{D}_2$ can be easily obtained as

$$d_{S_2R} = \sqrt{d_{S_1S_2}^2 + d_{S_1R}^2 - 2d_{S_1S_2}d_{S_1R}\cos(\alpha_1 - \alpha_R)}, \tag{47}$$

$$d_{RD_1} = \sqrt{d_{S_1D_1}^2 + d_{S_1R}^2 - 2d_{S_1D_1}d_{S_1R}\cos\alpha_R}, \tag{48}$$

$$d_{RD_2} = \sqrt{d_{S_2D_2}^2 + d_{S_2R}^2 - 2d_{S_2D_2}d_{S_2R}\cos\beta_R}, \tag{49}$$

respectively. Here, β_R denotes the angle $\widehat{\mathscr{D}_2\mathscr{S}_2\mathscr{R}}$, which can be computed by

$$\beta_R = \alpha_2 - \sin^{-1}\left(\frac{d_{S_1R}}{d_{S_2R}}\sin(\alpha_1 - \alpha_R)\right). \tag{50}$$

Let $\{d^*_{S_1R,\delta_{PNC}}, \alpha^*_{S_1R,\delta_{PNC}}\}$, $\{d^*_{S_1R,\delta_{ANC}}, \alpha^*_{R,\delta_{ANC}}\}$, $\{d^*_{S_1R,\varepsilon_{PNC}}, \alpha^*_{R,\varepsilon_{PNC}}\}$ and $\{d^*_{S_1R,\varepsilon_{ANC}}, \alpha^*_{R,\varepsilon_{ANC}}\}$ denote the optimized positioning parameters for the relay location using PNC and ANC protocols subject to minimizing δ_{PNC}, δ_{ANC}, ε_{PNC} and ε_{ANC}, respectively. The RP optimization problem can be formulated as

$$\{d^*_{S_1R,\delta_{PNC}}, \alpha^*_{R,\delta_{PNC}}\} = \arg\min_{d_{S_1R},\alpha_R} \delta_{PNC}, \tag{51}$$

$$\{d^*_{S_1R,\delta_{ANC}}, \alpha^*_{R,\delta_{ANC}}\} = \arg\min_{d_{S_1R},\alpha_R} \delta_{ANC}, \tag{52}$$

$$\{d^*_{S_1R,\varepsilon_{PNC}}, \alpha^*_{R,\varepsilon_{PNC}}\} = \arg\min_{d_{S_1R},\alpha_R} \varepsilon_{PNC}, \tag{53}$$

$$\{d^*_{S_1R,\varepsilon_{ANC}}, \alpha^*_{R,\varepsilon_{ANC}}\} = \arg\min_{d_{S_1R},\alpha_R} \varepsilon_{ANC}, \tag{54}$$

where δ_{PNC}, δ_{ANC}, ε_{PNC} and ε_{ANC} are generally given by Eqs. (13), (20), (14) and (21), respectively. Given the fixed location of the source and destination nodes (cf. Fig. 1), d_{S_1R} and α_R are bounded by the following ranges:

$$0 < d_{S_1R} < \max\left\{ \sqrt{d^2_{S_1D_1} + d^2_{S_1S_2} - 2d_{S_1D_1}d_{S_1S_2}\cos\alpha_1},\ \sqrt{d^2_{S_1D_1} + d^2_{D_1D_2} - 2d_{S_1D_1}d_{D_1D_2}\cos\alpha_4} \right\}, \tag{55}$$

$$0 < \alpha_R < \alpha_1. \tag{56}$$

The following remarks can be drawn:

Remark 4 (*ANC-based relay can be nearly located at the same location for minimizing both the delay and energy*) Given a compact set $\mathbb{S}$, $\arg\min_{x_1,x_2\in\mathbb{S}}\max\{f(x_1),f(x_2)\} \approx \arg\min_{x_1,x_2\in\mathbb{S}}(f(x_1)+f(x_2))$. Thus, from Eqs. (20) and (21), it can be approximated that $\arg\min_{d_{S_1R},\alpha_R}\delta_{ANC} \approx \arg\min_{d_{S_1R},\alpha_R}\varepsilon_{ANC}$, which means $\{d^*_{S_1R,\delta_{ANC}}, \alpha^*_{R,\delta_{ANC}}\} \approx \{d^*_{S_1R,\varepsilon_{ANC}}, \alpha^*_{R,\varepsilon_{ANC}}\}$.

Remark 5 (*Perspective transformation for a general setting of the node positions in an irregular quadrilateral*) Note that the nodes in a quadrilateral can be mapped to the nodes in a rectangle using spatial transformation approach which can be found in a book of Wolberg in 1990 [104] for digital image processing. The optimal relay position in the rectangular region, namely virtual relay positions, can be firstly found for minimizing either delay or energy. Then, the real relay position for the irregular quadrilateral node setting can be determined by an inverse mapping. Specifically, a perspective transformation or projective non-affine mapping with bilinear interpolation can be used to map a quadrilateral to a rectangle as follows: Given four 2-dimensional points A, B, C and D of a quadrilateral located at (x_A, y_A), (x_B, y_B), (x_C, y_C) and (x_D, y_D), and four 2-dimensional points A', B', C' and D' of a rectangle located at $(x_{A'}, y_{A'})$, $(x_{B'}, y_{B'})$, $(x_{C'}, y_{C'})$ and $(x_{D'}, y_{D'})$. $\{A, B, C, D\}$ can be mapped to $\{A', B', C', D'\}$ by finding an 4×4 mapping matrix $\mathbf{M}$ such that

$$
\begin{pmatrix}
1 & x_A & y_A & x_A y_A \\
1 & x_B & y_B & x_B y_B \\
1 & x_C & y_C & x_C y_C \\
1 & x_D & y_D & x_D y_D
\end{pmatrix}
\mathbf{M} =
\begin{pmatrix}
1 & x_{A'} & y_{A'} & x_{A'} y_{A'} \\
1 & x_{B'} & y_{B'} & x_{B'} y_{B'} \\
1 & x_{C'} & y_{C'} & x_{C'} y_{C'} \\
1 & x_{D'} & y_{D'} & x_{D'} y_{D'}
\end{pmatrix}
$$

Wolberg in 1990 [104] and Kim et al. in 2002 [105] showed that perspective transformation is planar mapping and thus both forward and inverse mapping are unique. Also, the lines connecting nodes are shown to be preserved in all orientations.

According to Remark 5, for simplicity, a specific scenario can be considered, where $\alpha_1 = \alpha_2 = \alpha_3 = \alpha_4 = \pi/2$, $d_{S_1 D_1} = d_{S_2 D_2}$ and $d_{S_1 S_2} = d_{D_1 D_2}$. The search range of the relay position given by Eqs. (55) and (56) is thus rewritten as

$$
0 < d_{S_1 R} < \sqrt{d_{S_1 D_1}^2 + d_{S_1 S_2}^2}, \tag{57}
$$

$$
0 < \alpha_R < \frac{\pi}{2}. \tag{58}
$$

With the total power constraint P and different power allocation at $\mathscr{S}_1$ and $\mathscr{S}_2$, there are three typical cases based on the relationship between P_1 and P_2 which are described as follows:

5.1 Equal Power Allocation at Sources

Due to the equal power allocation at $\mathscr{S}_1$ and $\mathscr{S}_2$, $\mathscr{R}$ is located on the median line between the pair nodes $\{\mathscr{S}_1, \mathscr{D}_1\}$ and $\{\mathscr{S}_2, \mathscr{D}_2\}$. Denote $d_R = \sqrt{d_{S_1 R}^2 - d_{S_1 S_2}^2/4}$. The RP optimization in Eqs. (51)–(54) can be determined through

$$
d_{R, \delta_X}^* = \arg \min_{0 < d_R < d_{S_1 D_1}} \delta_X, \tag{59}
$$

$$
d_{R, \varepsilon_X}^* = \arg \min_{0 < d_R < d_{S_1 D_1}} \varepsilon_X, \tag{60}
$$

where $X \in \{\mathtt{PNC}, \mathtt{ANC}\}$. Then, $\{d_{S_1 R, \delta_X}^*, \alpha_{R, \delta_X}^*\}$ and $\{d_{S_1 R, \varepsilon_X}^*, \alpha_{1R, \varepsilon_X}^*\}$ can be computed by

$$
d_{S_1 R, \delta_X}^* = \sqrt{d_{R, \delta_X}^{*2} + \frac{d_{S_1 S_2}^2}{4}}, \quad \alpha_{R, \delta_X}^* = \tan^{-1}\left(\frac{d_{S_1 S_2}}{2 d_{R, \delta_X}^*}\right), \tag{61}
$$

$$d^*_{S_1 R, \varepsilon_X} = \sqrt{d^{*2}_{R, \varepsilon_X} + \frac{d^2_{S_1 S_2}}{4}}, \quad \alpha^*_{R, \varepsilon_X} = \tan^{-1}\left(\frac{d_{S_1 S_2}}{2 d^*_{R, \varepsilon_X}}\right). \tag{62}$$

It can be observed that the search algorithms using Eqs. (59)–(62) require a lower complexity processing than an exhaustive search of all available relay positions in the whole region encompassing the four source and destination nodes with the constraints of Eqs. (57) and (58).

5.2 Unequal Power Allocation at Sources

Considering unequal power allocation at $\mathscr{S}_1$ and $\mathscr{S}_2$, i.e., $P_1 \neq P_2$, there are two cases including $P_1 > P_2$ and $P_1 < P_2$ as follows:

5.2.1 $P_1 > P_2$

In this scenario, $\mathscr{R}$ should be located in the neighborhood region of the pair node $\{\mathscr{S}_2, \mathscr{D}_2\}$. Thus, the search range for the optimal relay location in Eqs. (57) and (58) can be limited by two regions defined as follows:

$$\text{Region (I):} \begin{cases} \tan^{-1}\left(\frac{d_{S_1 S_2}}{2 d_{S_1 D_1}}\right) < \alpha_R < \tan^{-1}\left(\frac{d_{S_1 S_2}}{d_{S_1 D_1}}\right), \\ \frac{d_{S_1 S_2}}{2 \sin \alpha_R} < d_{S_1 R} < \frac{d_{S_1 D_1}}{\cos \alpha_R}. \end{cases} \tag{63}$$

$$\text{Region (II):} \begin{cases} \tan^{-1}\left(\frac{d_{S_1 S_2}}{d_{S_1 D_1}}\right) < \alpha_R < \frac{\pi}{2}, \\ \frac{d_{S_1 S_2}}{2 \sin \alpha_R} < d_{S_1 R} < \frac{d_{S_1 S_2}}{\sin \alpha_R}. \end{cases} \tag{64}$$

With various relay positions in regions (I) and (II), the optimal relay location $\{d^*_{S_1 R, \delta_X}, \alpha^*_{R, \delta_X}\}$ and $\{d^*_{S_1 R, \varepsilon_X}, \alpha^*_{R, \varepsilon_X}\}$, $X \in \{\text{PNC, ANC}\}$, subject to minimizing either δ_X or ε_X can be determined as in Eqs. (51)–(54). Regarding the search range in the context of $P_1 > P_2$, it can be observed that the search regions (I) and (II) are narrower than the region determined by Eqs. (57) and (58), and thus the complexity of the search for the optimal relay location is reduced.

5.2.2 $P_1 < P_2$

Similarly, in this scenario, $\mathscr{R}$ is located near the two nodes $\mathscr{S}_1$ and $\mathscr{D}_1$. The search range for the optimal relay location in Eqs. (57) and (58) can thus be limited by two regions defined as follows:

$$\text{Region (III):} \begin{cases} 0 < \alpha_R < \tan^{-1}\left(\dfrac{d_{S_1 S_2}}{2 d_{S_1 D_1}}\right), \\ 0 < d_{S_1 R} < \dfrac{d_{S_1 D_1}}{\cos \alpha_R}. \end{cases} \tag{65}$$

$$\text{Region (IV):} \begin{cases} \tan^{-1}\left(\dfrac{d_{S_1 S_2}}{2 d_{S_1 D_1}}\right) < \alpha_R < \dfrac{\pi}{2}, \\ 0 < d_{S_1 R} < \dfrac{d_{S_1 S_2}}{2 \sin \alpha_R}. \end{cases} \tag{66}$$

Then, the optimal relay location $\{d^*_{S_1 R, \delta_X}, \alpha^*_{R, \delta_X}\}$ and $\{d^*_{S_1 R, \varepsilon_X}, \alpha^*_{R, \varepsilon_X}\}$, $X \in \{\text{PNC, ANC}\}$, can be determined in regions (III) and (IV) so as to minimize either δ_X or ε_X. Additionally, it can be observed that the search regions (III) and (IV) for the scenario $P_1 < P_2$ are also narrower than the region determined by Eqs. (57) and (58), and again a low-complexity search algorithm is achieved.

Consider for illustration the following example of the RP optimization problem for minimum ED and EB in a typical WBN.

Example 3 A symmetric WBN as in Example 1 is investigated where $d_{S_1 D_1} = d_{S_2 D_2}$ $= 2$ m and $d_{S_1 S_2} = d_{D_1 D_2} = 1/2$ m. Figures 4 and 5 plot the optimal relay locations for minimizing ED and EB, respectively, as a function of power allocation at source nodes when HARQ-IR protocols are employed with PNC and ANC. The optimal relay locations in Figs. 4 and 5 are determined through $d_{S_1 R}$ and α_R. It is assumed that $R = 4$ b/s/Hz and $P = P_1 + P_2 + P_R = 5$ W. Both equal power allocation, i.e., $P_1 = P_2$, and unequal power allocation with $P_1 = 3P_2$ and $P_1 = 5P_2$, are considered. Note that the RP for the scenario $P_1 < P_2$ can be similarly observed to be symmetric with that for the scenario $P_1 > P_2$. The bisection search method is applied to find the optimal relay position in the search region. Investigating the optimal relay location for minimum ED, Fig. 4 shows that for the scenario $P_1 = P_2$, as P_1 and P_2 increase, the optimal location of both the ANC-based and PNC-based $\mathscr{R}$ move from the region near $\mathscr{S}_1$ and $\mathscr{S}_2$ to the region near $\mathscr{D}_1$ and $\mathscr{D}_2$. However, when P_1 and P_2 are small, the ANC-based $\mathscr{R}$ is closer to $\mathscr{S}_1$ and $\mathscr{S}_2$ than the PNC-based $\mathscr{R}$. For the case $P_1 = nP_2, n > 1$, as n increases, the optimal location of the ANC-based $\mathscr{R}$ is closer to $\mathscr{S}_2$, while that of the PNC-based $\mathscr{R}$ is farther away from $\mathscr{D}_2$.

For minimum EB, it can be observed in Fig. 5 that for the scenario $P_1 = P_2$, as P_1 and P_2 increase, the optimal location of the ANC-based $\mathscr{R}$ moves from the region near $\mathscr{S}_1$ and $\mathscr{S}_2$ to the region near $\mathscr{D}_1$ and $\mathscr{D}_2$, while that of the PNC-based $\mathscr{R}$ moves in the reverse direction. For the case $P_1 = nP_2, n > 1$, similar to Fig. 4, it is shown that, as n increases, the ANC-based $\mathscr{R}$ should be closer to $\mathscr{S}_2$, while that of the PNC-based $\mathscr{R}$ should be farther away from $\mathscr{D}_2$. Furthermore, the optimal locations for the ANC-based $\mathscr{R}$ are shown to be nearly similar for both objectives of minimum ED and minimum EB, while the optimised locations for the PNC-based $\mathscr{R}$ are different with respect to the objective functions. These nearly similar locations of the ANC-based $\mathscr{R}$ verify the statement in Remark 4.

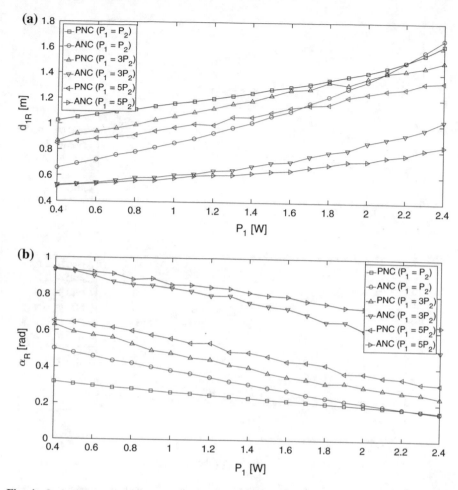

Fig. 4 Optimization of relay location subject to minimizing ED with $R = 4$ b/s/Hz, $P = 5$ W, $d_{S_1 D_1} = d_{S_2 D_2} = 2$ m, $d_{S_1 S_2} = d_{D_1 D_2} = 1/2$ m: **a** $d_{S_1 R}$ and **b** α_R

6 Relay Placement in Wireless Multicast Networks

The RP optimization in WBNs can be extended for a general wireless multicast network (WMN) consisting of N_s sources, N_r relays and N_d destinations. The positions of N_s sources and N_d destinations are assumed to be fixed in a two-dimensional plane while the positions of N_r relays vary in a convex set $\mathfrak{S}_T$ having its boundary formed by all the source and destination points. HARQ-IR protocol and NC techniques are also applied at the relays to assist the data transmission between the sources and destinations.

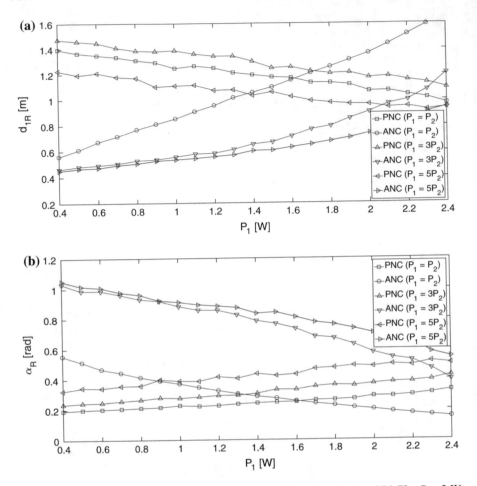

Fig. 5 Optimization of relay location subject to minimizing EB with $R = 4$ b/s/Hz, $P = 5$ W, $d_{S_1 D_1} = d_{S_2 D_2} = 2$ m, $d_{S_1 S_2} = d_{D_1 D_2} = 1/2$ m: **a** $d_{S_1 R}$ and **b** α_R

In a WMN, the k-th relay, $k = 1, 2, \ldots, N_r$, i.e., $\mathscr{R}_k$, assists the data transmission from a group of $N_{s,k}$ sources, i.e., $\{\mathscr{S}_{k,1}, \mathscr{S}_{k,2}, \ldots, \mathscr{S}_{k,N_{s,k}}\} \triangleq \mathscr{S}_k^{(N_{s,k})}$, to a group of $N_{d,k}$ destinations, i.e., $\{\mathscr{D}_{k,1}, \mathscr{D}_{k,2}, \ldots, \mathscr{D}_{k,N_{d,k}}\} \triangleq \mathscr{D}_k^{(N_{d,k})}$. The indices of nodes are determined based upon their vertical axis values in a decreasing order, i.e., the node located higher has a lower index. Denote $\mathfrak{S}_k$ as the convex set generated by points $\{\mathscr{S}_{k,1}, \mathscr{S}_{k,2}, \ldots, \mathscr{S}_{k,N_{s,k}}\}$ and $\{\mathscr{D}_{k,1}, \mathscr{D}_{k,2}, \ldots, \mathscr{D}_{k,N_{d,k}}\}$ which are in supporting region of $\mathscr{R}_k$, i.e.,

$$\mathfrak{S}_T \supseteq \bigcup_{k=1}^{N_r} \mathfrak{S}_k \tag{67}$$

The relay-aided transmission is realized in two time slots as follows: In the first time slot, $\mathscr{S}_{k,i_k}$, $i_k = 1, 2, \ldots, N_{s,k}$, sends data to $\mathscr{R}_k$ and the corresponding $\mathscr{D}_{k,i'_k}$, $i'_k = 1, 2, \ldots, N_{d,k}$ via direct links. Then, $\mathscr{R}_k$ carries out either PNC or ANC on the received signals before broadcasting the combined signal to all $\mathscr{D}_k^{(N_{d,k})}$ in the second time slot. For simplicity, it is assumed that there is no interference caused by non-intended nodes and there is no cooperation between relays, between sources and between destinations in the WMN.

Let (x_A, y_A), $A \in \{\{\mathscr{S}_i\}, \{\mathscr{R}_k\}, \{\mathscr{D}_j\}\}$, denote the coordinate values of a point $\mathscr{A}$. Exploiting the properties of perspective transformation (cf. Remark 5), the nodes in the irregularly-shaped WMN can be mapped to the nodes in a rectangle. The optimal placement of virtual relays can be found in the rectangular region to minimize either ED or EB. The real optimal positions of the relays can be thus determined by an inverse mapping.

Algorithm 1 Proposed relay placement algorithm

> **for** $k = 1$ to N_r **do**
>
> $\quad \mathfrak{S}_k \leftarrow \{\mathscr{S}_{k,1}, \mathscr{S}_{k,2}, \ldots, \mathscr{S}_{k,N_{s,k}}, \mathscr{D}_{k,1}, \mathscr{D}_{k,2}, \ldots, \mathscr{D}_{k,N_{d,k}}\}$
>
> $\quad$ *Step 1*: Map the boundary of $\mathfrak{S}_k$ to a rectangle $\mathfrak{S}'_k$:
>
> $\quad (\mathscr{S}'_{k,1} \mathscr{S}'_{k,N_{s,k}} \mathscr{D}'_{k,1} \mathscr{D}'_{k,N_{d,k}}) \leftarrow (\mathscr{S}_{k,1} \mathscr{S}_{k,N_{s,k}} \mathscr{D}_{k,1} \mathscr{D}_{k,N_{d,k}})$
>
> $\quad$ Find mapping matrix $\mathbf{M}$.
>
> $\quad$ *Step 2*: Find virtual positions of remaining nodes in $\mathfrak{S}'_k$:
>
> $\quad$ **for** $i = 2$ to $N_{s,k} - 1$ **do**
>
> $\quad\quad [1, x_{S'_{k,i}}, y_{S'_{k,i}}, x_{S'_{k,i}} y_{S'_{k,i}}] \leftarrow [1, x_{S_{k,i}}, y_{S_{k,i}}, x_{S_{k,i}} y_{S_{k,i}}] \mathbf{M}$
>
> $\quad$ **end for**
>
> $\quad$ **for** $i = 2$ to $N_{d,k} - 1$ **do**
>
> $\quad\quad [1, x_{D'_{k,i}}, y_{D'_{k,i}}, x_{D'_{k,i}} y_{D'_{k,i}}] \leftarrow [1, x_{D_{k,i}}, y_{D_{k,i}}, x_{D_{k,i}} y_{D_{k,i}}] \mathbf{M}$
>
> $\quad$ **end for**
>
> $\quad$ *Step 3*: Find virtual relay placement in $\mathfrak{S}'_k$ to either minimize ED or minimize EB: (x'_{R_k}, y'_{R_k}).
>
> $\quad$ *Step 4*: Find real relay placement in $\mathfrak{S}_k$:
>
> $\quad\quad [1, x_{R_k}, y_{R_k}, x_{R_k} y_{R_k}] \leftarrow [1, x'_{R_k}, y'_{R_k}, x'_{R_k} y'_{R_k}] \mathbf{M}^{-1}$
>
> **end for**

For convenience, the entire set $\mathfrak{S}_T$ is divided into N_r subsets with respect to N_r relays (cf. Eq. (67)) and consider a specific subset $\mathfrak{S}_k$, $k = 1, 2, \ldots, N_r$. The RP in the WMN can be carried out as in Algorithm 1, which consists of the following steps:

- *Step 1*: Map the boundary of $\mathfrak{S}_k$ to a rectangle, namely $\mathfrak{S}'_k$, by finding a mapping matrix $\mathbf{M}$.
- *Step 2*: Find virtual positions of remaining sources and destinations in $\mathfrak{S}'_k$.
- *Step 3*: Find virtual relay position (x'_{R_k}, y'_{R_k}) in $\mathfrak{S}'_k$ for either minimizing ED or minimizing EB.
- *Step 4*: Find real relay position in $\mathfrak{S}_k$ by inverse mapping.

It can be observed that the RP algorithm only requires the perspective transformation and determination of the optimal relay positions in a particular rectangle.

7   Conclusions

This chapter has provided an overview of cooperative communications with different diversity approaches and cooperative protocols along with NC techniques at the physical layer. In particular, WBN has been investigated as a typical application of the NC techniques. The EDT has been derived for HARQ-IR protocols with PNC and ANC in the WBN by taking into account the effects of both relay location and power allocation. In the high power regime, the use of the relay in both PNC and ANC schemes has been shown to have no advantage over the non-relay-aided DT scheme. In the low power regime, the PNC scheme is more energy efficient than both the ANC and DT schemes when the relay node is located either at the centre of the network or close to the destination nodes, while the DT scheme outperforms both the PNC and ANC schemes when the relay node is in the neighborhood of the source nodes. Furthermore, an RP algorithm for reducing the search region has been developed to find the optimal relay locations for the HARQ-IR protocols with PNC and ANC to minimize either the total delay or the total energy consumption in the WBN. The RP algorithm has also been discussed for a general WMN. For future work, the mobility of nodes as well as network infrastructures in the practical WMNs, such as mobile ad hoc networks, wireless sensor networks, vehicular networks and more generally wireless mesh networks, could be considered in the RP optimization problem subject to constraints on the limited power of nodes and their geographic locations.

References

1. Sendonaris, A., Erkip, E., Aazhang, B.: User cooperation diversity—Part I. System description. IEEE Trans. Commun. **51**(11), 1927–1938 (2003)
2. Sendonaris, A., Erkip, E., Aazhang, B.: User cooperation diversity—Part II. Implementation aspects and performance analysis. IEEE Trans. Commun. **51**(11), 1939–1948 (2003)
3. Laneman, J., Tse, D., Wornell, G.: Cooperative diversity in wireless networks: efficient protocols and outage behavior. IEEE Trans. Inf. Theory **50**(12), 3062–3080 (2004)
4. Loa, K., Wu, C.C., Sheu, S.T., Yuan, Y., Chion, M., Huo, D., Xu, L.: IMT-advanced relay standards [WiMAX/LTE update]. IEEE Commun. Mag. **48**(8), 40–48 (2010)
5. Sheng, Z., Leung, K., Ding, Z.: Cooperative wireless networks: from radio to network protocol designs. IEEE Commun. Mag. **49**(5), 64–69 (2011)
6. Sharma, S., Shi, Y., Hou, Y., Kompella, S.: An optimal algorithm for relay node assignment in cooperative ad hoc networks. IEEE/ACM Trans. Netw. **19**(3), 879–892 (2011)
7. Sun, L., Zhang, T., Lu, L., Niu, H.: Cooperative communications with relay selection in wireless sensor networks. IEEE Trans. Consum. Electron. **55**(2), 513–517 (2009)
8. Chen, Y., Teo, J., Lai, J., Gunawan, E., Low, K.S., Soh, C.B., Rapajic, P.: Cooperative communications in ultra-wideband wireless body area networks: channel modeling and system diversity analysis. IEEE J. Sel. Areas Commun. **27**(1), 5–16 (2009)
9. Dimakis, A., Ramchandran, K., Wu, Y., Suh, C.: A survey on network codes for distributed storage. Proc. IEEE **99**(3), 476–489 (2011)
10. Ahlswede, R., Cai, N., Li, S.Y., Yeung, R.: Network information flow. IEEE Trans. Inf. Theory **46**(4), 1204–1216 (2000)

11. Koetter, R., Medard, M.: An algebraic approach to network coding. IEEE/ACM Trans. Netw. **11**(5), 782–795 (2003)
12. Zhang, S., Liew, S.C., Lam, P.P.: Hot topic: Physical-layer network coding. In: Proceedings of ACM MobiCom'06, Los Angeles, CA, USA, September 2006, pp. 358–365
13. Katti, S., Gollakota, S., Katabi, D.: Embracing wireless interference: analog network coding. In: Proceedings of ACM SIGCOMM'07, Kyoto, Japan, August 2007, pp. 397–408
14. Louie, R., Li, Y., Vucetic, B.: Practical physical layer network coding for two-way relay channels: performance analysis and comparison. IEEE Trans. Wirel. Commun. **9**(2), 764–777 (2010)
15. Ju, M., Kim, I.M.: Error performance analysis of BPSK modulation in physical-layer network-coded bidirectional relay networks. IEEE Trans. Commun. **58**(10), 2770–2775 (2010)
16. Nguyen, D., Tran, T., Nguyen, T., Bose, B.: Wireless broadcast using network coding. IEEE Trans. Veh. Technol. **58**(2), 914–925 (2009)
17. Chen, Y., Kishore, S.: On the tradeoffs of implementing randomized network coding in multicast networks. IEEE Trans. Commun. **58**(7), 2107–2115 (2010)
18. Liu, J., Goeckel, D., Towsley, D.: Bounds on the throughput gain of network coding in unicast and multicast wireless networks. IEEE J. Sel. Areas Commun. **27**(5), 582–592 (2009)
19. Zhan, A., He, C., Jiang, L.: A channel statistic based power allocation in a butterfly wireless network with network coding. In: Proceedings of IEEE ICC 2010, Cape Town, South Africa, May 2010, pp. 1–5
20. Hu, J., Fan, P., Xiong, K., Yi, S., Lei, M.: Cooperation-based opportunistic network coding in wireless butterfly networks. In: Proceedings of IEEE GLOBECOM 2011, Houston, TX, USA, December 2011, pp. 1–5
21. Zheng, L., Tse, D.N.C.: Diversity and multiplexing: a fundamental tradeoff in multiple antenna channels. IEEE Trans. Inf. Theory **49**(5), 1073–1096 (2003)
22. Winters, J., Salz, J., Gitlin, R.: The impact of antenna diversity on the capacity of wireless communication systems. IEEE Trans. Commun. **42**(234) 1740–1751 (1994)
23. Foschini, G.J., Gans, M.J.: On limits of wireless communications in a fading environment when using multiple antennas. Wirel. Pers. Commun. **6**, 311–335 (1998)
24. Telatar, E.: Capacity of multi-antenna gaussian channels. Eur. Trans. Telecommun. **10**(6), 585–596 (1999)
25. Tarokh, V., Seshadri, N., Calderbank, A.: Space-time codes for high data rate wireless communication: performance criterion and code construction. IEEE Trans. Inf. Theory **44**(2), 744–765 (1998)
26. Guey, J.C., Fitz, M., Bell, M., Kuo, W.Y.: Signal design for transmitter diversity wireless communication systems over Rayleigh fading channels. IEEE Trans. Commun. **47**(4), 527–537 (1999)
27. Alamouti, S.: A simple transmit diversity technique for wireless communications. IEEE J. Sel. Areas Commun. **16**(8), 1451–1458 (1998)
28. Tarokh, V., Jafarkhani, H., Calderbank, A.: Space-time block codes from orthogonal designs. IEEE Trans. Inf. Theory **45**(5), 1456–1467 (1999)
29. Ganesan, G., Stoica, P.: Space-time block codes: a maximum SNR approach. IEEE Trans. Inf. Theory **47**(4), 1650–1656 (2001)
30. Tirkkonen, O., Hottinen, A.: Square-matrix embeddable space-time block codes for complex signal constellations. IEEE Trans. Inf. Theory **48**(2), 384–395 (2002)
31. Jafarkhani, H.: A quasi-orthogonal space-time block code. IEEE Trans. Commun. **49**(1), 1–4 (2001)
32. Su, W., Xia, X.G.: Signal constellations for quasi-orthogonal space-time block codes with full diversity. IEEE Trans. Inf. Theory **50**(10), 2331–2347 (2004)
33. Hughes, B.: Differential space-time modulation. IEEE Trans. Inf. Theory **46**(7), 2567–2578 (2000)
34. Hochwald, B., Marzetta, T., Richardson, T., Sweldens, W., Urbanke, R.: Systematic design of unitary space-time constellations. IEEE Trans. Inf. Theory **46**(6), 1962–1973 (2000)

35. Damen, M., Abed-Meraim, K., Belfiore, J.C.: Diagonal algebraic space-time block codes. IEEE Trans. Inf. Theory **48**(3), 628–636 (2002)
36. Du, J., Li, Y.: Parallel detection of groupwise space–time codes by predictive soft interference cancellation. IEEE Trans. Commun. **54**(12), 2150–2154 (2006)
37. Lindskog, E., Paulraj, A.: A transmit diversity scheme for channels with intersymbol interference. In: IEEE ICC'00, vol. 1, New Orleans, LA, USA, June 2000, pp. 307–311
38. Al-Dhahir, N.: Single-carrier frequency-domain equalization for space-time block-coded transmissions over frequency-selective fading channels. IEEE Commun. Lett. **5**(7), 304–306 (2001)
39. Zhou, S., Giannakis, G.: Space-time coding with maximum diversity gains over frequency-selective fading channels. IEEE Signal Process. Lett. **8**(10), 269–272 (2001)
40. Agrawal, D., Tarokh, V., Naguib, A., Seshadri, N.: Space-time coded OFDM for high data-rate wireless communication over wideband channels. In: Proceedings of IEEE VTC'98, vol. 3, Ottawa, Canada, May 1998, pp. 2232–2236
41. Lu, B., Wang, X.: Space-time code design in OFDM systems. In: Proceedings of IEEE GLOBECOM'00, vol. 2, San Francisco, USA, November 2000, pp. 1000–1004
42. Blum, R., Li, Y.G., Winters, J., Yan, Q.: Improved space-time coding for MIMO-OFDM wireless communications. IEEE Trans. Commun. **49**(11), 1873–1878 (2001)
43. Su, W., Safar, Z., Liu, K.: Full-rate full-diversity space-frequency codes with optimum coding advantage. IEEE Trans. Inf. Theory **51**(1), 229–249 (2005)
44. Gong, Y., Letaief, K.: Space-frequency-time coded OFDM for broadband wireless communications. In: Proceedings of IEEE GLOBECOM'01, vol. 1, San Antonio, Texas, USA, November 2001, pp. 519–523
45. Liu, Z., Xin, Y., Giannakis, G.: Space-time-frequency coded OFDM over frequency-selective fading channels. IEEE Trans. Signal Process. **50**(10), 2465–2476 (2002)
46. Molisch, A., Win, M., Winters, J.: Space-time-frequency (STF) coding for MIMO-OFDM systems. IEEE Commun. Lett. **6**(9), 370–372 (2002)
47. van der Meulen, E.C.: Three-terminal communication channels. Adv. Appl. Probab. **3**, 120–154 (1971)
48. Cover, T., Gamal, A.: Capacity theorems for the relay channel. IEEE Trans. Inf. Theory **25**(5), 572–584 (1979)
49. Kramer, G., Gastpar, M., Gupta, P.: Cooperative strategies and capacity theorems for relay networks. IEEE Trans. Inf. Theory **51**(9), 3037–3063 (2005)
50. Hunter, T., Nosratinia, A.: Diversity through coded cooperation. IEEE Trans. Wirel. Commun. **5**(2), 283–289 (2006)
51. Laneman, J., Wornell, G.: Distributed space-time-coded protocols for exploiting cooperative diversity in wireless networks. IEEE Trans. Inf. Theory **49**(10), 2415–2425 (2003)
52. Nabar, R., Bolcskei, H., Kneubuhler, F.: Fading relay channels: performance limits and space-time signal design. IEEE J. Sel. Areas Commun. **22**(6), 1099–1109 (2004)
53. Yiu, S., Schober, R., Lampe, L.: Distributed space-time block coding. IEEE Trans. Commun. **54**(7), 1195–1206 (2006)
54. Hassibi, B., Hochwald, B.: High-rate codes that are linear in space and time. IEEE Trans. Inf. Theory **48**(7), 1804–1824 (2002)
55. Jing, Y., Hassibi, B.: Distributed space-time coding in wireless relay networks. IEEE Trans. Wireless Commun. **5**(12), 3524–3536 (2006)
56. Jing, Y., Jafarkhani, H.: Using orthogonal and quasi-orthogonal designs in wireless relay networks. IEEE Trans. Inf. Theory **53**(11), 4106–4118 (2007)
57. Yi, Z., Kim, I.M.: Single-symbol ML decodable distributed STBCs for cooperative networks. IEEE Trans. Inf. Theory **53**(8), 2977–2985 (2007)
58. Scutari, G., Barbarossa, S.: Distributed space-time coding for regenerative relay networks. IEEE Trans. Wirel. Commun. **4**(5), 2387–2399 (2005)
59. Anghel, P., Kaveh, M.: Relay assisted uplink communication over frequency-selective channels. In: Proceedings of IEEE Workshop SPAWC'03, Rome, Italy, June 2003, pp. 125–129

60. Mheidat, H., Uysal, M., Al-Dhahir, N.: Equalization techniques for distributed space-time block codes with amplify-and-forward relaying. IEEE Trans. Signal Process. **55**(5), 1839–1852 (2007)
61. Tran, L.N., Vien, Q.T., Hong, E.K.: Training sequence-based distributed space-time block codes with frequency domain equalization. In: Proceedings of IEEE PIMRC'09, Tokyo, Japan, September 2009, pp. 501–505
62. Tran, L.N., Vien, Q.T., Hong, E.K.: Unique word-based distributed space-time block codes for two-hop wireless relay networks. IET Commun. **6**(7) 715–723 (2012)
63. Vien, Q.T., Tran, L.N., Hong, E.K.: Design of distributed space-time block code for two-relay system over frequency selective fading channels. In: Proceedings of IEEE GLOBECOM'09, Honolulu, Hawaii, USA, November 2009, pp. 1–5
64. Vien, Q.T., Tran, L.N., Hong, E.K.: Distributed space-time block code over mixed Rayleigh and Rician frequency selective fading channels. EURASIP J. Wirel. Commun. Net. **2010**, Article ID 385872, 9 p (2010)
65. Vien, Q.T.: Distributed Space-Time Block Codes for Relay Networks: Design for Frequency-Selective Fading Channels. LAP LAMBERT Academic Publishing (2011)
66. Vien, Q.T., Hong, E.K.: Design of quasi-orthogonal space-time block codes for cooperative wireless relay networks over frequency selective fading channels. In: Proceedings of IEEE ATC'09, Hai Phong, Vietnam, October 2009, pp. 275–278
67. Vien, Q.T., Nguyen, D.T.H., Tran, L.N., Hong, E.K.: Design of DSTBC and HARQ schemes for turbo-coded cooperative wireless relay networks over frequency selective fading channels. In: Proceedings of ITC-CSCC'09, Jeju, Korea, July 2009, pp. 584–587
68. Vien, Q.T., Nguyen, H.X., Gemikonakli, O., Barn, B.: Performance analysis of cooperative transmission for cognitive wireless relay networks. In: Proceedings of IEEE GLOBECOM 2013, Atlanta, Georgia, USA, December 2013, pp. 4186–4191
69. Vien, Q.T., Stewart, B.G., Nguyen, H.X., Gemikonakli, O.: Distributed space-time-frequency block code for cognitive wireless relay networks. IET Commun. **8**(5), 754–766 (2014)
70. Katti, S., Rahul, H., Hu, W., Katabi, D., Medard, M., Crowcroft, J.: XORs in the air: practical wireless network coding. IEEE/ACM Trans. Netw. **16**(3), 497–510 (2008)
71. Rankov, B., Wittneben, A.: Spectral efficient protocols for half-duplex fading relay channels. IEEE J. Sel. Areas Commun. **25**(2), 379–389 (2007)
72. Verdu, S.: Multiuser Detection. Cambridge University Press, UK (1998)
73. Popovski, P., Yomo, H.: Physical network coding in two-way wireless relay channels. In: Proceedings of IEEE ICC'07, Glasgow, Scotland, June 2007, pp. 707–712
74. Zhang, R., Liang, Y.C., Chai, C.C., Cui, S.: Optimal beamforming for two-way multi-antenna relay channel with analogue network coding. IEEE J. Sel. Areas Commun. **27**(5), 699–712 (2009)
75. Song, L., Hong, G., Jiao, B., Debbah, M.: Joint relay selection and analog network coding using differential modulation in two-way relay channels. IEEE Trans. Veh. Technol. **59**(6), 2932–2939 (2010)
76. Wang, H.M., Xia, X.G., Yin, Q.: A linear analog network coding for asynchronous two-way relay networks. IEEE Trans. Wirel. Commun. **9**(12), 3630–3637 (2010)
77. Vien, Q.T., Tran, L.N., Nguyen, H.X.: Network coding-based ARQ retransmission strategies for two-way wireless relay networks. In: Proceedings of IEEE SoftCOM 2010, Split, Croatia, September 2010, pp. 180–184
78. Vien, Q.T., Tran, L.N., Nguyen, H.X.: Efficient ARQ retransmission schemes for two-way relay networks. J. Commun. Softw. Syst. **7**(1), 9–15 (2011)
79. Vien, Q.T., Nguyen, H.X.: CQI reporting strategies for nonregenerative two-way relay networks. In: Proceedings of IEEE WCNC 2012, Paris, France, April 2012, pp. 974–979
80. Vien, Q.T., Nguyen, H.X.: Network coding-based channel quality indicator reporting for two-way multi-relay networks. Wiley J. Wirel. Commun. Mob. Comput. **14**(15), 1471–1483 (2014)
81. Vien, Q.T.: Cooperative diversity techniques for high-throughput wireless relay networks. Ph.D. Thesis, Glasgow Caledonian University (2013)

82. Vien, Q.T., Tran, L.N., Hong, E.K.: Network coding-based retransmission for relay aided multisource multicast networks. EURASIP J. Wirel. Commun. Net. **2011**, Article ID 643920, 10 p (2011)
83. Vien, Q.T., Tianfield, H., Stewart, B.G., Nguyen, H.X., Choi, J.: An efficient retransmission strategy for multisource multidestination relay networks over Rayleigh flat fading channels. In: Proceedings of IEEE WPMC 2011, Brest, France, October 2011, pp. 171–175
84. Vien, Q.T., Stewart, B.G., Nguyen, H.X.: Outage probability of regenerative protocols for two-source two-destination networks. Springer J. Wirel. Pers. Commun. **69**(4), 1969–1981 (2013)
85. Vien, Q.T., Stewart, B.G., Tianfield, H., Nguyen, H.X., Choi, J.: An efficient network coded ARQ for multisource multidestination relay networks over mixed flat fading channels. Elsevier AEU Int. J. Electron. Commun. **67**(4), 282–288 (2013)
86. Vien, Q.T., Nguyen, H.X., Shah, P., Ever, E., To, D.: Relay selection for efficient HARQ-IR protocols in relay-assisted multisource multicast networks. In: Proc. IEEE VTC 2014-Spring, Seoul, Korea (May 2014) 1–5
87. Vien, Q.T., Tu, W., Nguyen, H.X., Trestian, R.: Cross-layer optimisation for topology design of wireless multicast networks via network coding. In: Proceedings of IEEE LCN 2014, Edmonton, Canada, September 2014, pp. 466–469
88. Vien, Q.T., Tu, W., Nguyen, H.X., Trestian, R.: Cross-layer topology design for network coding based wireless multicasting
89. Vien, Q., Nguyen, H., Barn, B., Tran, X.: On the perspective transformation for efficient relay placement in wireless multicast networks. IEEE Commun. Lett. **19**(2), 275–278 (2015)
90. Vien, Q.T., Nguyen, H.X., Choi, J., Stewart, B.G., Tianfield, H.: Network coding-based block ACK for wireless relay networks. In: Proceedings of IEEE VTC 2011-Spring, Budapest, Hungary, May 2011, pp. 1–5
91. Vien, Q.T., Nguyen, H.X., Choi, J., Stewart, B.G., Tianfield, H.: Network coding-based block acknowledgement scheme for wireless regenerative relay networks. IET Commun. **6**(16) 2593–2601 (2012)
92. Vien, Q.T., Stewart, B.G., Tianfield, H., Nguyen, H.X.: An efficient cooperative retransmission for wireless regenerative relay networks. In: Proceedings of IEEE GLOBECOM 2012, Anaheim, California, USA, December 2012, pp. 4417–4422
93. Vien, Q.T., Stewart, B.G., Tianfield, H., Nguyen, H.X.: Cooperative retransmission for wireless regenerative multirelay networks. IEEE Trans. Veh. Technol. **62**(2), 735–747 (2013)
94. Vien, Q.T., Nguyen, H.X., Tu, W.: Optimal relay positioning for green wireless network-coded butterfly networks. In: Proceedings of IEEE PIMRC 2013, London, UK, September 2013, pp. 286–290
95. Vien, Q.T., Stewart, B.G., Choi, J., Nguyen, H.X.: On the energy efficiency of HARQ-IR protocols for wireless network-coded butterfly networks. In: Proceedings of IEEE WCNC 2013, Shanghai, China, April 2013, pp. 2559–2564
96. Vien, Q.T., Nguyen, H.X., Stewart, B.G., Choi, J., Tu, W.: On the energy-delay tradeoff and relay positioning of wireless butterfly networks. IEEE Trans. Veh. Technol. **64**(1), 159–172 (2015)
97. Wicker, S.B.: Error Control Systems for Digital Communication and Storage. Prentice-Hall (1995)
98. Caire, G., Tuninetti, D.: The throughput of hybrid-ARQ protocols for the Gaussian collision channel. IEEE Trans. Inf. Theory **47**(5), 1971–1988 (2001)
99. Choi, J., To, D.: Energy efficiency of HARQ-IR for two-way relay systems with network coding. In: Proceedings of EW 2012, Poznan, Poland, April 2012
100. Zhang, S., Liew, S.C.: Channel coding and decoding in a relay system operated with physical-layer network coding. IEEE J. Sel. Areas Commun. **27**(5), 788–796 (2009)
101. Cover, T.M., Thomas, J.A.: Elements of Information Theory, 2nd edn. Wiley, NJ (2006)
102. Chen, X., Song, S.H., Letaief, K.: Relay position optimization improves finite-SNR diversity gain of decode-and-forward mimo relay systems. IEEE Trans. Commun. **60**(11), 3311–3321 (2012)

103. Han, L., Huang, C., Shao, S., Tang, Y.: Relay placement for amplify-and-forward relay channels with correlated shadowing. IEEE Wirel. Commun. Lett. **2**(2), 171–174 (2013)
104. Wolberg, G.: Digital Image Warping, 1st edn. Wiley-IEEE Computer Society Press, Los Alamitos (1990)
105. Kim, D.K., Jang, B.T., Hwang, C.J.: A planar perspective image matching using point correspondences and rectangle-to-quadrilateral mapping. In: Proceedings of IEEE SSIAI'02, Sante Fe, New Mexico, April 2002, pp. 87–91

The Bat Algorithm, Variants and Some Practical Engineering Applications: A Review

T. Jayabarathi, T. Raghunathan and A.H. Gandomi

Abstract The bat algorithm (BA), a metaheuristic algorithm developed by Xin-She Yang in 2010, has since been modified, and applied to numerous practical optimization problems in engineering. This chapter is a survey of the BA, its variants, some sample real-world optimization applications, and directions for future research.

Keywords Algorithm · Bat algorithm · Engineering application · Optimization · Swarm intelligence · Metaheuristics

1 Introduction

Real-world optimization problems do not conform to the requirements of calculus or gradient based optimization methods that require functions to be continuous, smooth and unimodal, with ever present derivatives and ideal constraints. In reality, functions can be noisy, filled with discontinuities, have multiple optimums, and their derivatives may be non-existent [1]. To solve such problems, researchers have been increasingly looking to nature as the ultimate expert on optimization. A pioneer in the field who conclusively demonstrated that practical problems of significant complexity could be solved by nature inspired algorithms is Goldberg, who solved an oil transportation problem that was not amenable to gradient based optimization, and thought to be even less amenable to nature inspired methods, for his PhD in the year 1983 [2]. Given their enormous popularity, nature inspired algorithms are known by many names, including evolutionary computation, the oldest name. Since an element of learning or heuristics is involved, and this

T. Jayabarathi · T. Raghunathan (✉)
School of Electrical Engineering, VIT University, Vellore 632014, India
e-mail: raghunathan.t@vit.ac.in; nathraghu@yahoo.com

A.H. Gandomi
School of Business, Stevens Institute of Technology, Hoboken, NJ 07030, USA

© Springer International Publishing AG 2018
X.-S. Yang (ed.), *Nature-Inspired Algorithms and Applied Optimization*,
Studies in Computational Intelligence 744, https://doi.org/10.1007/978-3-319-67669-2_14

313

heuristics is of higher level, they are also popularly known as metaheuristic algorithms at present.

The genetic algorithm (GA) used by Goldberg belongs to the family of evolutionary algorithms (EAs), which loosely model evolution in biology as an optimization process. The 1990s onwards saw the development of another family of nature based algorithms that modeled the collective behavior of social animals as an optimization process. The ant colony optimization (ACO) [3], particle swarm optimization (PSO) [4] and krill herd [5] algorithms belong to this family of swarm intelligence (SI) algorithms. Nature inspired algorithms continued to be developed at an even faster pace in the 2000s, and the year 2010 saw the emergence of the bat algorithm (BA), an SI algorithm developed by Xin-She Yang [6]. This was followed by the application of the BA to engineering applications in 2012 [7], and to constrained optimization problems in 2013 [8]. A good reference to swarm intelligence methods of the period is [9]. Since then, the BA has seen numerous variants being developed, and these applied to solve many real-world problems.

In the search for an optimum in multimodal search or solution space, two conflicting objectives need to be catered to: exploration or diversification, and exploitation or intensification. Striking the right balance these two objectives can be the difference between successfully locating the global optimum and not doing so. In the initial stages of the search, all areas of the solution space have to be explored, if the global optimum is not to be missed. On the other hand, once the most promising areas have been located, further exploration would lead the search to meander about aimlessly. Instead, the requirement at this later stage of the search is to exploit or intensify the search in the narrowed down promising areas, so that the optimum can be located. A good search algorithm must thus have operators for exploration and exploitation.

1.1 Exploration or Diversification

Reference [6] contains a detailed description of the food location behavior of bats in nature, and the modeling of this behavior to form the BA metaheuristic. Hence the motivation here is to recapture the most essential details of the BA in the language of general optimization theory. In summary, the BA mimics the collective behavior of a colony of bats that use a phenomenon known as echolocation. Echolocation by bats is characterized by the emission of pulses of some frequency f, loudness A and emission rate R. The exploration capability of the BA is provided by its velocity and position update equations, given by

$$V^i(t+1) = V^i(t) + f^i \{X^i(t) - X^{best}(t)\} \tag{1}$$

$$f^i(t) = f^{min} + \text{rand}() \, (f^{max} - f^{min}) \tag{2}$$

$$X^i(t+1) = X^i(t) + V^i(t+1) \tag{3}$$

where t is the current iteration number, V^i, f^i and X^i are the velocity, frequency, and position, respectively, of the ith bat in the population of N_b bats, and X^{best} is the bat location (solution) that has the best fitness in the current population. $\text{rand}() \in [0, 1]$ is a randomly generated number from a uniform distribution in the interval $[0,1]$. f^{max} and f^{min} are the allowable maximum and minimum frequencies, which can assume the default values of 0 and 100, but can be varied to suit the problem being solved. At initialization $(t=0)$, V^i is assumed to be 0.

1.2 Exploitation or Intensification

The exploitation or local search is by means of a random walk. Two parameters, the loudness $A^i(t)$ and the pulse emission rate $R^i(t)$ are updated at every iteration, for every bat in the population. Depending on $R^i(t)$, a local search is conducted, either around the best solution or a randomly chosen solution:

$$X^{i,\text{new}} = \begin{cases} X^{i,\text{old}}(t) + r_2 A^i(t) & \text{if } \text{rand}() > R^i(t) \\ X^r(t) + r_2 A^i(t) & \text{else} \end{cases} \tag{4}$$

where $\text{rand}() \in [0, 1]$ and $r_2 \in [-1, 1]$ are uniformly distributed random numbers, and $r \in [1, 2, \ldots, N_b], r \neq i$ is a randomly chosen integer. In other words, $X^r(t)$ is a randomly chosen solution in the current iteration, and different from the ith solution.

The right balance between exploration and exploitation as the search progresses is provided by adjusting the pulse emission rate $R^i(t)$ and loudness $A^i(t)$ dynamically:

$$R^i(t+1) = R^i(0)[1 - \exp(-\gamma t)] \tag{5}$$

$$A^i(t+1) = \alpha A^i(t) \tag{6}$$

where $R^i(0) \in [0, 1]$ and $A^i(t) \in [1, 2]$, both randomly generated, within their respective limits. As a first choice, the default values that can be used are $\gamma = \alpha = 0.9$, as in [6].

1.3   Selection

In order to improve the solutions over the iterations, a fitness based, tournament type of solution, in which the competitors are the old and new solutions is implemented. The fitter solution replaces the less fit one, with a probability

$$A^i(t): X^i(t) = X^{i,\text{new}}(t) \text{ if } F\left(X^{i,\text{new}}(t)\right) < F\left(X^i(t)\right) \text{ and rand}() < A^i(t) \forall i, i \in [1, 2, \ldots, N_b] \tag{7}$$

where $\text{rand}() \in [0, 1]$ is a uniformly distributed random number, and $F(X(t))$ is the fitness or cost function to be minimized.

The BA is shown to perform well on some benchmark unimodal and multimodal functions in comparisons against the PSO and GA in [6].

2   Variants of the Bat Algorithm

Once the basic BA showed initial promise as a good metaheuristic algorithm, the focus shifted towards improving it further and applying it to real-world optimization problems for better results. Some of the modifications proposed to improve the performance of the BA are outlined next.

2.1   Chaotic Bat Algorithm

The basic BA outlined in Sect. 1 used default values of the algorithm parameters, and exponentially decreasing values for loudness A^i. For reasons yet to be studied, using chaotic maps or sequences to update the loudness has been shown to produce better performance on the complicated real-world problem solved in [10]. Instead of using Eq. (6) to update the loudness, Ref. [10] updated the loudness using

$$A^i(t+1) = a\left[A^i(t)\right]^2 \sin\left[\pi A^i(t)\right] \tag{8}$$

where a is set to 2.3 and $A^i(0) \in [0, 1]$, and is randomly generated. Equation (8) describes the sinusoidal chaotic map or sequence of the variable A^i. Other chaotic maps are given in [11], which proposed the chaotic BA (CBA) used in [10]. Another version of the CBA was proposed by Jordehi [12].

2.2　Directional Bat Algorithm

Instead of updating the bat positions using Eqs. (1)–(3), the directional BA proposed by Chakri et al. [13] probes the search space in two different directions: one in the direction of the best bat, and the other a randomly selected one. The one with better fitness is used to update the current bat position. When used along with a few other modifications, this directional BA is claimed to improve the performance when tested on a suite of benchmark functions.

2.3　θ-Modified Bat Algorithm

The central idea of the θ-modified BA is to use a polar framework, instead of the Cartesian coordinate framework used by Eqs. (1)–(3) [14]. A second modification is to dynamically update of the parameter α, instead of using a fixed value, as in the basic BA described in Sect. 1. This is claimed to have produced the best results on the stochastic multi-objective problems solved in [14].

2.4　The BA with Mutation

One of the earliest ideas was to equip the BA, which is an SI algorithm, with operators from EAs.

In nature, mutation introduces new genetic material into the existing gene pool, thereby helping to maintain the diversity of the population. The mutation operator in EAs plays the same role of maintaining the diversity of the search space. Its operation can be explained quite simply with the help of the binary genetic algorithm (BGA) in [1], since the solutions or chromosomes or strings in a BGA consist of just 1's and 0's. If the string length of the individual solution or chromosome or string is len_{str} and the population consists of N_p number of individuals, the total number of bits in the population is $N_p \times str_{len}$. If the probability of mutation is equal to $1/(N_p \times str_{len})$, applying the mutation operator to this population involves flipping one randomly chosen bit in the population: if this bit is a 1, it is changed to 0 and vice versa.In multimodal search space, mutation applied correctly can help the search break out of being trapped in local minima. In real valued or continuous GA, mutation is slightly more complex to implement. The interested reader is referred to [15] for more details.

On some problems, the BA could be lacking in exploratory capability, [16] introduced the BA with mutation, to successfully produce better results on a variant of the basic economic dispatch problem, which is to be discussed in Sect. 3 of this chapter. The BA with mutation for a different application can be found in [17]. An enhanced BA with four different types of mutation for self-adaptive learning mechanism (SALM) can be found in [18].

2.5   The BA with Mutation and Crossover

As a logical next step, another evolutionary operator of crossover was also introduced in [19]. Crossover between two parent individuals produces a new offspring or solution that could be potentially fitter than the parents, thereby progressing the search towards the optimum.

2.6   The BA with DE Mutation and Crossover

Differential evolution (DE) is an evolutionary algorithm whose performance was found to be superior to the simple GA on complicated real-world problems [20, 21], by virtue of the kind of mutation and crossover that the DE uses, and known as DE mutation and crossover. Reference [22] explored the hybridization of the bat algorithm with differential evolution, to obtain better results.

2.7   The BA with DE Mutation and Lévy Flights Trajectory (DLBA)

The basic BA is equipped with differential evolution (DE) mutation, and Lévy flights trajectory, to improve its performance. The DLBA is shown to perform better than the basic BA on a suite of unimodal and multimodal benchmark functions in [23].

2.8   The Double-Subpopulation Lévy Flight Bat Algorithm

In this variant [24], the bat population is divided into two subpopulations, internal and external. The internal subpopulation aims at better exploitation, by employing the current speed and global optimal values for updating the speed, and a Lévy flight model. The external subpopulation aims at better exploration, employing DE mutation and crossover, whenever the population diversity reaches a minimum value called the diversity threshold.

2.9 BA with Habitat Selection and Self-adaptive Compensation for Doppler Effect in Echoes

In this variant [25], the position of a bat in the population can display both quantum and mechanical behavior. In quantum behavior, it can appear anywhere in the whole search space with a certain probability. In physical behavior, the frequency update equation has a term that compensates the Doppler effect of change of frequency as the target moves relative to its source.

3 Application of the Bat Algorithm to the Economic Dispatch Problem

3.1 Problem Formulation

Whenever a new algorithm is proposed, its performance is first tested by applying it on unconstrained, unimodal and multimodal benchmark test functions. In contrast, most real-world problems are constrained, and often, even the existence of a solution that satisfies all the constraints is not known beforehand. Details like how violations of the constraints are handled during the solution process too can often be far from clear to beginners. Hence it is instructive to go through the solution procedure for solving a complicated real-world problem by applying the BA. We choose to demonstrate this by going through the steps involved in solving the economic dispatch (ED) problem in electrical power systems engineering [10]. The cost function in this problem is nonlinear, multimodal, with discontinuities, subject to numerous inequality and equality constraints as outlined below.

3.1.1 Objective Function

The cost function is assumed to be quadratic, and minimization of the fuel cost of N_g number of power plants in the system is the objective here:

$$\min_{P \in R^{N_g}} F = \sum_{j=1}^{N_g} F_j(P_j) = \sum_{j=1}^{N_g} \left(a_j + b_j P_j + c_j P_j^2 \right) \tag{9}$$

where $F_j(P_j)$ is the fuel cost of the jth generating unit in \$/hr, Pj is the power generated by the jth generating unit in MW, and a_j, bj and c_j are cost coefficients of the jth generator.

However, the valve point effect superimposes ripples on this quadratic cost curve, thereby making it multimodal, and the cost curve becomes

$$\min_{P \in R^{N_g}} F = \sum_{j=1}^{N_g} F_j(P_j) = \sum_{j=1}^{N_g} (a_j + b_j P_j + c_j P_j^2) + \left| e_j \sin(f_j(P_j^{\min} - P_j)) \right| \quad (10)$$

where e_j and f_j are the constants of the valve-point effect of generators.

If there are multiple fuel options, the fuel cost of the jth generator is given by

$$F_j(P_j) = \begin{cases} a_{j1} + b_{j1}P_j + c_{j1}P_j^2, \text{fuel } 1, P_j^{\min} < P_j \le P_{j1} \\ a_{j2} + b_{j2}P_j + c_{j2}P_j^2, \text{fuel } 2, P_{j1} < P_j \le P_{j2} \\ \vdots \\ a_{jk} + b_{jk}P_j + c_{jk}P_j^2, \text{fuel } k, P_{jk-1} < P_j \le P_j^{\max} \end{cases} \quad (11)$$

A generator with k fuel options has k discrete regions.

3.1.2 Optimization Constraints

The power generated has to obviously satisfy the minimum and maximum power generation limits:

$$P_j^{\min} \le P_j \le P_j^{\max} \quad (12)$$

While the above inequality constraints are relatively easier to satisfy, the more difficult-to-satisfy equality constraint is that the solution or total power generated P_G must satisfy the total load demand P_D plus the total losses P_L in the system:

$$\sum_{j=1}^{N_g} P_j = P_D + P_L \quad (13)$$

where P_L represents the line losses which is calculated using B-coefficients, given by

$$P_L = \sum_{j=1}^{N_g} \sum_{i=1}^{N_g} P_j B_{ji} P_i + \sum_{j=1}^{N_g} B_{0j} P_j + B_{00} \quad (14)$$

where P_i and P_j are the real power injection at ith and jth buses, respectively, and the B_{ij}'s are the loss coefficients which can be assumed to be constant under normal operating conditions.

3.1.3 Practical Operating Constraints of Generators

3.1.4 Prohibited Operating Zones (POZ)

The prohibited zones arise due to practical operational constraints on generators. The feasible operating zones of unit j can be described as follows:

$$P_j \in \begin{cases} P_j^{\min} \leq P_j \leq P_{j,1}^l \\ P_{j,k-1}^u \leq P_j \leq P_{j,k}^l, \quad k=2,3,\ldots n_j, j=1,2,\ldots n \\ P_{j,n_j}^u \leq P_j \leq P_j^{\max} \end{cases} \tag{15}$$

where n_j is the number of prohibited zones of the jth generator. $P_{j,k}^l, P_{j,k}^u$ are the lower and upper power outputs of the kth prohibited zone of the jth generator, respectively.

3.1.5 Ramp Rate Limits

The physical limitations of starting up and shutting down of generators imposeramp rate limits, which are modeled as follows. The increase in generation is limited by

$$P_j - P_j^0 \leq UR_j \tag{16}$$

Similarly, the decrease is limited by

$$P_j^0 - P_j \leq DR_j \tag{17}$$

where P_j^0 is the previous output power, UR_j and DR_j are the up-ramp limit and the down-ramp limit, respectively, of the jth generator.

Combining (16) and (17) with (12) results in the change of the effective operating or generation limits to

$$\underline{P_j} \leq P_j \leq \overline{P_j} \tag{18}$$

where

$$\underline{P_j} = \max(P_j^{\min}, P_j^0 - DR_j) \tag{19}$$

$$\overline{P_j} = \min(P_j^{\max}, P_j^0 + UR_j) \tag{20}$$

Combining this with (15), the ED problem can be formulated as

$$\min_{P \in R^{N_g}} F = \sum_{j=1}^{N_g} F_j(P_j) = \sum_{j=1}^{N_g} (a_j + b_j P_j + c_j P_j^2)$$

$$+ \left| e_j \sin \left[f_j (P_j^{\min} - P_j) \right] \right|$$

(21a)

$$\text{s.t.} \quad \sum_{j=1}^{N_g} P_j = P_D + P_L$$

$$\max(P_j^{\min}, P_j^0 - DR_j) \le P_j \le P_{j,1}^l$$

(21b)

$$P_{j,k-1}^u \le P_j \le P_{j,k}^l, \quad k = 2, 3, \ldots n_j, j = 1, 2, \ldots N_g$$

$$P_{j,n_j}^u \le P_j \le \min(P_j^{\max}, P_j^0 + UR_j)$$

3.2 Implementation of the BA to ED Problem

Step 0: The 0th or the first step consists of initialization, which is executed as follows:

- For every solution (or bat or generating unit), generate randomly within the specified limits the generation values.
- For units with POZ, if the randomly generated value falls within the POZ, fix it at the nearest limit that is violated.
- If a unit has ramp-rate limits, the power output is uniformly distributed between the effective lower and upper limits.

Generate N_b number of bats or solutions, each comprising N_g number of generating units:

$$\begin{bmatrix} p_1^1 & p_2^1 & \cdots & p_{N_g}^1 \\ p_1^2 & p_2^2 & \cdots & p_{N_g}^2 \\ \vdots & & & \\ p_1^{N_b} & p_2^{N_b} & \cdots & p_{N_g}^{N_b} \end{bmatrix} = \begin{bmatrix} P^1 \\ P^2 \\ \vdots \\ P^{N_b} \end{bmatrix}$$

(22)

Step 1: Calculate the fitness values of all the bats using the objective or fitness function F, in (21a).

Step 2: For ith bat, define pulse frequency f^i using (2).

Step 3: Update the velocity and position (which is a vector of generation values) of each bat using (1) and (3), respectively.

Step 4: Generate a new solution by random walk using (4).

Step 5: Select the fitter of the old and new solutions, with a probability $A^i(t)$, using (7).

Step 6: Update the values of R^i and A^i using (5) and (6), respectively.

Step 7: Check if the effective generation limits and POZ limits are violated. Fix the generation at the limit that is violated. This takes care of the inequality constraints. After this is done, any violation of the power balance equality constraint (13) is dealt with by using a penalty factor approach. By this approach, (21a) is modified to

$$\min_{P \in R^{N_g}} \ L = F + \lambda \left| \sum_{j=1}^{N_g} P_j - (P_D + P_L) \right| \tag{23}$$

where λ is the penalty coefficient, and a fixed large, positive real number.

Step 8: Repeat steps 1–7 until the maximum number of iterations is reached.

4 Application of the Bat Algorithm to Real-World Problems

Given that the BA is a good metaheuristic algorithm capable of solving complicated optimization problems, it was not long before it was applied to practical real-world problems. These are classified as falling into one of the areas below.

4.1 Structural Optimization

One of the first problems solved by using the BA is the welded beam design problem [26]. The problem has four design variables and two objectives: to minimize both the overall fabrication cost and the end deflection. The noteworthy aspect of the problem is that, since multiple solutions of the same objective function value exist, and an efficient algorithm or optimizer must discover all the solutions that form the Pareto front. The BA successfully solved this problem. Probably the next real-world problem to be solved by the BA is the design of a brushless DC wheel motor [27]. The objective here is to maximize the efficiency, with five design variables, six constraints and seventy-eight nonlinear equations. Other problems in this category that were successfully solved are (a) the design of steel truss structures, a problem of medium to high dimensionality [28], (b) optimal design of steel frames for minimum weight [29], and (c) the design of a shell and tube heat exchanger to optimize the bi-objective fitness function comprising cost and effectiveness [30].

4.2 Classification and Feature Selection

Reference [31] solves the problem of classification of high dimensional microarray data sets by a functional link artificial neural network (FLANN) classifier. The BA is used to optimize the weights of the FLANN. The BA has also been used to find both the optimal structure as well as the weights and biases of a neural network used for data classification [32]. In [33], a neural network used for data classification is trained using a BA with multiple co-operative sub-populations, and a chaotic map to preserve the diversity of solutions. Such co-operation, in the place of the usual competition between solutions, is claimed to perform better, for this particular application.

In [34], the multispectral satellite image classification using the BA is shown to be a better performer than the bat-K-means clustering, GA, and PSO methods, both in terms of classification efficiency and time complexity.

Feature selection methods aim to provide as simple a classification model as possible, since feature selection is a high dimensional problem affected by the curse of dimensionality. Using a wrapper approach that consists of a heuristic search of a subspace of all possible feature combinations, and a fitness function that is the classifier's performance, [35] uses a binary BA to maximize classifier performance. Reference [36] proved the ability of the BA to solve a high dimensional, combinatorial optimization problem involving feature selection by modifying the continuous valued basic BA into a binary BA. The solution space comprised Boolean hypercube, and the most informative features had to be selected.

Reference [37] considers image thresholding as a constrained optimization problem, and determines the optimal one- or multi-level, fuzzy entropy based thresholds that are maximized using the BA. The BA is used to optimize the parameters of support vector machine, to reduce classification errors in classification problems in [38].

4.3 Electrical Power Systems

The BA has been applied to solve a large number of problems in electrical power systems. Reference [10] solves the ED problem in electrical power systems engineering, using the chaotic BA (in Sect. 2.1), an enhanced version of the basic BA. The problem herein has a cost function that is nonlinear, multimodal, has discontinuities, and has to satisfy numerous inequality and equality constraints as outlined in Sect. 3. Reference [39] solves the same ED problem using chaotic BA as in [10], but using a pseudo-code based algorithm to deal with the power balance equality constraint, instead of the penalty factor approach in [10]. Reference [16] solved an advanced version of the ED problem using the BA with mutation, described in Sect. 2.4. Another application of the BA to the ED problem can be found in [40].

The optimal power flow problem whose objective is to minimize the losses in the presence of power balance equality constraints, and inequality constraints like limits on voltages, real and reactive powers at buses is solved using the BA, without and with an unified power flow controller, a flexible ac transmission system device, in [41].

The objective of load frequency control (LFC) in an electrical power system is to minimize frequency oscillations and tie-line flows due to sharp load changes. Reference [19] solves the multi-area LFC problem using the BA with mutation and crossover, described in Sect. 2.5. A PD-PID controller tuned by the BA is used for solving the multi-area LFC in [42]. The LFC of a two-area system is achieved by using a dual mode gain scheduling of PI controllers, which are tuned by using the BA in [43]. Reference [44] contains the LFC of a two-area system with superconducting magnetic energy storage (SMES) units using model predictive control (MPC) is solved by using the BA, to choose the parameters of the MPC and SMES.

The problem of tuning the parameters of the power system stabilizer (PSS) to minimize system oscillations due to load changes and disturbances is solved using the BA, and compared against the other approaches of the conventional PSS (CPSS) and GA based PSS (GAPSS) in [45]. It found the BA based PSS to be the best performer. The same problem of tuning the parameters of the PSS to minimize oscillations due to disturbances, but for a nonlinear model of the power system comprising single machine connected to an infinite-bus through a transmission line is solved in [46]. It uses the integral of time weighted errors as the cost function.

The optimal phasor measurement unit (PMU) placement to ensure observability of the power system is an NP-hard, combinatorial optimization problem. This problem is solved using a binary bat algorithm hybridized with Taguchi method (TBBA) in [47]. Reference [48] solves the problem of parameter estimation in a power system based on PMU recorded data using a hybridized algorithm comprising the BA and DE, and found that this hybridized algorithm performs better than the other algorithms therein.

Reference [49] uses the binary bat algorithm (in which the solution vector comprises just 1's and 0's) to extract wavelet based fault features for predicting low speed bearing faults. The maximum power point tracking (MPPT) problem, whose objective is to maximize the power output of a photovoltaic array solar panel, is solved by employing a PI controller that is tuned using the BA in [50]. Reference [14] solved the distribution feeder reconfiguration problem, using θ-modified BA (in Sect. 2.3), another enhanced version of the basic BA.

The problem of improvement of power quality has been successfully solved using the BA. The cost function comprising multiple objectives of minimization of total harmonic distortion, initial investment cost, and total fundamental power losses, subject to multiple inequality constraints is minimized by optimal design of a passive power filter, using the BA, in [51].

For the optimal speed control of a brushless dc motor using an online adaptive neuro-fuzzy inference system (ANFIS) controller, the learning parameters of the

ANFIS controller are tuned using the BA in [52]. The position control of a piezoelectric actuator is complicated by the nonlinear and multi-valued mapping between the input and output of the actuator, due to the presence of highly non-linear hysteresis. Hence a neural network (NN) controller trained by using the BA is proposed in the place of conventional controllers and claimed to perform better in [53].

Using the BA, [54] solves the problem of optimal sizing of the battery energy storage (BES) in a microgrid, taking the fixed and running cost of the distributed generators and running cost of the BES per day as the cost function, subject to power balance equality constraint and a number of inequality constraints including charging and discharging rates of the BES, and operating reserve constraint.

Using the BA, [55] solves the problem of optimal spot pricing in a deregulated electricity market, using the fuel cost as the cost function, and power flows as inequality constraints that have to be satisfied optimally.

Using four different types of mutation for self-adaptive learning mechanism (SALM) along with the basic BA to produce the enhanced BA, for finding the linear supply function equilibrium of generating companies in a competitive electricity market is solved in [18].

4.4 Applications in Other Areas

In the field of arrays in electronic engineering, the design of a linear array antenna to minimize the side lobe level, mutual coupling effect, and null control is solved using the BA in [56]. In the field of process control, the NP-hard problem of minimizing the total production cost, of a process with five different tasks with their own costs is solved using the BA with mutation to improve the exploratory capability in [57].

In aerospace engineering, the challenging, high dimensional optimization problem of three-dimensional path planning for an unmanned combat air vehicle (UCAV), whose objective is to find the minimum cost path, subject to numerous constraints, is solved using the BA with DE mutation and crossover in [58].

In the field of petroleum engineering, the problem of minimizing drilling costs by predicting the rate of penetration (ROP) is solved using the BA. First, the data of simultaneous effect of six variables on the ROP is used to develop a mathematical relationship between the ROP and these variables. Next, the BA is used to determine the optimal values of these variables in [59]. Maximization of net present value (NPV) by optimal placement of wells for oil production is solved by the BA, and compared with the other algorithms of GA and PSO, in [60].

In the field of nuclear engineering, a fitness factor that involves maximization of the multiplication factor and minimization of the power peaking factor of a nuclear reactor core is maximized using the BA in [61].

Reference [62] solves the problem of minimizing the energy consumption of a heating, ventilation, and air conditioning (HVAC) system. Using a cubic cost curve to define the total power consumption of the daily optimal chiller loading (DOCL), the minimum DOCL subject to cooling load balance equality constraint, and lower/upper limits on chillers as inequality constraints is solved by the BA.

The examples and case studies cited herein are merely a typical sample of the real-world problems solved by the BA. Given the popularity of the BA, the interested reader of any background is probably quite likely to find applications of interest within their own fields.

5 Conclusion

As one of the better performing metaheuristic algorithms, the bat algorithm has been applied to solve numerous challenging real-world problems that are not that easily solvable by conventional calculus-based methods.

There is immense scope for further research on the bat algorithm. The BA has been applied to numerous continuous optimization problems. However, its application to solving combinatorial optimization problems like the traveling salesman problem have been far fewer in number. Another area in which the BA has been relatively untested is the solution of large scale optimization problems.

At present, the user has to tune the parameters of loudness $A^i(t)$ and pulse emission rate $R^i(t)$ to suit the problem being solved. An alternate, ideal solution would be to equip the algorithm with self-tuning capabilities so that these parameters are automatically adjusted to suit the problem being solved. This too has not attracted enough research at present.

Since metaheuristic algorithms work on the Darwinian principle of selection of the fittest, other types of selection like rank based selection could be experimented with, instead of the knockout or tournament type of selection described in this chapter.

Given that numerous metaheuristic algorithms exist at present, another direction for further research would be the hybridization of the bat algorithm with the operators of other metaheuristic algorithms, so that the hybrid would combine the best features of the algorithms being combined.

Another kind of hybridization that could be explored profitably is that between the bat algorithm and gradient based methods, eliminating the limitations and combining the strengths of these two contrasting types of optimization methods.

Some steps have been taken in many of the directions indicated here. However, fundamental research that addresses these difficult, core issues in optimization theory requires research at a different level altogether, rather than simple comparisons of performance between some x and y algorithms on some specific problem or set of problems. In that sense, it is hoped that this chapter can inspired more research in the foreseeable future.

References

1. Goldberg, D.E.: Genetic Algorithms in Search, Optimization and Machine Learning. Pearson Education, India (1989)
2. Goldberg, D.E.: Computer-aided gas pipeline operation using genetic algorithms and rule learning (Doctoral dissertation, University of Michigan). Dissertation Abstracts International, 44(10), 3174B (University Microfilms No. 8402282) (1983)
3. Dorigo, M.: Optimization, Learning and Natural Algorithms, PhD thesis, Politecnico di Milano, Italy (1992)
4. Kennedy, J., Eberhart R.: Particle swarm optimization. In: Proceedings of IEEE International Conference on Neural Networks, pp. 1942–1948 (1995)
5. Gandomi, A.H., Alavi, A.H.: Krill herd: a new bio-inspired optimization algorithm. Commun. Nonlinear Sci. Numer. Simul. 17(12), 4831–4845 (2012)
6. Yang, X.-S.: A new metaheuristic bat-inspired algorithm. Nature Inspired Cooperative Strategies for Optimization (NICSO) 284, 65–74 (2010)
7. Yang, X.S., Gandomi, A.H.: Bat algorithm: a novel approach for global engineering optimization. Eng. Comput. 29(5), 464–483 (2012)
8. Gandomi, A.H., Yang, X.S., Alavi, A.H., Talatahari, S.: Bat algorithm for constrained optimization tasks. Neural Comput. Appl. 22(6), 1239–1255 (2013)
9. Yang, X.S., Cui, Z., Xiao, R., Gandomi, A.H., Karamanoglu, M. (eds.): Swarm Intelligence and Bio-Inspired Computation: Theory and Applications. Elsevier, Waltham, MA (2013)
10. Adarsh, B.R., Raghunathan, T., Jayabarathi, T., Yang, X.-S.: Economic dispatch using chaotic bat algorithm. Energy 96, 666–675 (2016)
11. Gandomi, A.H., Yang, X.S.: Chaotic bat algorithm. J. Comput. Sci. 5(2), 224–232 (2014)
12. Jordehi, A.R.: Chaotic bat swarm optimisation (CBSO). Appl. Soft Comput. 26, 523–530 (2015)
13. Chakri, A., Kehlif, R., Benouaret, M., Yang, X.-S.: New directional bat algorithm for continuous optimization problems. Expert Syst. Appl. 69, 159–175 (2017)
14. Kavousi-Fard, A., Niknam, T., Fotuhi-Firuzabad, M.: A novel stochastic framework based on cloud theory and θ-modified bat algorithm to solve the distribution feeder reconfiguration. IEEE Trans. Smart Grid 7(2), 740–750 (2016)
15. Haupt, R.L., Haupt, S.E.: Practical genetic algorithms. Wiley, (2004)
16. Niknam, T., Azizipanah-Abarghooee, R., Zare, M., Bahmani-Firouzi, B.: Reserve constrained dynamic environmental/economic dispatch: a new multiobjective self-adaptive learning bat algorithm. IEEE Syst. J. 7(4), 763–776 (2013)
17. Wang, G., Guo, L,. Duan, H., Liu, L., Wang, H.: A bat algorithm with mutation for UCAV path planning. Sci. World J. (2012)
18. Niknam, T., Sharifinia, S., Azizipanah-Abarghooee, R.: A new enhanced bat-inspired algorithm for finding linear supply function equilibrium of GENCOs in the competitive electricity market. Energy Convers. Manag. 76, 1015–1028 (2013)
19. Khooban, M.H., Niknam, T.: A new intelligent online fuzzy tuning approach for multi-area load frequency control: self adaptive modified bat algorithm. Int. J. Electr. Power Energy Syst. 71, 254–261 (2015)
20. Raghunathan, T., Ghose, D.: An online-implementable differential evolution tuned all-aspect guidance law. Control Eng. Prac. 18(10), 1197–1210 (2010)
21. Raghunathan, T., Ghose, D.: Differential evolution based 3-D guidance law for a realistic interceptor model. Appl. Soft Comput. 16, 20–33 (2014)
22. Fister Jr., I., D. Fister, and X.-S. Yang. A hybrid bat algorithm. arXiv:1303.6310 (2013)
23. Xie, J., Zhou Y., Chen, H.: A novel bat algorithm based on differential operator and Lévy flights trajectory. Computat. Intell. Neurosci. (2013)
24. Jun, L., Liheng, L., Xianyi, W.: A double-subpopulation variant of the bat algorithm. Appl. Math. Comput. 263, 361–377 (2015)

25. Meng, X.B., Gao, X.Z., Liu, Y., Zhang, H.: A novel bat algorithm with habitat selection and Doppler effect in echoes for optimization. Expert Syst. Appl. **42**(17), 6350–6364 (2015)
26. Yang, X.S.: Bat algorithm for multi-objective optimisation. Int. J. Bio-Inspired Comput. **3**(5), 267–274 (2012)
27. Bora, T.C., Coelho, L.D.S., Lebensztajn, L.: Bat-inspired optimization approach for the brushless DC wheel motor problem. IEEE Trans. Magn. **48**(2), 947–950 (2012)
28. Hasançebi, O., Teke, T., Pekcan, O.: A bat-inspired algorithm for structural optimization. Comput. Struct. **128**, 77–90 (2013)
29. Hasançebi, O., Carbas, S.: Bat inspired algorithm for discrete size optimization of steel frames. Adv. Eng. Softw. **67**, 173–185 (2014)
30. Tharakeshwar, T.K., Seetharamu, K.N., Prasad, B.D.: Multi-objective optimization using bat algorithm for shell and tube heat exchangers. Appl. Therm. Eng. **110**, 1029–1038 (2017)
31. Mishra, S., Shaw, K., Mishra, D.: A new meta-heuristic bat inspired classification approach for microarray data. Procedia Technol. **4**, 802–806 (2012)
32. Jaddi, N.S., Abdullah, S., Hamdan, A.R.: Optimization of neural network model using modified bat-inspired algorithm. Appl. Soft Comput. **37**, 71–86 (2015)
33. Jaddi, N.S., Abdullah, S.S., Hamdan, A.R.: Multi-population cooperative bat algorithm-based optimization of artificial neural network model. Inf. Sci. **294**, 628–644 (2015)
34. Senthilnath, J., Kulkarni, S., Benediktsson, J.A., Yang, X.S.: A novel approach for multispectral satellite image classification based on the bat algorithm. IEEE Geosci. Remote Sens. Lett. **13**(4), 599–603 (2016)
35. Rodrigues, D., Pereira, L.A., Nakamura, R.Y., Costa, K.A., Yang, X.S., Souza, A.N., Papa, J. P.: A wrapper approach for feature selection based on bat algorithm and optimum-path forest. Expert Syst. Appl. **41**(5), 2250–2258 (2014)
36. Nakamura, R.Y., Pereira, L.A., Costa, K.A., Rodrigues, D., Papa, J.P., Yang X.-S.: BBA: a binary bat algorithm for feature selection. In: 2012 25th SIBGRAPI Conference on Graphics, Patterns and Images (SIBGRAPI), pp. 291–297. IEEE (2012, August)
37. Ye, Z.W., Wang, M.W., Liu, W., Chen, S.B.: Fuzzy entropy based optimal thresholding using bat algorithm. Appl. Soft Comput. **31**, 381–395 (2015)
38. Tharwat, A., Hassanien, A.E., Elnaghi, B.E.: A BA-based algorithm for parameter optimization of support vector machine. Pattern Recogn. Lett. **93**, 13–22 (2017)
39. Shukla, A., Singh S.N.: Pseudo-inspired CBA for ED of units with valve-point loading effects and multi-fuel options. IET Gener. Transm. Distrib. **11**(4), 1039–1045 (2017)
40. Hosseini, S.S.S., Yang, X.S., Gandomi, A.H., Nemati, A.: Solutions of non-smooth economic dispatch problems by swarm intelligence. In: Fister, I., Fister Jr., I. (eds.) Adaptation and hybridization in computational intelligence, pp. 129–146. Springer International Publishing, Switzerland (2015)
41. Rao, B.V., Kumar, G.N.: Optimal power flow by BAT search algorithm for generation reallocation with unified power flow controller. Int. J. Electr. Power Energy Syst. **68**, 81–88 (2015)
42. Dash, P., Saikia, L.C., Sinha, N.: Automatic generation control of multi area thermal system using Bat algorithm optimized PD–PID cascade controller. Int. J. Electr. Power Energy Syst. **68**, 364–372 (2015)
43. Sathya, M.R., Ansari, M.M.T.: Load frequency control using Bat inspired algorithm based dual mode gain scheduling of PI controllers for interconnected power system. Int. J. Electr. Power Energy Syst. **64**, 365–374 (2015)
44. Elsisi, M., Soliman, M., Aboelela, M.A.S., Mansour, W.: Optimal design of model predictive control with superconducting magnetic energy storage for load frequency control of nonlinear hydrothermal power system using bat inspired algorithm. J. Energy Storage **12**, 311–318 (2017)
45. Ali, E.S.: Optimization of power system stabilizers using BAT search algorithm. Int. J. Electr. Power Energy Syst. **61**, 683–690 (2014)

46. Sambariya, D.K., Prasad, R.: Robust tuning of power system stabilizer for small signal stability enhancement using metaheuristic bat algorithm. Int. J. Electr. Power Energy Syst. **61**, 229–238 (2014)
47. Basetti, V., Chandel, A.K.: Optimal PMU placement for power system observability using Taguchi binary bat algorithm. Measur. **95**, 8–20 (2017)
48. Rashidi, F., Abiri, E., Niknam, T., Salehi, M.R.: On-line parameter identification of power plant characteristics based on phasor measurement unit recorded data using differential evolution and bat inspired algorithm. IET Sci. Meas. Technol. **9**(3), 376–392 (2015)
49. Kang, M., Kim, J., Kim, J.M.: Reliable fault diagnosis for incipient low-speed bearings using fault feature analysis based on a binary bat algorithm. Inf. Sci. **294**, 423–438 (2015)
50. Oshaba, A.S., Ali, E.S., Elazim, S.A.: MPPT control design of PV system supplied SRM using BAT search algorithm. Sustain. Energy, Grids and Netw. **2**, 51–60 (2015)
51. Yang, N.C., Le, M.D.: Optimal design of passive power filters based on multi-objective bat algorithm and pareto front. Appl. Soft Comput. **35**, 257–266 (2015)
52. Premkumar, K., Manikandan, B.V.: Speed control of Brushless DC motor using bat algorithm optimized adaptive neuro-fuzzy inference system. Appl. Soft Comput. **32**, 403–419 (2015)
53. Svečko, R., Kusić, D.: Feedforward neural network position control of a piezoelectric actuator based on a BAT search algorithm. Expert Syst. Appl. **42**(13), 5416–5423 (2015)
54. Bahmani-Firouzi, B., Azizipanah-Abarghooee, R.: Optimal sizing of battery energy storage for micro-grid operation management using a new improved bat algorithm. Int. J. Electr. Power Energy Syst. **56**, 42–54 (2014)
55. Murali, M., Kumari, M.S., Sydulu, M.: Optimal spot pricing in electricity market with inelastic load using constrained bat algorithm. Int. J. Electr. Power Energy Syst. **62**, 897–911 (2014)
56. Das, A., Mandal, D., Ghoshal, S.P., Kar, R.: An efficient side lobe reduction technique considering mutual coupling effect in linear array antenna using BAT algorithm. Swarm Evol. Comput. (2017)
57. Wang, J., Fan, X., Zhao, A., Yang, M.: A hybrid bat algorithm for process planning problem. IFAC-PapersOnLine **48**(3), 1708–1713 (2015)
58. Wang, G.G., Chu, H.E., Mirjalili, S.: Three-dimensional path planning for UCAV using an improved bat algorithm. Aerosp. Sci. Technol. **49**, 231–238 (2016)
59. Moraveji, M.K., Naderi, M.: Drilling rate of penetration prediction and optimization using response surface methodology and bat algorithm. J. Nat. Gas Sci. Eng. **31**, 829–841 (2016)
60. Naderi, M., Khamehchi, E.: Well placement optimization using metaheuristic bat algorithm. J. Petrol. Sci. Eng. **150**, 348–354 (2017)
61. Kashi, S., Minuchehr, A., Poursalehi, N., Zolfaghari, A.: Bat algorithm for the fuel arrangement optimization of reactor core. Ann. Nuc. Energy **64**, 144–151 (2014)
62. dos Santos Coelho, L., Askarzadeh, A.: An enhanced bat algorithm approach for reducing electrical power consumption of air conditioning systems based on differential operator. Appl. Therm. Eng. **99**, 834–840 (2016)

Printed in the United States
By Bookmasters